U0172334

第九届结构工程新进展论坛文集

韧性结构与结构减隔震技术

周云　谭平　主编

中国建筑工业出版社

图书在版编目（CIP）数据

韧性结构与结构减隔震技术/周云，谭平主编. —北京：中国建筑工业出版社，2020.11

（第九届结构工程新进展论坛文集）

ISBN 978-7-112-25572-6

Ⅰ.①韧… Ⅱ.①周… ②谭… Ⅲ.①隔震－建筑结构－文集 Ⅳ.①TU352.12-53

中国版本图书馆 CIP 数据核字(2020)第 203898 号

本书为“第九届结构工程新进展论坛”特邀报告人论文集。本届论坛主题为“韧性结构与结构减隔震技术”，主要关注韧性结构与结构减隔震技术的理论、试验、计算、设计、施工和计算机应用等方面。特邀报告人的论文涵盖了以下主要议题：减隔震技术的最新研究成果；基于抗震韧性的新型结构体系研究；高性能结构的实现方法；减隔震组合技术在大跨结构、复杂高层及高烈度地震区中的应用；工程结构的韧性提升理论及方法。

责任编辑：赵梦梅　刘婷婷

责任校对：芦欣甜

第九届结构工程新进展论坛文集

韧性结构与结构减隔震技术

周云　谭平　主编

*

中国建筑工业出版社出版、发行（北京海淀三里河路9号）

各地新华书店、建筑书店经销

北京红光制版公司制版

北京圣夫亚美印刷有限公司印刷

*

开本：787毫米×1092毫米　1/16　印张：19½　字数：470千字

2020年11月第一版　2020年11月第一次印刷

定价：**88.00**元

ISBN 978-7-112-25572-6

(36503)

前　言　Preface

“结构工程新进展论坛”自2006年首次举办以来，近十几年间已经打造成为行业内一个颇具影响的交流平台。论坛旨在促进我国结构工程界对学术成果和工程经验的总结及交流，汇集国内外结构工程各方面的最新科研信息，追随和关注着结构工程学科与领域的前沿与热点，提高专业学术水平，推动我国建筑行业科技发展。

论坛原则上以两年一个主题的形式轮流举办，前八届的主题分别为：

- 新型结构材料与体系（第一届，2006，北京）
- 结构防灾、监测与控制（第二届，2008，大连）
- 钢结构研究和应用的新进展（第三届，2009，上海）
- 混凝土结构与材料新进展（第四届，2010，南京）
- 钢结构研究和应用的新进展（第五届，2012，深圳）
- 结构抗震、减震新技术与设计方法（第六届，2014，合肥）
- 工业建筑与特种结构新进展（第七届，2016，西安）
- 可持续结构与材料（第八届，2018，上海）

“结构工程新进展论坛”已作为结构工程领域重要的学术会议在国内外产生了重要影响，历届论坛都吸引了众多专家学者、工程设计人员、青年学生等参会。第九届论坛由中国建筑工业出版社、同济大学《建筑钢结构进展》编辑部和香港理工大学《结构工程进展》编委会联合主办，由广州大学和广东省工程勘察设计行业协会联合承办，由中国地震学会工程隔震与减震控制专业委员会、中国建筑学会抗震防灾分会结构减震控制委员会、中国土木工程学会防震减灾工程技术推广委员会和中国灾害防御协会城乡韧性与防灾减灾专业委员会共同协办，于2020年12月3日～4日在广州大学召开。

本次论坛的主题是“韧性结构与结构减隔震技术”。工程结构是人类生存和发展所必须具备的工程性基础设施，也是人类生命与财产安全的天然屏障。建立能够凭借自身能力抵御灾害、迅速从灾害中恢复的“韧性”工程结构以及研发提高结构“韧性”的减隔震技术，既是当今全人类面临的共同需求，也是当前结构工程科技发展热点和科技前沿。在本次论坛中我们荣幸地邀请到近20位特邀报告人，他们的报告主题涵盖了近年来与“韧性结构与结构减隔震技术”相关的最新学术思想、研究成果和设计方法；阐述了在这些领域内的最新发展动态；同时也向与会者提供了一个与专家互动并获取宝贵经验的机会。

感谢特邀报告人，他们不仅在大会上作了精彩的主题报告，而且还奉献了精心准备的论文，使得本书顺利出版。

感谢参加本次论坛的所有代表，正是大家的积极参与，才使得本次论坛能够顺利进行。还要特别感谢为本书出版辛勤工作的广州大学刘彦辉老师和郝霖霏老师。

感谢中华人民共和国住房和城乡建设部执业资格注册中心、中国建筑工业出版社、同济大学《建筑钢结构进展》编辑部、香港理工大学《结构工程进展》编辑部对本次论坛的指导、支持和帮助。

目 录 Contents

1 Recent Progress and Application on Seismic Isolation, Energy Dissipation and Control for Structures in China *

Fulin Zhou, Ping Tan

(Guangzhou University 248 Guang Yuan Zhong Lu,

Guangzhou 510405, China ptan@gzhu. edu. cn)

Abstract: China is a seismic country which 100% of territory is located in seismic zone. Most of strong earthquakes are over prediction. Most people dead caused by structures collapse. Earthquakes not only cause severe damages for structures, but also cause non-structural elements and inside facilities, cause to stop the city's life, like hospital, airport, bridge and power plant, etc.. Designers need to use the new techniques to protect the structures and inside facilities. So the Isolation, energy dissipation and Control are more and more widely used in recent years in China. There are nearly 6500 structures with isolation and about 3000 structures with passive energy dissipation or hybrid control in China now. The fields of application includes house buildings, large or complex structures, bridges, immersed tunnel under sea or river, historical or cultural relic protection, industries facilities and retrofit for existed structures. Paper also introduces the design rules and some new innovating devices of seismic isolation, energy dissipation and hybrid control for civil and industrial structures. Paper also makes discussion for the tendency of development on seismic resistance、seismic isolation、passive and active control technique in the future in China and in the world.

Keywords: seismic isolation, energy dissipation, passive control, hybrid control

1. Earthquake Tragedy and Lessons in China

A tragedy strong earthquake, Tangshan Earthquake M7. 8 happened at 3:15 on 26 July 1976. The epicenter depth was only 13KM. The broken faults run through the city. The whole city become ruin, 240000 people died, 96% buildings collapsed, including houses, schools, hospitals, office block and any kinds of other buildings (Fig. 1, Fig. 2).

Another tragedy strong earthquake, Wenchuan Earthquake M8. 0 happened at 14:28 on 12 May 2008. The epicenter depth was only 17KM. The broken faults also run through the city. The whole city become ruin, 90000 people died, 80% buildings collapsed (Fig. 3).

* 论文发表于《EARTHQUAKE ENGINEERING AND ENGINEERING VIBRATIO》2018 年第 1 期。

Fig. 1 One school in Tangshan

Fig. 2 One hospital in Tangshan

Fig. 3 Buildings collapsed in Wenchuan

In the same area of Wenchuan Earthquake, just after 5 years, a another strong Lushan earthquake M7.0 happened in 20 April 2013, over 75% buildings were damaged or collapsed in Lushan County, So call: A Standing Ruin!

But after Lushan Earthquake, A surprise message comes: one of building of Lu Shan County Hospital used Isolation Technique, it is no any damaged, and It performances very well during and after earthquake!

There are 3 buildings in Lu Shan County Hospital, all are reinforce concrete structures with 7 stories and 1 story basement (Fig. 4). One building used Isolation, no any damaged for structure and decoration, no any fall down for facilities & equipment inside the building. The persons inside do no have any feeling during earthquake, This Isolation hospital building become unique Rescue Center in whole county after earthquake, Thousands of injured people accept first aid rescue in it (Fig. 6). And the other two hospital buildings with traditional anti-seismic designed without isolation, persons inside the building feel shaking very severely, building damaged for structure, wall and ceilings, facilities and equipment all fall down, the hospital was all break down (Fig. 5)

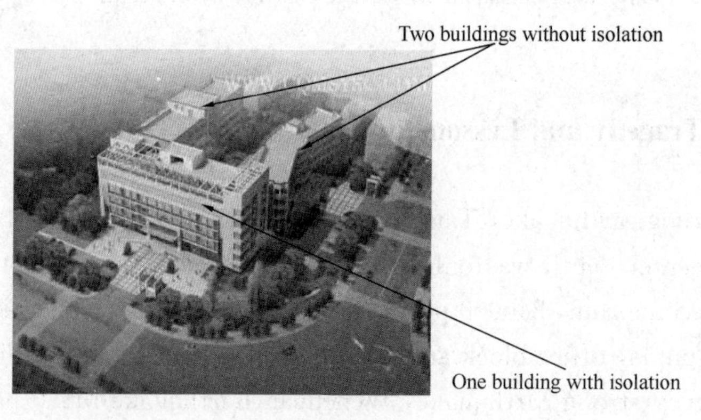

Fig. 4 Lu Shan County Hospital, including three buildings

Many lessons have been learnt from strong earthquakes in China [ZHOU 2016]:

Fig. 5 Two buildings without isolation in Lu Shan County Hospital, break down

Fig. 6 One building with isolation in Lu Shan County Hospital, everything are perfect

1.1 Most of strong earthquakes are over prediction in China.

Over prediction Strong earthquakes in China **Tab. 1**

Earthquake	Design ground acceleration	recorded ground acceleration	times
2008.05.12Wenchuan earthquake	0.10g	0.96g	10
2013.04.20Lushan earthquake	0.15g	0.80g	6

So, the designers need to consider the over prediction strong earthquake.

1.2 Most people dead caused by structures collapse.

Depend on the statistics, 90% of people dead caused by building collapse in earthquake. The earthquake not only causes severe damages or collapse of structures, but also causes severe damages of non-structural elements and inside facilities. It will cause to stop the city life. Like hospital, power plant, and so on.

1.3 Designers need to use the new technique

The isolation, energy dissipation and structural control are effective technique for using. It could protect people life, protect both the building structures and inside facilities in over prediction strong earthquake.

2. Technical Rules for Seismic Isolation & Energy Dissipation in China

Technical rules of seismic isolation and energy dissipation consists three different sets in China [ZHOU 2014]:

2.1 Technical specifications:

- Technical specification for seismic isolation with laminated rubber bearing isolators (CECS 2001). This is the national technical specification for design and construction of buildings and bridges with seismic isolation issued on 2001 in China.
- Technical specification for Energy Dissipation (CECS 2013). This is the national technical specification for design and construction of buildings and bridges with energy dissipation issued on 2013 in China.

2.2 Design codes:

- Seismic design code of buildings with isolation and energy dissipation (Chapter 12 in national code for seismic design of buildings, GB 50011—2001 and 2010). This is a part of national code in China for seismic design of buildings issued on 2001and 2010 in China.
- Seismic Design Code for Isolation Structures (GB 2017). This is the national code for seismic isolation design of buildings structures which is issued on 2017 in China.

2.3 Standards for isolators and energy dissipation dampers

- Standard of laminated rubber bearing isolators (GB 20688—2006). This is the national standard of isolators for laminated rubber bearings issued on 2006 in China.
- Standard of energy dissipation dampers (JG/T 209—2011). This is the national standard of energy dissipation dampers issued on 2011 in China.

3. Testing and Design of Seismic Isolation System

There are five kinds of material have been used for isolators in China, including Sand layer, Graphite lime mortar layer, Slide friction layer, Roller and Rubber bearing, and the Rubber bearing which is the most popular to be used [ZHOU 1997].

Many tests have been finished and whole sets of computation theory of seismic isolationrubber bearings system have been established in China now. The tests include three kinds of work:

3.1 Test of mechanical characteristics for isolator, includes compression tests (capacities, stiffness), compression with shear cycle loading tests (stiffness, damping radio and maximum horizontal displacement). The results of testing showed that, the failure compression stress reach 90MPa., when horizontal strain is zero, the failure compression stress reach 35MPa., when horizontal strain is 400%, so, choose designing compression stress ≤15MPa. is more enough for safety (Fig. 7).

3.2 Tests and investigation of Durability for isolator, includes low cycle fatigue failure tests, creep and ozone aging tests. From the investigation for rubber aging show that, the maximum thickness of rubber ozone is 5 mm exposed to sunlight and air for 105 years , so if make the cover layer 10mm for rubber bearing to isolate the air and sunlight (Fig. 8), the rubber bearing inside the cover layer could make sure the working life of rubber bearing to reach over 100 years!

Fig. 7 compression with shear loading tests

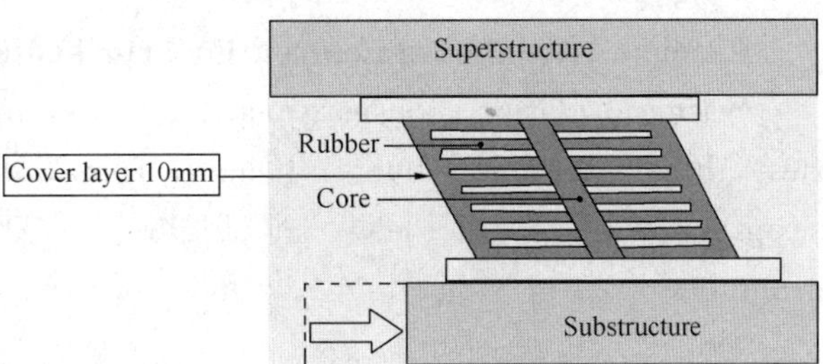

Fig. 8 Cover layer for rubber bearing

3.3 Shaking table tests for large scale structural model, including structure with different location of isolation bearings layers are tested on the shaking table (Fig. 9). The testing results show that the isolation effect depends on the ratio of mass of superstructure to the substructure, the base isolation (isolator on the base) is the best, the acceleration of isolation structure could reduce to be 1/4-1/12 comparing with the no-isolation structure. For middle isolation (isolator on the story), the isolation effect also is significant [HUANG 2003] .

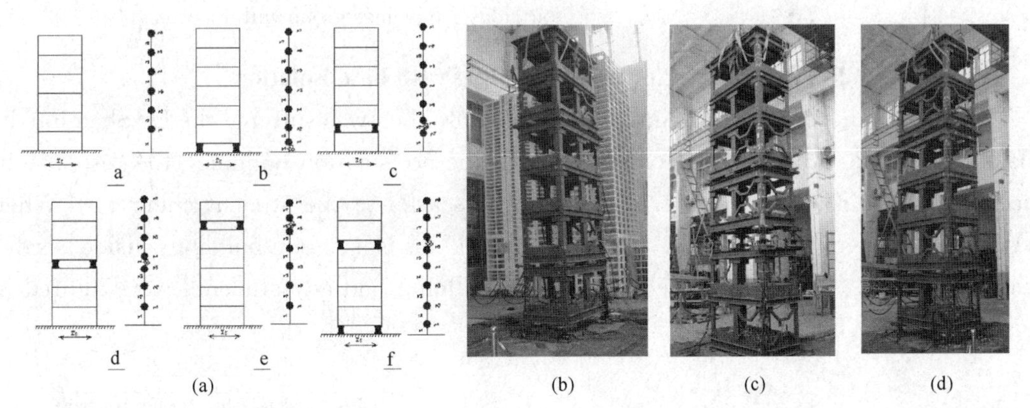

(a) (b) (c) (d)

Fig. 9 Shaking table test for different location of isolation bearings layers

(a) Different location of isolation layers; (b) On the base;

(c) On the Story; (d) Multi-layers

4. Examples of Application on Isolation in China

There are over 6500 buildings with isolation rubber bearings built in China until 2016. These buildings include houses, hospital, school, museum and library, etc.. The buildings are with 3～31 stories. The most of structural types of buildings are concrete, steel frame and brick structures. Some railway bridges and highway bridges with seismic isolation have been built also in China. It has become a very strong tendency to widely using seismic isolation rubber bearings system in China now [ZHOU 2016]

4.1 House Buildings

Example 1 Reinforced Concrete high rise house building with basement isolation.

Seismic Isolation Houses group buildings, including 24 high rise buildings (31 stories), Floor area 980.000 m² in Beijing, China, which is the largest area of Isolation Houses group buildings in the world (Fig. 10). There are many similar Isolation Houses group buildings in Xinjian, Yunan and other areas in China.

Fig. 10 24 high rise (31 stories) Seismic Isolation Houses buildings group with Floor area 980.000 m²

Example 2 RC frame 6 stories school building with base isolation

This isolation school building was built in 2008, then was suffered the Lu Shan Earthquake M7.0 in 20 April 2013. Comparing with the no-isolation building, the response of acceleration was reduced to be 1/6, The teachers asks to the student children: "When earthquake happen, keep in room, don' t run out! this is isolation building, inside is safer than outside!". No any damage for isolation building, no any students were injured in this earthquake (Fig. 11, Fig. 12)

Fig. 11 Seismic Isolation school in Wenchuan County

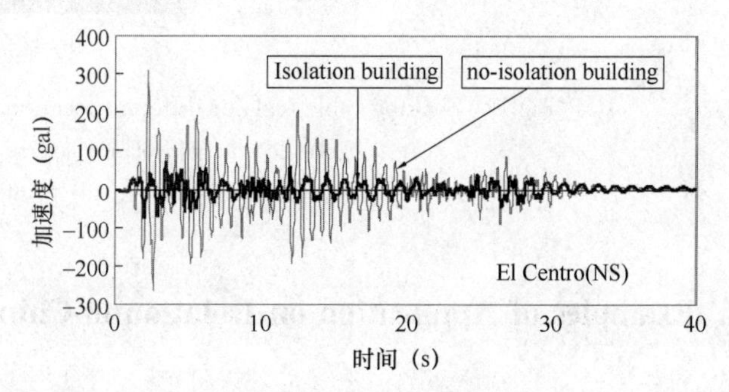

Fig. 12 The records of Acceleration response in earthquake

Example 3 RC frame 2 stories platform+9 stories house with story isolation.

The Seismically isolated artificial ground which is the largest area with 3D isolation in the world (Fig. 13). There is a very large platform (2 stories RC frame) with 1500m wide and 2000m long to cover a railway area in Beijing City. There are 48 isolation buildings (7～9 stories RC frame) with floor area 240000 M2 built on the top floor of the platforms The rubber bearings layer is located on the top floor of the platform to isolate the seismic motion also to isolate the railway vibration which is 3D isolation.

Fig. 13 All 48 isolation buildings using Stories 3D Isolation on RC Platform

4.2 Large span or complex structures

Example 4 Isolation Kunming New Airport (2007-2012) Terminal

The floor Area of Isolation Kunming New Airport is 500000 m^2 which is the largest floor Area of isolation building in the world at recent. Because the location of Airport is near the seismic faults (10km), it is needed to protect the complex structure of airport, column with curve shape, large wall's glass, large ceiling and the important facilities inside the building during earthquake, only use the isolation! This project uses 1892 Rubber Bearings (ϕ1200mm) to isolate the ground motion and 108 Oil Dampers to reduce the displacement during earthquake (Fig. 14). This airport terminal was suffered a earthquake M4.5, Chonmin earthquake in 9 March 2015, the acceleration response records has been got: terminal floor F3/base= 1/4 (Fig. 15). It is very effect to reduce the structural response for Large Area Airport Terminal. The Beijing new airport terminal with isolation floor Area 700000m^2 (2015～2019) and Hainan new airport terminal with isolation floor Area 300000m^2 (2016～2019) are constructing now. Many new airports with isolation are planning or designing in China now.

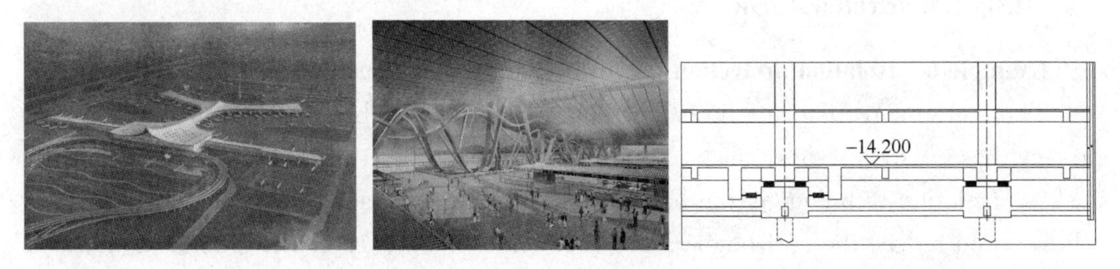

Fig. 14 Isolation Kunming New Airport (2007～2012) Terminal

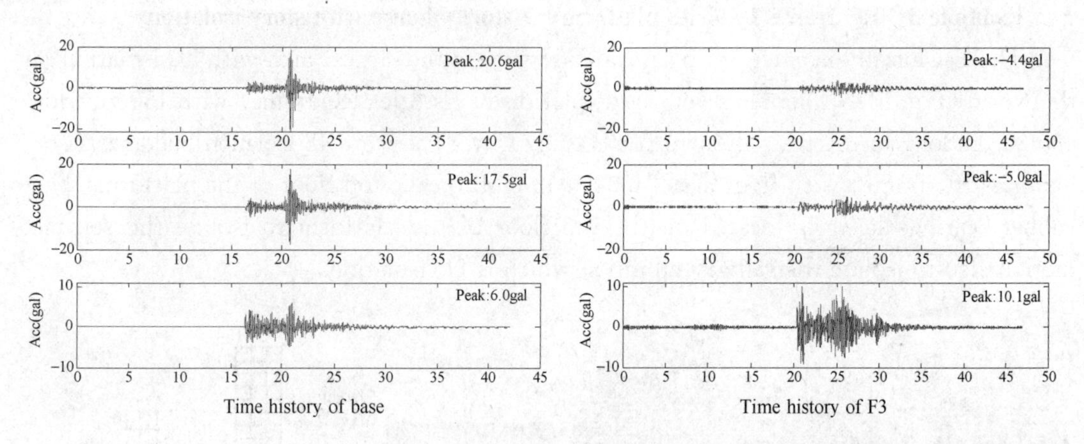

Fig. 15 Acceleration response records : Airport floor F3/base=1/4.

4.3 Bridges

Example 5 Isolation Bridge (HK-MACAO-ZHUHAI) Crossing the South China sea

This 26KM bridge crossing the see uses the isolation technique (Fig. 16). It is very effective to avoid the damage cracks at the bottom of piers in the sea water, to move the nonlinear area from the bottom of pier to the isolators on the top of piers, keep the pier on elastic state in earthquake, also to reduce the seismic response to be 1/4 response for traditional bridge.

Fig. 16 Seismic Isolation Bridge (HK-MACAO-ZHUHAI) 26km

4.4 Historical or cultural relic

Example 6 Isolation protection for history statue and stone tablet (1200ys history)

There are many history relics with thousands years in China. It is very important to protect these history relics, such as statues, tablets or pictures, etc. in strong earthquakes. Isolation technique is the best option! Putting one isolation layer on the base of relics, could reduce the horizontal response of relics to be 1/12 of the ground motion and make sure the safety of relics in strong earthquake (Fig. 17).

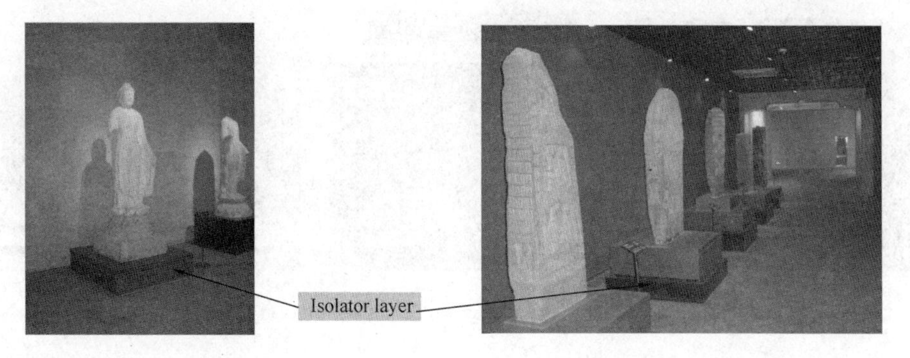

Fig. 17 Isolation protection for history statue and stone tablet (1200ys history)

4. 5 Seismic Retrofit

Example 7 Seismic isolation retrofits for old Buildings

Millions of old buildings of schools, hospital or other lifeline engineering are lack of capacity of seismic resistance in China, which need to retrofit to be satisfied the safety in strong earthquake. Shanxi Province successfully use Isolation to retrofit school buildings, the Government held a National Meeting to extend this demonstrate (Fig. 18, Fig. 19) .

Fig. 18 base Isolation retrofit In Shanxi Province

Fig. 19 1st story Isolation retrofit in Shanxi Province

4. 6 Low rise house

Example 8 Low rise house building with "Elastic Isolation Brick"

There are many houses in rural areas in China. These houses always are low rise, low cost simple design and simple construction, but most of them are lack of capacity for seismic safety. A new isolation system called "Elastic Isolation Brick" has been invented for this requirement and began to be popularly used in large countryside areas in China (Fig. 20) . The cost of this "Elastic Isolation Brick" is only 1/4 of the general rubber bearing, its design is very simple, and its construction is able to be handmade. The shaking table tests showed that, the low rise house with "Elastic Isolation Brick" could suffer very strong earthquake with ground acceleration 0. 80*g* and no any damage, it could become the "safety island" in very strong earthquake.

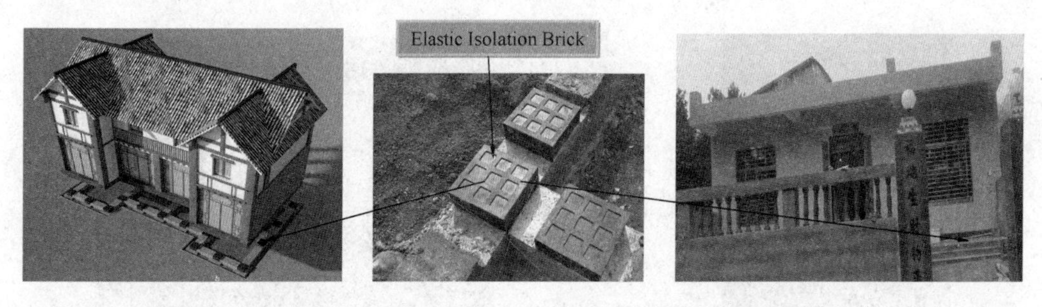

Fig. 20 low rise house with "Elastic Isolation Brick"

5 Energy Dissipation

There are over 3500 buildings with energy dissipation dampers until 2016. Energy dissipation system is formed by adding some energy dissipaters into the structure. The energy dissipaters provide the structure with large amounts of damping which will dissipate most vibration energy from ground motion previous to the structural damage, then to avoid the structure to go in to the severe damage or collapse states in earthquake or to satisfy the requirement in wind. The energy dissipaters may be set on the bracing, walls, joints, connection parts, non-structural elements or any suitable spaces in structures, which may reduce 20%～50% of the structural response comparing the traditional structure without energy dissipaters. This technique is very reliable and simple, suitable to be used for general or important new or existed buildings in seismic regions.

Six kinds of dampers have been used in China now:

- Steel yielding
- Lead yielding
- Oil Damper
- BRB Bracing
- Smart Material (SMA)
- Eddy-Current Damper (ECD)

Example 9 Tunnel with Smart Material (SMA) Energy Dissipation

There are some tunnel crossing the river or sea, need to control the open width of tunnel joints to avoid the water going to the tunnel to kill the people in the cars or trains inside the tunnel in strong earthquake. One shield tunnel with length 6 km crossing the sea, which is located in seismic area with seismic intensity 0.40g in Southern China (Fig. 21). This tunnel follows the soft design concept to create some soft joints then uses Smart Material (SMA) to control the open width of tunnel joints during earthquake and recover the close state after earthquake, make sure the tunnel becomes watertight tunnel experienced the strong earthquake. Depend on the theory analysis and testing result for this new soft design tunnel system show that, this soft design tunnel could reduce the response about 30%～50%, and could make sure the tunnel becomes watertight in very strong earthquakes (Fig. 22) .

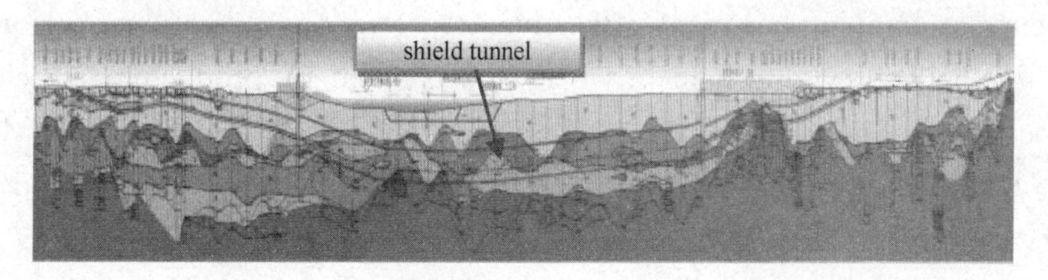

Fig. 21 shield tunnel with length 6 km crossing the sea in Southern China

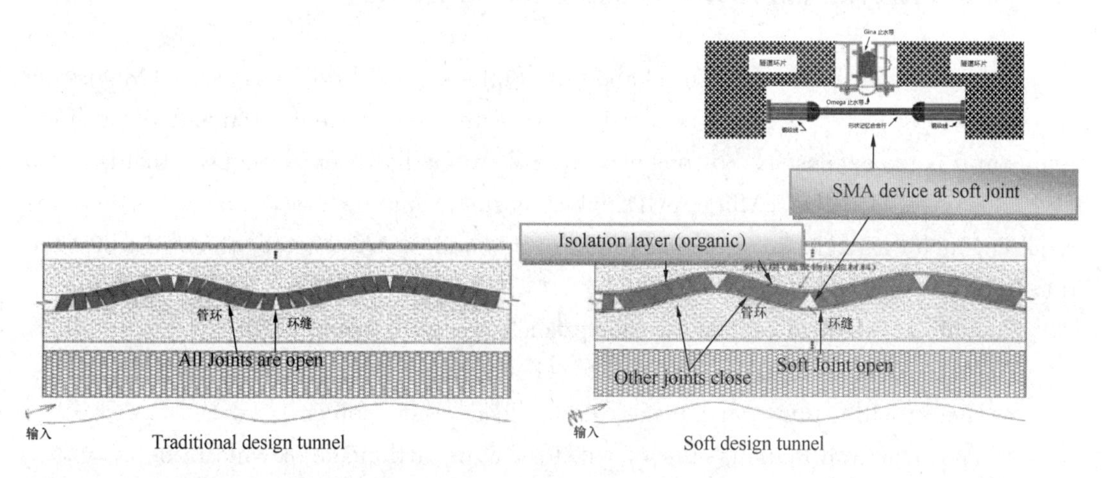

Fig. 22 New soft design concept for Tunnel with Smart Material (SMA)

Example 10 Eddy-Current Damper (ECD)

A novel Damper has been developed by Z. Q Chen, China. The concept is that when the iron plate put in magnetic field will induce large damping. This is a new Generation of Energy Dissipater, which could provide large damping for structures, it is more precise, more durable because no touch between material parts. This new damping device has been used in Shanghai Center tower building with 606m height (Fig. 23) . The building with

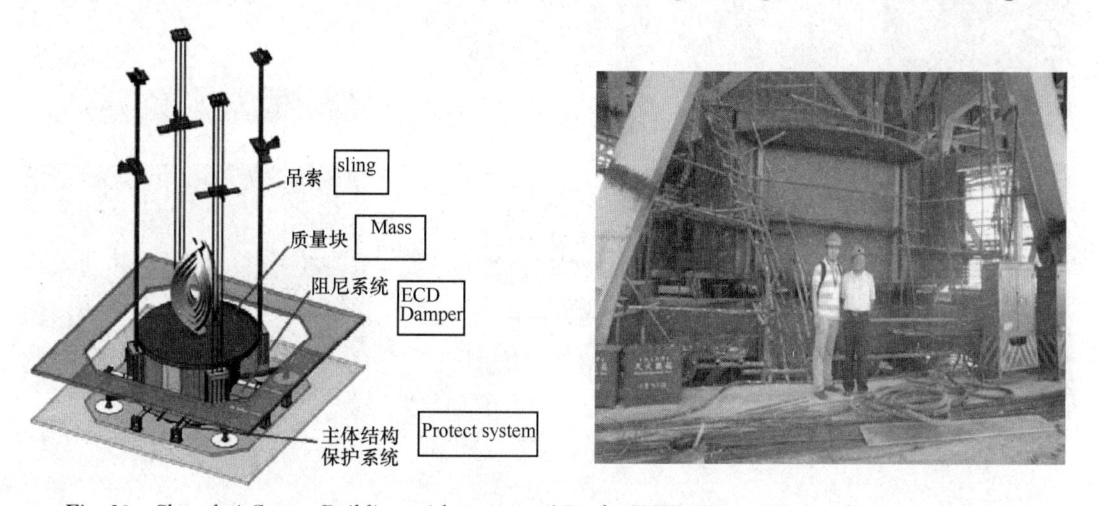

Fig. 23 Shanghai Center Building with 606m and Prof. CHEN Zeng Qing in the construction site

this new system could reduce 30%~50% response during strong wind.

Eddy-current damping has many attractive properties and operational performance. By employing the ball screw mechanism, which is capable of producing sufficiently large damping coefficients and apparent mass. This new device has been used on TMD's damper in high-rise buildings, bridges other project. It is more effective, durable and economical comparing with other types of Dampers above mentioned.

6. Hybrid (Passive Add Active) Control for Structures

Structural control has been used about 30 buildings and bridges. PassiveTMD is low cost and stable, but it could not be satisfied for earthquake or wind, Active AMD is effective, but it is too expensive. So, one new control system has been developed, that is called Hybrid Control HTMD (TMD+AMD) which is stable and low cost, and could satisfy the requirement for earthquake or strong wind, so this new system become one of the best option for structures in China.

Example 11 Hybrid control for Guangzhou Tower with height 645m

The reasons of using hybrid control for Guangzhou Tower are:

- Height with 610m, earthquake & wind load is big problem!
- The structural Plane is ellipse, will torsion in earthquake or wind load
- The slender conformation is not be satisfied in strong wind (Fig. 24) .
- TMD is low cost, but can not satisfy the safe requirements. AMD is effective, but too expensive

TMD+AMD=HTMD may be the best balance (Fig. 25, Fig. 26)

Using two water tanks as mass of TMD, effectively to reduce the cost of new system.

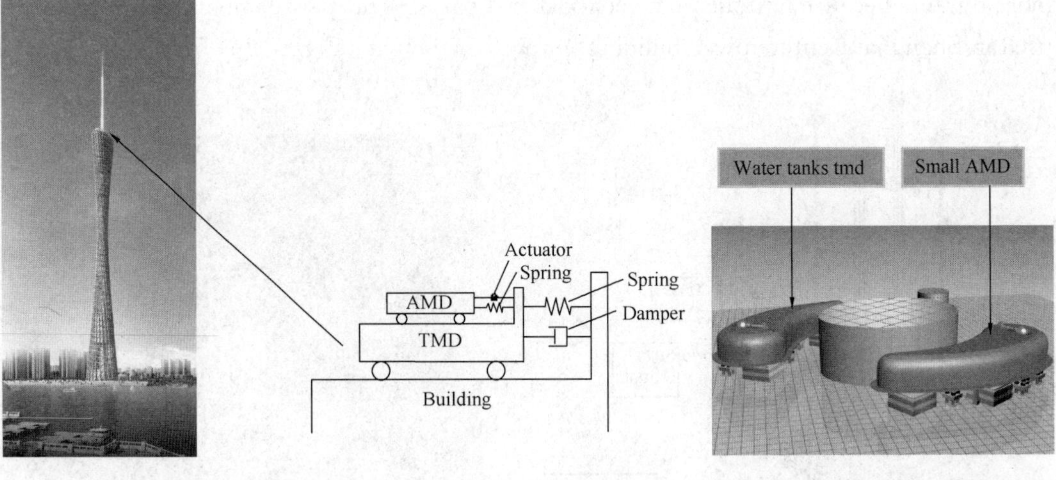

Fig. 24 Guangzhou Tower

Fig. 25 HIBRID (TMD+AMD)

Fig. 26 Using 2 Water tanks as Mass of TMD

7. Conclusion

The low damage and resilient structures has become the seeking goal for seismic design of buildings structures, and the seismic isolation, energy dissipation and structural control have become one of the best options at present in China!

There are many advantages for seismic isolation, energy dissipation & structural control system is:

- More safe, even in over prediction earthquakes.
- More effective protection for both structures & facilities inside in earthquakes.
- More effective to keep elastic state of supper structures in earthquakes..
- More economical, inexpensive, about add cost 5%~10% of structural cost.
- More Satisfied for irregular architectural design for anti-seismic or wind load.

So, it could say that, the coming years:

- Traditional anti-seismic structures still is main structure System.
- Seismic isolation, energy dissipation & structural control will be one of main system for anti-seismic structures, it will be more and more widely used in coming years in China !

References

[1] Zhou Fu Lin[1997]. "Seismic Control of Structures". Chinese Seismic Publishing House.

Huang X. Y. [2003]. "Earthquake Engineering Research & Test Centre. Technical Reports". Guangzhou University. 2003.

[2] Zhou Fu Lin[2014]. "Discussion on Compiling Chinese Design Code of Seismic Isolation and Energy Dissipation". Guangzhou, China.

[3] Zhou Fu Lin [2016]. "Seismic isolation, energy dissipation and structural control system". Beijing, CITY AND DISASTER REDUCTION, No. 5, 2016.

2 大位移摩擦摆隔震韧性结构*

欧进萍[1,2]，武沛松[1,2]，关新春[1,2]

（1. 哈尔滨工业大学结构工程灾变与控制教育部重点实验室单位，哈尔滨；
2. 哈尔滨工业大学土木工程智能防灾减灾工业和信息化部重点实验室，哈尔滨）

摘　要：随着抗震性能需求的提高，根据结构在极罕遇地震下的表现定义结构韧性性能的基本性能目标是：在极罕遇地震作用时，结构不发生严重破坏；震后结构回复到震前的结构状态，进而恢复原有的建筑功能。隔震结构隔震层的地震响应超过允许值，引发隔震支座破坏，导致减震基本性能失效，成为制约隔震结构韧性性能目标实现的关键。本文提出并建立大位移摩擦摆支座隔震结构体系，以整层作为滑动面、框架作为滑块，构建其统一分析模型，阐述其减震机理及工作性能，论证该类隔震结构体系在地震响应大位移情况下具有良好的韧性结构基本性能。从韧性结构目标和指标评价角度，研究该结构体系的摩擦摆支座隔震层位移、主体结构的变形损伤、加速度敏感型的非结构构件损伤以及相应震后可恢复能力的设计实现。

关键词：大位移摩擦摆支座；隔震结构；罕遇地震；极罕遇地震；隔震韧性

中图分类号：TU352.1　文献标志码：As

Seismic-Isolated Resilience Structures with Super-Large Displacement Friction Pendulum Bearings

Ou Jinping[1,2], Wu Peisong[1,2], Guan Xinchun[1,2]

(1. Key Lab of Structures Dynamic Behavior and Control of the Ministry of Education, Heilongjiang, Harbin, China;
2. Key Lab of Smart Prevention and Mitigation of Civil Engineering Disaster of the Ministry of Industry and Information Technology, Heilongjiang, Harbin, China)

Abstract: With increasing of seismic demand, according to the performance of structures subjected to very rare earthquakes, the basic performance objectives of structure resilience are: Structures can survive very-rare earthquakes; the state of structures and building functions are recoverable. The response of isolation layer exceeds its limit, cause the damage of isolation bearing and failure of seismic performance. This is the key to realize structure resilience. Super-Large Displacement Friction Pendulum Bearings (SLDFPB) whose whole layer acts as sliding surface and frames act as sliding blocks is studied. Its resilience is analyzed through deformation damage of isolation layer and superstructure, damage of acceleration sensitive non-structural component and functional recovery of each component. This study provides a new structural sys-

* 本文得到国家重点研发计划项目课题（2017YFC0703603）资助。

tem and development direction for resilience structures.

Keywords: Super-Large Displacement Friction Pendulum Bearings; seismic isolation resilience; very-rare earthquakes; safety of isolation bearings; equivalent radius of Friction Pendulum Bearings

1. 引言

韧性（Resilience）概念最早来源于生态系统的研究，用于衡量某生态系统受到外部扰动后，具有维持其基本性能，继续修复损伤，并逐渐恢复其预期功能的基本性能[1]。Lance 等[2]认为衡量生态系统韧性指标是恢复系统功能速度和承受扰动能力。韧性概念符合建筑结构性能，特别是抗灾防灾，发展的需求，韧性结构已经成为国内外研究热点，结构发展的目标。根据结构构成，可分为材料韧性[3]和结构韧性，两者关注的焦点和研究的内容不同。混凝土材料的自修复[4]、纳米混凝土[5]等都属于材料韧性关键内容。但在结构韧性方面，有些基本概念、关键问题还有待深入研究。Chang 等人[6]认为结构韧性决定于某一灾害下结构功能损失、恢复时间不超过容许预期的概率。Cimellaro 等[7]强调了结构全寿命过程中的韧性。吕西林等[8]提出韧性指地震后不需修复，或稍加修复即可恢复其使用功能。Filiatrault 等[9]指出结构韧性时将结构分为结构构件、建筑构件、服务型构件和结构内容物，分别针对其抗震需求制定韧性标准，通过结构各部分功能函数随时间的变化量化韧性评价[10]。针对特殊功能的结构，应该有针对性的韧性评价指标，API 规范固定式平台中强调杆件或节点的韧性，避免结构关键环节韧性不足，储备刚度不足，导致结构发生倒塌[11]；而医疗系统建筑更重视医疗功能的保全，即关键设备等非结构构件的韧性性能[12,13]。上述研究成果都有利地推动了韧性结构发展。同时，也应该看出，由于上述研究存在区分韧性结构基本性能与建筑功能的不同，导致韧性结构基础理论体系和研究内容存在概念模糊，甚至自相矛盾。

隔震（减震）结构具有隔离（减小）地震动对结构基本性能的影响，减小结构地震响应，是实现韧性结构的有效途径之一。各种隔震支座、减震器以及对应结构体系的研究和实践，也有力地推动了韧性结构发展。但由于地震具有很强的随机性、极大的破坏性，以及地震破坏机理的多样性，导致韧性结构发展面临巨大的挑战。特别是极罕遇，超大地震作用下，由于传统的橡胶隔震支座的抗震性能具有一定局限性[14]，导致现有隔震支座变形超过允许值，导致隔震支座破坏，甚至引发整体结构倒塌的风险[15]。为了提高隔震层变形能力，而增加橡胶支座尺寸，又将导致每个支座的侧向刚度增加，隔震层的总水平刚度增大，整体隔震效果降低。而摩擦摆支座[16,17]作为一种能够自动复位的平面滑移系统，水平刚度与上部结构质量成正比，其水平隔震能力能够避开竖向承载力、支座尺寸的限制，可以实现更柔的隔震层，但过大的变形会产生结构倾覆的问题。这个客观存在的矛盾成为制约隔震结构韧性性能提升的关键问题。

本文提出的大位移摩擦摆支座根据其在不增加刚度的同时可以具备更大水平变形能力的特点，充分利用隔震层的变形能力，极大增加了隔震结构的安全性。摩擦摆支座的变形来自滑动面间的位移，不同于由材料塑性产生的非线性，极罕遇地震下支座大位移可恢复，不出现明显损伤，隔震层韧性大幅增加；并由于其远大于传统摩擦摆支座的隔震半径，上部结构在地震下的响应也得到极大的降低，结构整体隔震韧性大幅增加。

2. 大位移摩擦摆隔震韧性结构概念与构造

从整体提高隔震层的水平变形能力的角度出发，放弃采用每根柱下独立的隔震支座，参考摩擦摆隔震支座的原理，将结构隔震层区域底面整体设计为大跨度大曲率半径球壳，上部结构柱在同一个或几个球壳中滑动，实现隔震效果，并可以极大幅度的增加隔震层整体水平变形能力。

依此思路设计了底部整体式大位移摩擦摆支座、底部组合式大位移摩擦摆支座，并分析其各项性能。

2.1 底部整体式大位移摩擦摆隔震结构

图1为底部整体式大位移摩擦摆支座示意图，将隔震层底面整体设计为大跨度大曲率半径球壳，上部设为可滑动框架，取代传统隔震层中互相独立的隔震支座；大跨度球壳上表面采用低摩擦材料球面层，上部设置滑块若干，滑块间通过连接构件、连梁、承重柱相连形成可滑动的球面框架，通过滑动框架整体在底面球壳上的滑动，保证隔震层具有充足的水平变形能力。

大位移摩擦摆支座的核心目标在于：通过大跨度底面球壳和整体滑动框架形成的整体隔震层，解决隔震层允许变形难以满足极罕遇地震需求的问题。同时球壳曲率半径也远大于传统摩擦摆支座，隔震层刚度及结构自振频率都远小于传统隔震结构自振频率，隔震效果更明显。地震作用的明显降低也有助于高层隔震结构抗倾覆。隔震层允许整体扭转保证了地震下结构的抗扭能力。

图2为支座柱底构造，滑块整体近似为扁圆柱体，下表面弧度与滑动面相同，加工时按对应位置焊接连接梁柱的接头；滑块底面喷涂聚四氟乙烯；连系梁与工字钢接头可通过螺栓连接，滑块箱型钢接头通过栓钉、锚栓等方式与上部混凝土框架柱浇筑。下部滑动面采用混凝土整浇，参考筏板基础的配筋原则，并将上端通长钢筋沿滑动面弯折。

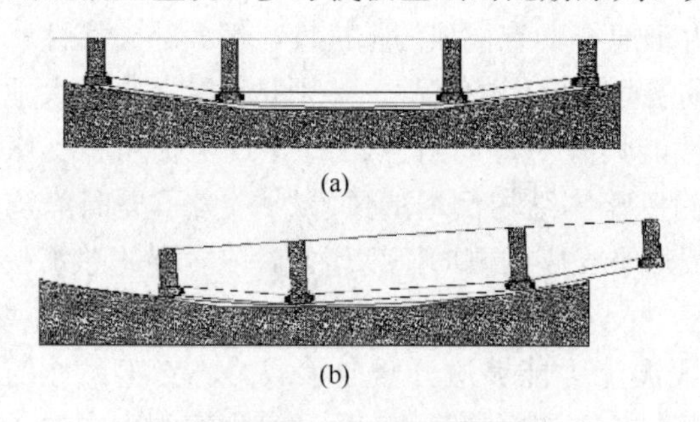

图1　底部整体式大位移摩擦摆支座

Fig. 1　Super-Large Displacement Translation Friction Pendulum Bearings

(a) 变形前隔震结构示意图；(b) 变形后隔震结构示意图

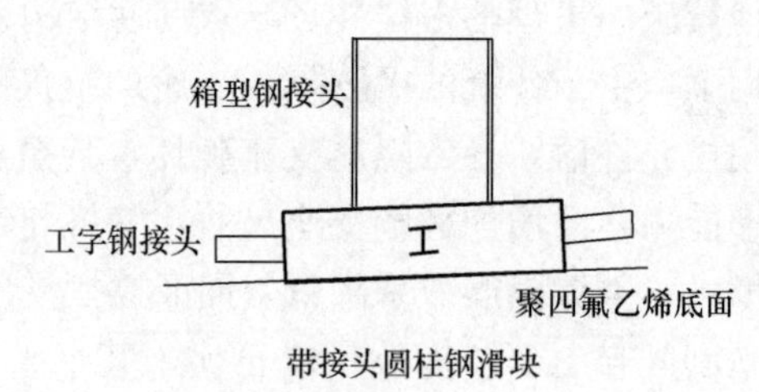

图2　大位移摩擦摆支座柱底构造

Fig. 2　Bottom of column and its detail

大位移摩擦摆支座和柱下摩擦摆隔震支座在计算时具有不同的力学性能。对于柱下独立的摩擦摆支座，地震来临时各支座位移及反力方向均相同；而在大位移摩擦摆支座中，各柱底支座反力均指向轨道圆心，方向不同。摩擦摆支座的上部结构在地震中沿支座轨道运动并保持水平，轴线始终垂直于地面；大位移摩擦摆支座对应的上部结构在地震下沿轨道发生整体转动，结构轴线发生倾斜。地震荷载不仅会由于不同的运动方向而激发横向加速度。角加速度分解和转动动能使力学性能更加复杂。上部结构倾斜导致隔震层与上部结构各层之间的刚度不为零。

根据以往的研究[18]，底部整体式大位移摩擦摆支座等效半径小于滑动轨道半径，大于上部结构重心到轨道球心的距离；隔震层等效摩擦系数大于滑动轨道摩擦系数，放大倍数约为按结构底部计算的轨道半径与按结构重心计算的轨道半径之比。

地震下底部整体式大位移摩擦摆支座结构与其他隔震结构最明显的区别在于上部结构发生倾斜，虽然转角不大，但这种新的运动形式可能使上部结构、内部非结构构件及内容物的安置产生特殊的要求。

2.2 底部组合式大位移摩擦摆支座隔震结构

为充分发挥整体式大位移摩擦摆支座很好的隔震能力，并回避上部结构的倾斜，本文提出底部组合式大位移摩擦摆支座。提取隔震层的框架作为摩擦摆滑块部分，并采用上下双曲面的构造将上部结构的转动转化为平动。单一双曲面支座形成的系统是不稳定的，通过若干大位移摩擦摆支座组合，在确保滑动面各点接触良好的前提下，相邻支座由于整体变形协调运动保持一致，在重力作用下具有良好的复位能力，形成稳定的系统。底部组合式大位移摩擦摆支座示意图见图 3。

底部组合式大位移摩擦摆支座的具体构造为：两组大跨度球壳分别固定于基础（或下部结构）顶部及上部结构底部，每组四个或多个球壳开口处于同一水平面，两组球壳开口相对并留有足够间隙，球壳内共形成四个或多个扁腔，内部设置可滑动框架，形成建筑地下室（或设备层）；可滑动框架上下分别设置若干滑块，整体作为一个大位移摩擦摆支座，在地震下可以发生相对滑动。

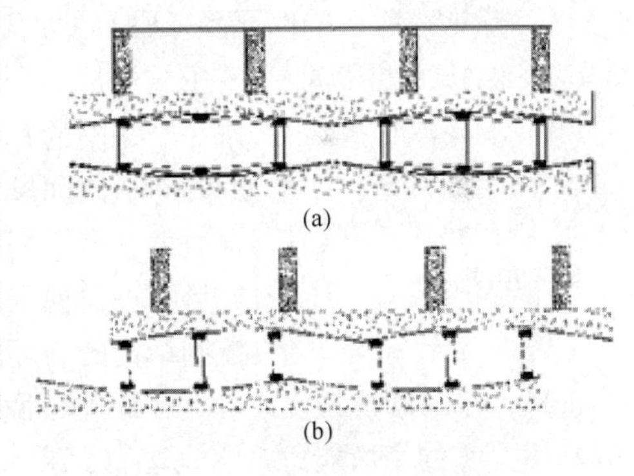

图 3　底部组合式大位移摩擦摆支座

Fig. 3　Super-Large Displacement Rotation Friction Pendulum Bearings

(a) 变形前隔震层示意图；(b) 变形后隔震层示意图

柱底摩擦摆支座与底部整体式、底部组合式大位移摩擦摆支座的静力特性差异主要体现在支座反力方向、弯矩方向和摩擦力方向。传统摩擦摆支座支反力总是竖直向上，整体式大位移摩擦摆支座的支反力始终指向球面轨道的球心；在组合式大位移摩擦摆支座中，同一个滑动框架的一组支座反力分别指向上、下球壳的两个球心，同一组柱底支座反力在不同滑动框架中方向相同。因此，三种隔震装置提供的回复力不同。三种摩擦摆支座滑块与滑动面的接触点数量、位置均不相同，提供的摩擦力也有区别。

三种隔震装置之间的上部结构惯性矩也因运动形式的不同而起不同的作用。对于传统摩擦摆支座隔震结构，在地震荷载作用下，上部结构近似发生平移运动，竖向振动小。整体式大位移摩擦摆支座的整个上部结构在地震作用下绕滑动面球心转动，上部结构的惯性矩明显改变了结构整体的运动。在组合式大位移摩擦摆支座中，滑动框架相对于地面及上部结构在轨道中转动，上部结构类似摩擦摆支座重心沿圆弧轨道移动的同时保持轴线竖直，其每个滑动框架的惯性矩会在运动方程中得到体现。

2.3 大位移摩擦摆支座高层结构多层面隔震体系

大位移摩擦摆支座不仅能够适用于底层隔震技术，对于高层建筑，层间隔震形成的多层隔震也具有很好的效果。

高层建筑在地震作用下两侧柱子竖向变形不能忽略，成为弯曲型或弯曲-剪切混合型结构。在地震下各层已存在倾斜的现象，并且梁柱转角受层间倾斜的差异而放大，不利于结构抗震。

高层隔震技术的重要难点在隔震层的倾覆，其主要原因在于橡胶支座、摩擦摆支座均不宜受拉。

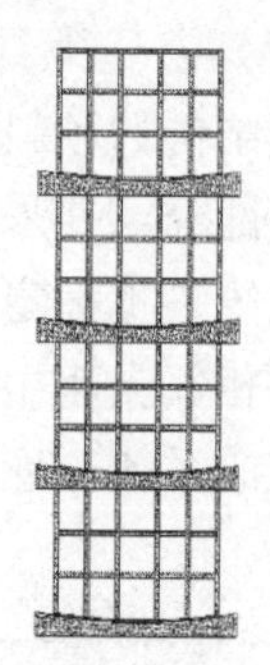

图 4 整体式大位移摩擦摆支座多层面隔震结构示意图

Fig. 4 Multilayer structural system of SLDRFPB

底面整体式大位移摩擦摆支座形成的多层隔震体系可以很好地解决上述两个问题。整体式大位移摩擦摆支座具备极优秀的抗倾覆能力，其突出的隔震效果在降低了上部结构响应的同时也大幅降低倾覆弯矩，并且在倾覆较大的时刻地震荷载产生的倾覆弯矩与上部结构整体倾斜产生的弯矩分配方向相反，支座柱底很难出现拉力。

图 4 显示了具有多个底面整体式大位移摩擦摆支座的高层隔震结构。高层建筑通过多个整体式大位移摩擦摆支座分成若干子结构，每个子结构由若干楼层组成。每个子结构都可以看作一个完整的整体大位移摩擦摆支座底层隔震结构。上部结构的倾斜从层间倾斜转化为整体倾斜，大幅降低梁柱转角。将多自由度、多振型的复杂结构简化为自由度数少、振型明显的结构形式不仅便于分析与设计，其明确的变形形式能够确保人为控制结构的破坏路径、有针对性的加强结构的抗震能力。

选择大位移摩擦摆支座作为子结构的隔震支座正是将变形主要集中在每个隔震层，并充分利用支座的变形能力，明显增加了结构的安全储备。

3. 底部组合式大位移摩擦摆支座底层隔震结构分析模型

本文重点介绍底部组合式大位移摩擦摆支座性能及隔震韧性。大位移摩擦摆支座与传统柱底摩擦摆支座具有相同的隔震原理，二者的主要区别在于大位移支座的滑动框架各方面远大于摩擦摆支座中的滑块部分，滑动框架运动及几何尺寸产生的影响在计算中应得到考虑。

3.1 底部组合式大位移摩擦摆支座力学模型建立

图 5 为底部组合式大位移摩擦摆支座的隔震结构示意图。经研究和试验表明，摩擦系

数、质量比和曲率半径都会对结构的抗震性能有明显影响，而上部结构质量是否偏心对结构抗震性能的影响很小[19]，分析时只考虑对称框架结构模型。

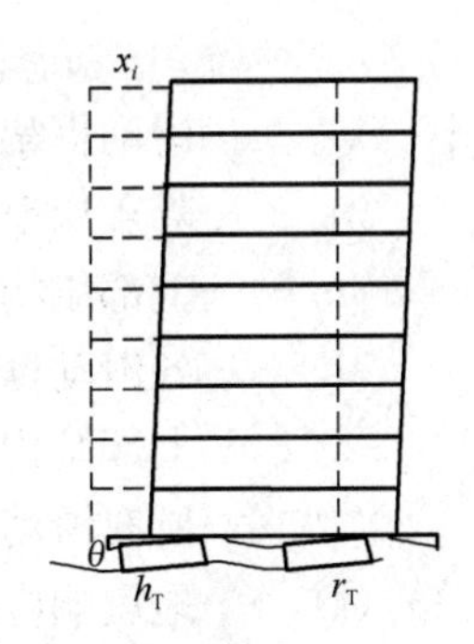

图 5　底部组合式大位移摩擦摆隔震结构示意图

Fig. 5　Seismic isolated structure of SLDTFPB

假设：

1. 每层楼的质量分布均匀。

2. 上部结构满足刚性楼板假设。

3. 框架柱具有无限的竖向刚度，即上部结构为“剪切型”。

4. 计算时考虑一维水平地震作用，地震荷载通过质量乘以地面加速度以惯性力的形式来表示。

5. 各滑块与滑动面为点接触，支反力指向轨道球心。

6. 计算时忽略上部结构的阻尼。

对于摩擦摆类的隔震支座，地震下摩擦力的大小应尽可能准则模拟。有研究表明，具有黏性阻尼的等效线性模型预测摩擦摆隔震结构响应时误差较大[20]，本文在分析时不采用双线性刚度模型或 Boc-Wen 模型处理摩擦力能够避免其不易分析静摩擦力的限制，将根据库仑摩擦理论计算各时刻摩擦力的大小。

滑动面上各接触位置的摩擦力大小可能不同，但每个时刻运动状态判断的结果是相同的。摩擦力由支反力和运动状态决定。在大位移摩擦摆支座中，滑动框架上下表面均会产生摩擦力，各接触点的支反力会随着外荷载改变而重新分配。各滑动框架和上部整体滑动面可以分别看作刚体，因此同一滑动框架表面各接触点的相对运动状态保持一致。为满足变形协调，不同滑动框架运动状态一致。

分析中采用了滑动框架的倾角 θ 和结构各层水平位移 x_i 作为结构自由度。

从滑动框架中心到隔震层顶部和底部的相对位移相同，因此隔震层的位移响应表示为

$$x_{T0}=2\left(r_T-\frac{1}{2}h_T\right)\sin\theta \tag{1}$$

$$y_{T0}=2\left(r_T-\frac{1}{2}h_T\right)(1-\cos\theta) \tag{2}$$

式中　r_T——球面轨道半径；

　　　h_T——滑动框架中轴线高度。

可以看出，组合隔震层中上部框架柱底的运动轨迹是一个圆心高 $2r_T-h_T$、半径为 $r_{T0}=2r_T-h_T$ 的圆。在地震作用下，上部结构将沿其弧线移动并保持竖直。等效质心位置不同对摩擦摆支座的性能具有一定影响，这一点在 Xia 的研究中也有所体现[21]。

滑动框架边柱底与对应轨道圆心连线的偏转角度为 α，每个滑动框架下表面的运动轨迹为一个圆心高 $r_T\cos\alpha$，半径为 $r_T\cos\alpha$ 的圆，每个滑动框架整体绕相应的圆心转动。

对于组合隔震层隔震体系，根据 Lagrange 原理建立隔震层倾角 θ 和隔震层与上部结构各层水平位移 x_i 的运动方程。考虑 $\sin\theta\approx\theta$，并忽略高阶项，采用 $x_0\approx r_{T0}\theta$ 表示隔震层水平位移，运动方程整理为

$$\left(\frac{m_b(r_T\cos\alpha)^2+I_b}{2r_T-h_T}+m_0(2r_T-h_T)\right)\ddot{x}_0+\left(\frac{r_T}{2r_T-h_T}m_b\cos\alpha+m_0+\sum_{i=1}^{n}m_i\right)gx_0+$$
$$k_1(2r_T-h_T)(x_0-x_1)=[m_br_T\cos\alpha+m_0(2r_T-h_T)]\ddot{x}_g(t)-r_T\mathrm{sgn}(\dot{x}_0)\Sigma f \tag{3}$$

$$m_i\ddot{x}_i + k_i x_i - k_i x_{i-1} + k_{i+1} x_i - k_{i+1} x_{i+1} = m_i \ddot{x}_g(t) \tag{4}$$

式中　m_b——滑动框架即地下室区域总质量；

m_0——隔震层上表面该层建筑总质量；

m_i——上部结构第 i 层质量；

I_b——各滑动框架总惯性矩；

k_i——上部结构第 i 层水平刚度；

Σf——所有滑动框架上下表面摩擦力代数和；

$\ddot{x}_g(t)$——地震响应加速度；

g——重力加速度；

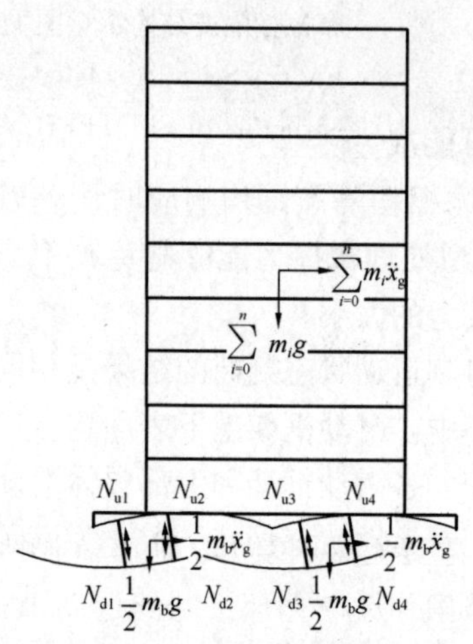

图 6　底部组合式大位移摩擦摆支座受力分析示意图

Fig. 6　Force analysis of SLDTFPB

可以看出，对于大位移摩擦摆支座的滑动框架，地震荷载不仅会由于不同的运动方向而引起横向加速度。角加速度分解和转动动能使力学性能复杂。

运动方程中摩擦力的大小仍需确定。静止状态下静摩擦力的大小受其余荷载的合力影响，理论分析时主要考虑支座运动状态。以运动状态下摩擦力的大小描述摩擦作用，摩擦力的代数和可以通过支反力的代数和表示。

假设每个滑动框架与轨道为点接触，由于摩擦系数较小，摩擦力的变化对支持力的重分布影响不大，求解各柱底支反力时不考虑摩擦力，组合隔震层的受力分析示意图如图 6 所示。

根据达朗贝尔原理，每个滑动框架及上部结构的合力矩与含惯性力的合外力均为零，考虑各位置支反力的重分配，

$$\Sigma f = \mu \Sigma N \tag{5}$$

$$N_{u1}\cos\alpha + N_{u2}\cos\alpha + N_{u3}\cos\alpha + N_{u4}\cos\alpha - \sum_{i=0}^{n} m_i g\cos\theta - \sum_{i=0}^{n} m_i\ddot{x}_g\sin\theta = \sum_{i=0}^{n} m_i\dot{\theta}^2 r_{T0} \tag{6}$$

$$N_{d1}\cos\alpha + N_{d2}\cos\alpha - N_{u1}\cos\alpha - N_{u2}\cos\alpha - \left(\frac{1}{2}m_b\right)g\cos\theta - \left(\frac{1}{2}m_b\right)\ddot{x}_g\sin\theta = \left(\frac{1}{2}m_b\right)\dot{\theta}^2\left(\frac{1}{2}r_{T0}\right) \tag{7}$$

$$N_{d3}\cos\alpha + N_{d4}\cos\alpha - N_{u3}\cos\alpha - N_{u4}\cos\alpha - \left(\frac{1}{2}m_b\right)g\cos\theta - \left(\frac{1}{2}m_b\right)\ddot{x}_g\sin\theta = \left(\frac{1}{2}m_b\right)\dot{\theta}^2\left(\frac{1}{2}r_{T0}\right) \tag{8}$$

经计算可得

$$\Sigma N=\frac{m_{\mathrm{b}}g\cos\theta+m_{\mathrm{b}}\ddot{x}_{\mathrm{g}}\sin\theta+2\sum_{i=0}^{n}m_{i}g\cos\theta+2\sum_{i=0}^{n}m_{i}\ddot{x}_{\mathrm{g}}\sin\theta+\frac{1}{2}m_{\mathrm{b}}\dot{\theta}^{2}r_{\mathrm{T0}}+2\sum_{i=0}^{n}m_{i}\dot{\theta}^{2}r_{\mathrm{T0}}}{\cos\alpha} \tag{9}$$

计算时将支反力代数和化简为$\Sigma N=\dfrac{m_{\mathrm{b}}g+2\sum_{i=0}^{n}m_{i}g}{\cos\alpha}$，结果有所减小，最大误差在1%左右。

进行隔震初步设计时，可将隔震层以上的结构整体简化为刚体进行分析，将支反力代数和代入隔震层运动方程，此时隔震结构整体运动方程可写作

$$\left(\frac{m_{\mathrm{b}}\,(r_{\mathrm{T}}\cos\alpha)^{2}+I_{\mathrm{b}}}{(2r_{\mathrm{T}}-h_{\mathrm{T}})^{2}}+\sum_{i=0}^{n}m_{i}\right)\ddot{x}_{0}+\mu g\,\frac{r_{\mathrm{T}}}{2r_{\mathrm{T}}-h_{\mathrm{T}}}\,\frac{m_{\mathrm{b}}+2\left(\sum_{i=0}^{n}m_{i}\right)}{\cos\alpha}\mathrm{sgn}(\dot{x}_{0})+$$

$$\frac{\dfrac{r_{\mathrm{T}}}{2r_{\mathrm{T}}-h_{\mathrm{T}}}m_{\mathrm{b}}\cos\alpha+\sum_{i=0}^{n}m_{i}}{2r_{\mathrm{T}}-h_{\mathrm{T}}}gx_{0}=\left(m_{\mathrm{b}}\,\frac{r_{\mathrm{T}}\cos\alpha}{2r_{\mathrm{T}}-h_{\mathrm{T}}}+\sum_{i=0}^{n}m_{i}\right)\ddot{x}_{\mathrm{g}}(t) \tag{10}$$

3.2 底部组合式大位移摩擦摆支座力学性能分析

对比传统的摩擦摆支座隔震结构运动方程，惯性力、摩擦力、回复力、地震荷载相关项前系数均不相同，差异主要来自滑块的转动以及上部结构和滑动框架不同的运动半径。将运动方程写成如下形式：

$$m\ddot{x}_{0}+\mu_{\mathrm{Te}}mg\,\mathrm{sgn}(\dot{x}_{0})+\frac{mg}{r_{\mathrm{Te}}}x_{0}=m_{\mathrm{Te}}\ddot{x}_{\mathrm{g}}(t) \tag{11}$$

其中，

$$\frac{r_{\mathrm{Te}}}{2r_{\mathrm{T}}-h_{\mathrm{T}}}=\frac{m_{\mathrm{b}}\,(r_{\mathrm{T}}\cos\alpha)^{2}+I_{\mathrm{b}}+(2r_{\mathrm{T}}-h_{\mathrm{T}})^{2}\sum_{i=0}^{n}m_{i}}{m_{\mathrm{b}}r_{\mathrm{T}}\cos\alpha(2r_{\mathrm{T}}-h_{\mathrm{T}})+(2r_{\mathrm{T}}-h_{\mathrm{T}})^{2}\sum_{i=0}^{n}m_{i}} \tag{12}$$

$$\frac{\mu_{\mathrm{Te}}}{\mu}=\frac{r_{\mathrm{T}}(2r_{\mathrm{T}}-h_{\mathrm{T}})\left(m_{\mathrm{b}}+2\sum_{i=0}^{n}m_{i}\right)}{\left[m_{\mathrm{b}}\,(r_{\mathrm{T}}\cos\alpha)^{2}+I_{\mathrm{b}}+(2r_{\mathrm{T}}-h_{\mathrm{T}})^{2}\sum_{i=0}^{n}m_{i}\right]\cos\alpha} \tag{13}$$

$$\frac{m_{\mathrm{Te}}}{m}=\frac{m_{\mathrm{b}}r_{\mathrm{T}}(2r_{\mathrm{T}}-h_{\mathrm{T}})\cos\alpha+(2r_{\mathrm{T}}-h_{\mathrm{T}})^{2}\sum_{i=0}^{n}m_{i}}{m_{\mathrm{b}}\,(r_{\mathrm{T}}\cos\alpha)^{2}+I_{\mathrm{b}}+(2r_{\mathrm{T}}-h_{\mathrm{T}})^{2}\sum_{i=0}^{n}m_{i}} \tag{14}$$

可以看出，组合式大位移摩擦摆支座表现出的力学性能与支座自身的物理参数均有些区别，具体分析如下。

图 7～图 9 所示分别为不同轨道深度的大位移摩擦摆支座隔震的框架结构的等效半径、等效摩擦系数以及等效质量。选择一个跨度为 25m，层高 3m 的 20 层框架作为上部

结构进行比较，底面组合式大位移摩擦摆支座每个球壳的深度 h_{T0} 取 0.1～1 m。滑动框架的侧柱为 3m 高，与上部结构的层高相同。

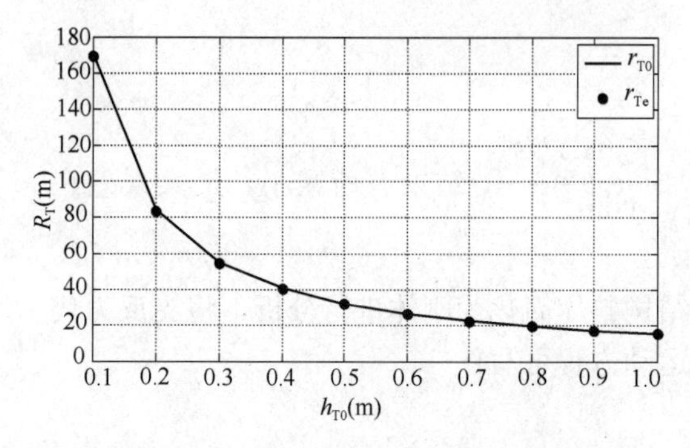

图 7　组合式大位移摩擦摆支座运动半径与等效半径

Fig. 7　Motivation radius and equivalent radius of SLDTFPB

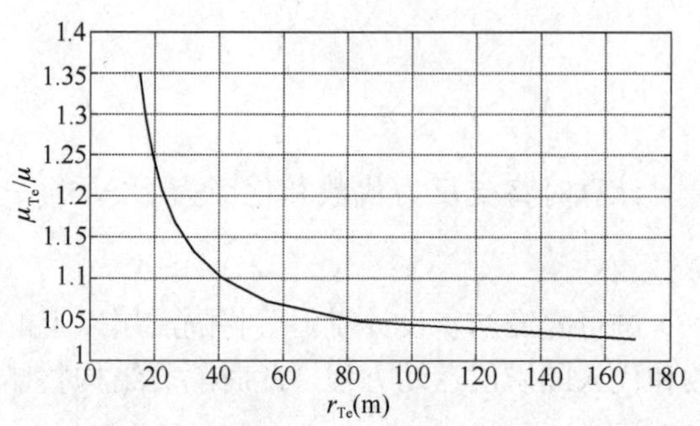

图 8　组合式大位移摩擦摆支座等效摩擦系数随等效半径变化

Fig. 8　Equivalent friction coefficient and equivalent radius of SLDTFPB

图 7 中大位移平动摩擦摆支座运动半径与等效半径吻合很好，根据球壳轨道自身的几何关系，在相同跨度下，球壳轨道半径和等效半径随隔震层深度的减小而减小。大位移摩擦摆支座的等效半径可以与上部结构实际运动轨迹的半径相匹配，设计时采用 $r_{T0}=2r_T-h_T$ 计算隔震频率可以保证足够的准确性。图 8 中，组合式大位移摩擦摆支座的等效摩擦系数随着支座等效半径的增加逐渐接近滑动面摩擦系数，选择较小等效半径的支座需要考虑摩擦系数的放大作用。图 9 中，惯性力对应质量与地震力对应质量比较接近，由于支座滑动框架的转动对于整个结构占比很小，对地震荷载的影响较小。

图 10～图 12 所示分别为不同层数的上部结构的大位移平动摩擦摆基础隔震的等效半径、等效摩擦系数以及等效质量，其每个球壳的深度 h_{T0} 取 0.2m。

随着上部结构层数的增加，隔震层在整个结构中的占比越来越小，等效半径增加，等效摩擦系数、等效质量均在减小，其中等效半径接近上部结构真实运动半径，等效摩擦系数接近滑动面摩擦系数，地震荷载等效质量接近惯性力对应质量。

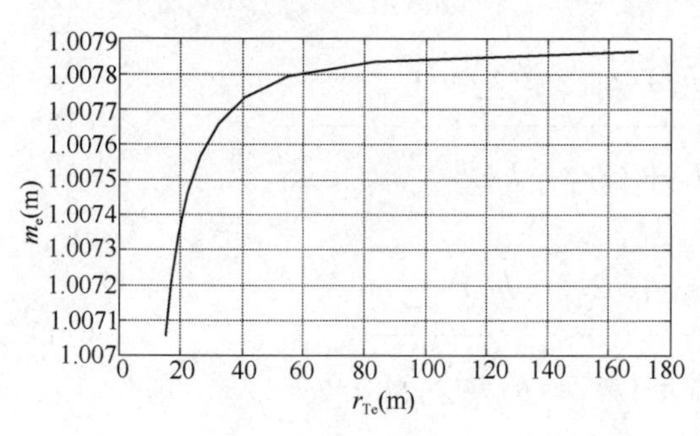

图 9　组合式大位移摩擦摆支座等效质量随等效半径变化

Fig. 9　Equivalent mass and equivalent radius of SLDTFPB

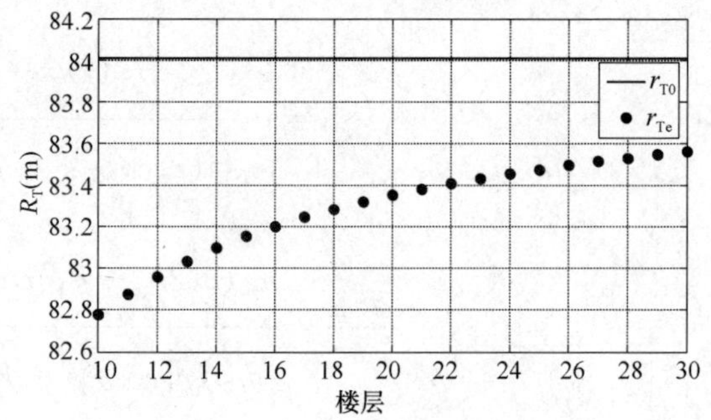

图 10　组合式大位移摩擦摆支座运动半径与等效半径随上部结构层数比较

Fig. 10　Comparison of motivation radius, equivalent radius of SLDTFPB and story of superstructure

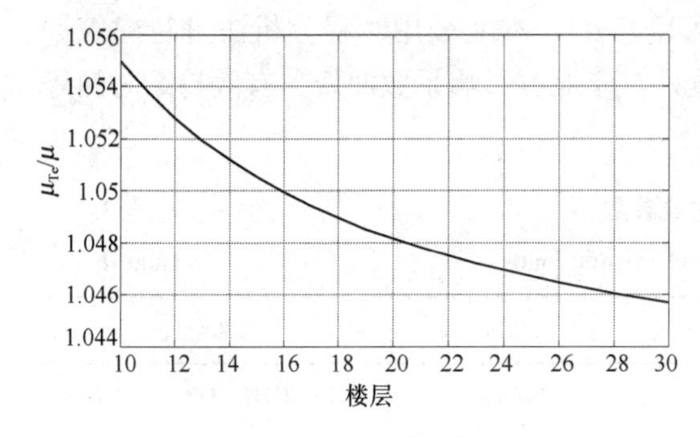

图 11 组合式大位移摩擦摆支座等效摩擦系数随上部结构层数变化

Fig. 11 Comparison of equivalent friction coefficient of SLDTFPB and stories of superstructure

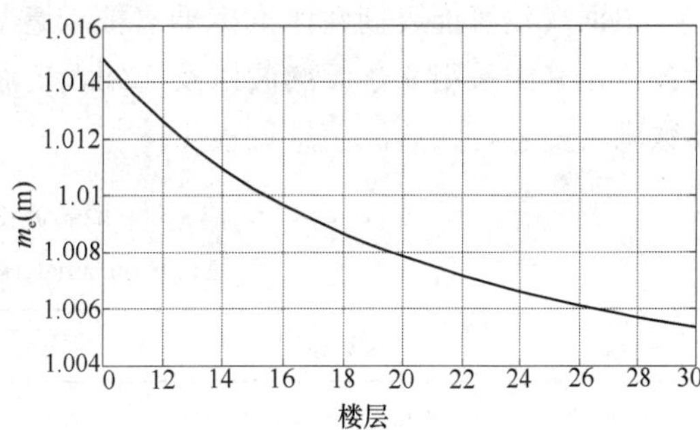

图 12 组合式大位移摩擦摆支座等效质量随上部结构层数变化

Fig. 12 Comparison of equivalent mass of SLDTFPB and stories of superstructure

简化计算时采用上部结构运动半径和惯性力对应质量代替运动方程中的等效半径和等效质量不会引入很大误差，等效摩擦系数需要单独计算。

4. 底部组合式大位移摩擦摆支座底层隔震结构韧性性能分析

对隔震结构的韧性分析主要包括上部结构的变形损伤、加速度敏感型的非结构构件损伤、隔震层稳定的工作性能以及隔震层的变形损伤，分别根据上部结构最大层间位移角、各层绝对加速度、隔震支座最大位移及隔震支座残余位移进行判断。

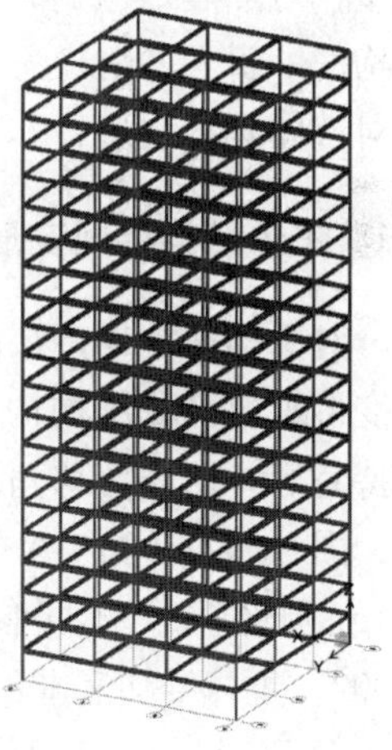

计算时选择某 20 层框架结构模型，见图 13。其平面尺寸为 25m× 25m，高度为 60m，结构自振周期为 1.48s，阻尼比为 0.05。按 8 度 2 类场地设计，设计分组为第二组，设计基本地震加速度为 0.15g。

对于三维结构，在单向地震作用下，每层上部结构绕球体中心旋转，半径为球体中心到各层中心的距离。由于每层自身厚度远小于长度和宽度，围绕纵轴的转动惯量近似等于平面上两个垂直轴上的转动惯量之和。如式 15，对于规则四边形，平面内任意一组垂直轴的转动惯量相同，水平面内地震的入射方向不会造成该方向对应转动惯量的改变，隔震层的运动方程不随地震入射方向改变。质量均匀分布在各层的三维结构的单向地震作用可以简化为二维平面问题。

$$I_z = I_x + I_y = I_a + I_b \tag{15}$$

图 13 某 20 层框架模型

Fig. 13 Frame structure with 20 layers

常见的各材料间摩擦系数中，液体或石墨润滑剂的摩擦系数约为 0.01、钢与混凝土的摩擦系数接近 0.02，常规摩擦摆支座的摩擦系数一般在 0.02～0.1 之间。本文分析时采用的摩擦系数为 0.01、0.02、0.03；轨道深度取 0.2m、0.3m，对应等效轨道半径为 159m、106m。

根据《建筑抗震韧性评价标准（征求意见稿）》[22]，本文采用时程分析法进行建筑韧性评价，其中采用 9 条天然波以及 2 条人工波，对应地震影响系数曲线、有效持续时间等参数满足要求，天然波信息见表 1。

地震波主要参数　　表 1

Main parameters of seismic loads　　Table 1

	地震动概况	矩震级	基站	具体组成
天然地震波 1	Chi-Chi- Taiwan，1999	7.62	HWA043	CHICHI/HWA043-FN
天然地震波 2	Chi-Chi- Taiwan，1999	7.62	TCU038	CHICHI/TCU038-FP
天然地震波 3	Kocaeli- Turkey，1999	7.51	Atakoy	KOCAELI/ATK-FP
天然地震波 4	LomaPrieta，1989	6.93	Salinas-John & Work	LOMAP/SJW-FP
天然地震波 5	LomaPrieta，1989	6.93	Agnews State Hospital	LOMAP/AGW-FP
天然地震波 6	Chi-Chi-Taiwan，1999	7.62	TCU038	CHICHI/TCU038-FP
天然地震波 7	Imperial Valley-06，1979	6.53	ElCentro Array #13	IMPVALL/H-E13-FN
天然地震波 8	LomaPrieta，1989	6.93	Salinas -John & Work	LOMAP/SJW-FP
天然地震波 9	LomaPrieta，1989	6.93	Sunnyvale-Colton Ave.	LOMAP/SVL-FP

基于隔震层运动方程公式 10、上部结构运动方程公式 4，在 Matlab 中建立多自由度模型，对隔震结构的时程分析进行仿真。采用隔震结构的质量矩阵和刚度矩阵，以及上部结构的一阶和二阶模态频率来模拟上部结构的有效阻尼矩阵。采用经典的 Runge-Kutta 方法对整体隔震层进行非线性数值计算，并判断隔震层的黏滑运动。

4.1 主体结构层间变形损伤

对于框架结构，层间位移角小于 1/550 时可以认为上部结构主体结构基本完好。对层间变形敏感的非结构构件中，层间位移角小于 0.0021 时填充墙、隔墙饰面、玻璃幕墙等保持完好，层间位移角小于 0.005 时楼梯保持完好。

图 14 为不同 PGA 的地震波下 20 层隔震结构最大层间位移角。在设计罕遇地震 PGA=0.4g 的地震下，不同参数的底面组合式大位移摩擦摆支座隔震结构层间位移角远未达到 0.002，主体结构和非结构构件具有充足的安全储备。

在 PGA=0.62g 的极罕遇地震下，该隔震结构也能满足主体结构和非结构构件的弹性变形要求，不出现无法恢复的塑性变形，具备最高等级的隔震韧性。

图中可以看出，随地震动的增加上部结构变形增加趋势变缓，隔震效果越明显；这是由于地震强度越大隔震支座受摩擦阻碍越小，隔震支座的等效阻尼降低，隔震效果越好。

图中相同摩擦系数不同支座半径的隔震结构响应曲线基本重合，具有相同的隔震效果，与反应谱曲线在较大周期处趋于平稳的规律吻合。隔震效果受摩擦系数的影响明显，支座摩擦系数为 0.01 时隔震效果最好。设计时从降低上部结构响应的角度，选择较小的摩擦系数更有利。

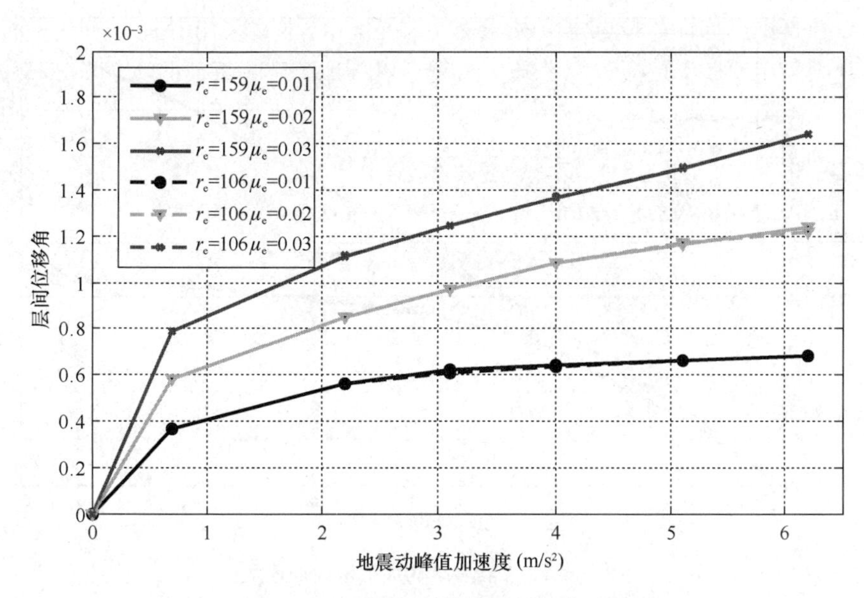

图 14　20 层隔震结构最大层间位移角

Fig. 14　Inter story draft ratio of seismic isolated structure

4.2　加速度敏感型非结构构件损伤

据统计，非结构系统的投资可以占到建筑总投资的 75%～85%；而在历次震害中，非结构系统震害也占到了总震害的 80%以上；同时，造成非结构系统损伤的地震强度远小于造成结构损伤的地震强度。

加速度敏感型非结构系统，主要包括设备、家具等非结构构件，其在地震作用下的破坏主要是由结构的楼面加速度反应导致的。该类非结构构件大多是独立浮放式自然搁置，是非结构系统中最不稳定的部分，在地震中易发生滑移、摇摆、倾覆等现象，造成构件损坏和功能丧失[23]。

当上部结构各层峰值加速度小于 0.5g 时，常规加速度敏感型非结构构件如吊顶、管道、电梯，有锚固或隔震的各类设备均能正常工作；小于 0.2g 时，能够保证无锚固、无隔震的冷水机组、空气压缩机、空气处理机组等设备正常使用。

图 15 为不同 PGA 的地震波下 20 层隔震结构最大楼面加速度。在罕遇地震下，小摩擦系数的大位移摩擦摆支座能够充分降低上部结构的楼面加速度，各类加速度敏感型设备均能正常使用；在极罕遇地震下，对加速度最敏感的冷水机组等设备可能出现无法工作需要修复的情况，对于 100～1000t 的冷水机组修复工时在 9～31h 不等。将对应某一损伤状态下的某类构件的修复费用与经济损失的比值作为修复系数，无锚固，无隔振的冷水机组修复系数约为 1.02。地震下其他构件可以正常使用无需修复，结构总体修复系数很小，具备最高等级的韧性。

降低上部结构的楼面加速度与层间位移角可以共同归纳为增加隔震效果，在设计中都可通过选择小摩擦系数的滑动面实现。在实际工程中注重重要结构非结构构件的安装措施、抗震能力能够使大位移摩擦摆隔震结构成为较为完美的韧性结构。

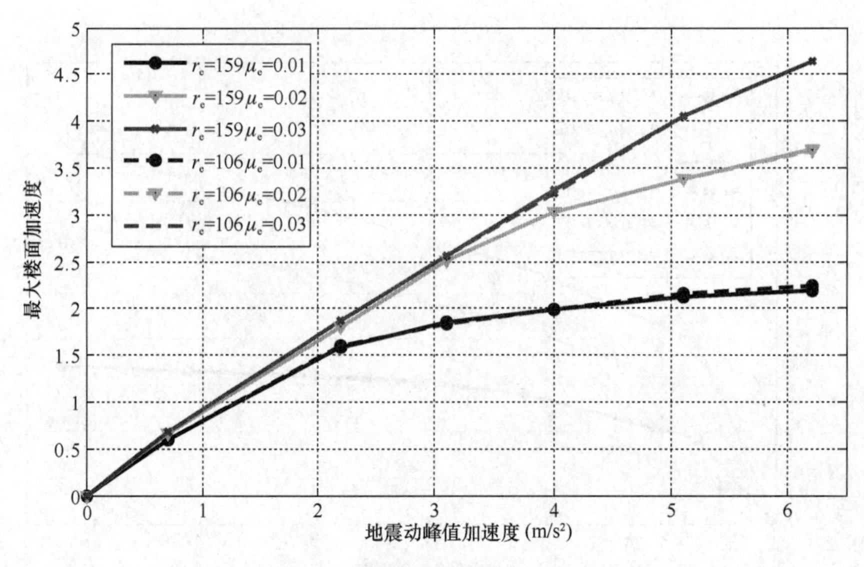

图 15　20 层隔震结构最大楼面加速度

Fig. 15　maximum acceleration of seismic isolated structure

4.3　隔震层性能及残余变形

隔震层的抗震韧性同样根据其在地震下是否能够正常使用以及震后是否需要修复判断。具体为地震中隔震支座力学性能的稳定性、隔震支座的水平变形、隔震支座的抗倾覆能力以及震后残余位移是否需要修复。

根据公式 10，系统的刚度与摩擦系数在地震中不发生变化，隔震层工作性能稳定。对于高宽比小于 3 的隔震结构，一般也不会出现倾覆问题。

图 16～图 18 为 20 层隔震结构隔震支座最大位移以及残余位移。个别罕遇地震下支座最大位移可超过 1m，极罕遇地震下支座最大位移可能接近 2m，仅有大位移摩擦摆支座

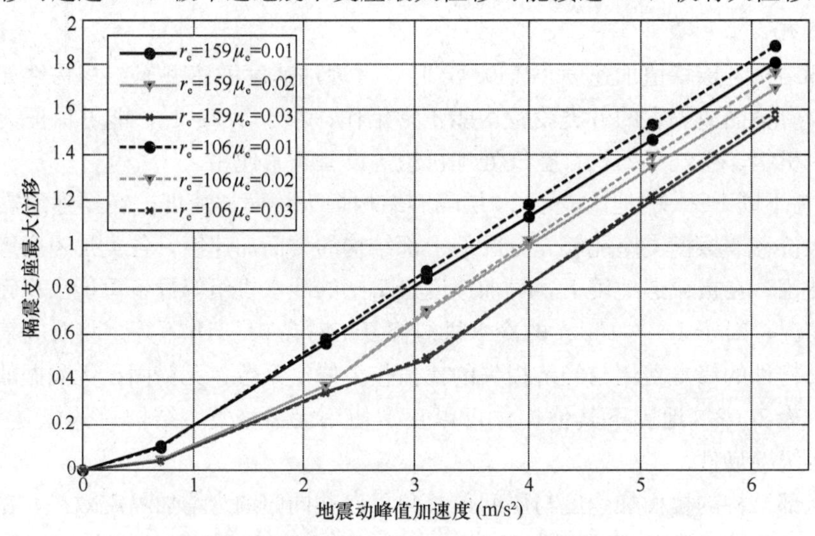

图 16　20 层隔震结构隔震支座最大位移

Fig. 16　Maximum displacement of isolation bearing

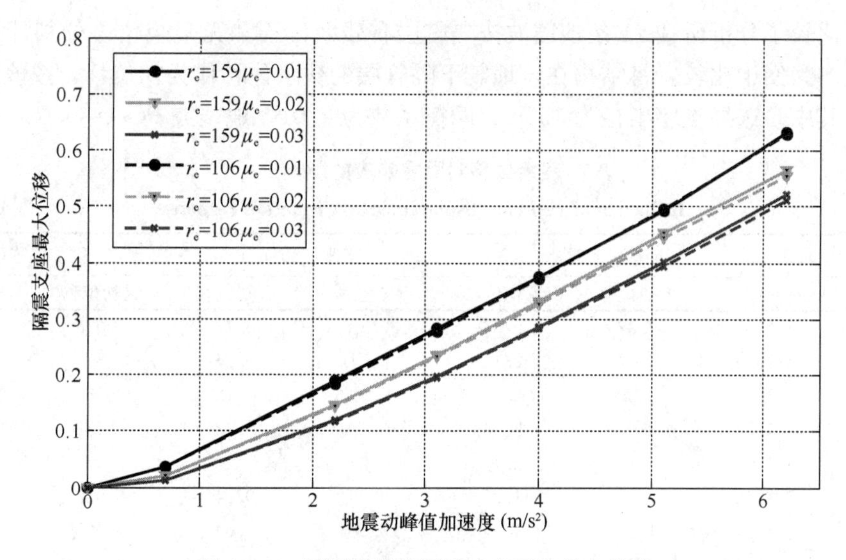

图 17　20 层隔震结构隔震支座平均最大位移

Fig. 17　Mean of 11 earthquakes for maximum displacement of isolation bearing

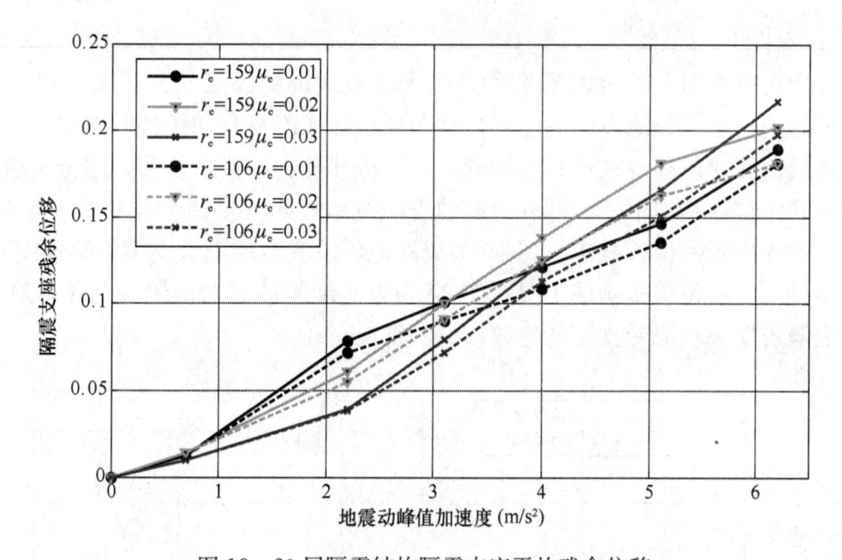

图 18　20 层隔震结构隔震支座平均残余位移

Fig. 18　Mean of 11 earthquakes for residual displacement of isolation bearing

能够具备足够的变形能力，避免了限位装置对隔震支座的损伤以及隔震效果降低。

不同地震下支座响应差异明显。图 17 中，11 条地震动的支座峰值位移平均值约为最大值的 1/3，以结构安全作为首要目标时宜预留充足的安全储备。

极罕遇地震下支座残余位移平均可达 0.2m，若将隔震支座整体设计为结构基础部分，不考虑其他使用功能，0.2m 的残余位移并不影响上部结构的继续使用。对于大位移摩擦摆支座 0.2m 的残余位移下支座各方面性能也相差不大。

大位移摩擦摆支座的残余位移随机性较大，同时受支座半径、支座摩擦系数以及不同地震作用的影响，从统计的角度，平均残余位移近似平均支座最大位移的 1/3，残余位移服从 Rayleigh 分布。

表 2 比较了分析所选 11 条地震波按隔震层有残余位移和无残余位移分别计算的地震下的响应。为便于比较，将结构在该地震下隔震层的残余位移作为初位移，按该地震动重新加载。其中，选择支座半径为 159m，摩擦系数为 0.01，地震波 PGA＝0.62g。

残余位移对隔震响应的影响 **表 2**

Influence of residual displacement on isolation response **Table 2**

层间位移角		层加速度(m/s^2)		支座位移(m)		残余位移(m)	
无初位移	有初位移	无初位移	有初位移	无初位移	有初位移	无初位移	有初位移
1/1400	1/1405	2.32	2.33	0.78	0.77	0.078	0.078
1/1580	1/1580	2.14	2.14	0.34	0.34	0.026	0.026
1/1576	1/1592	2.30	2.19	0.39	0.48	0.387	0.470
1/1350	1/1374	2.24	2.34	0.49	0.33	0.188	0.214
1/1493	1/1505	2.16	2.16	0.23	0.22	0.023	0.027
1/1212	1/1216	1.97	1.99	0.93	0.94	0.086	0.087
1/1910	1/1901	1.90	1.94	0.56	0.64	0.105	0.116
1/1227	1/1247	2.40	2.39	1.81	1.74	0.376	0.376
1/1685	1/1695	2.18	2.18	0.25	0.30	0.089	0.126
1/1463	1/1443	2.29	2.29	0.63	0.73	0.156	0.157
1/1570	1/1445	2.12	2.46	0.57	0.63	0.569	0.629

从表 2 中可以看出，11 条地震波中，绝大多数地震动的层间位移角和层加速度基本不受初位移影响，小部分地震动的层间位移角或层加速度在有初位移时增加，个别地震动的层加速度反而在初位移下减小；对于支座最大位移和支座残余位移，部分地震动下支座最大位移和残余位移基本不变，部分地震动受初位移影响支座最大位移和残余位移增加，个别地震动的支座最大位移减小。图 19 为天然地震波 8 对应隔震支座位移响应时程曲线，代表几乎不受支座初位移影响的工况；图 20 为天然地震波 3 对应隔震支座位移响应时程曲线，代表隔震支座响应受支座初位移影响的工况。

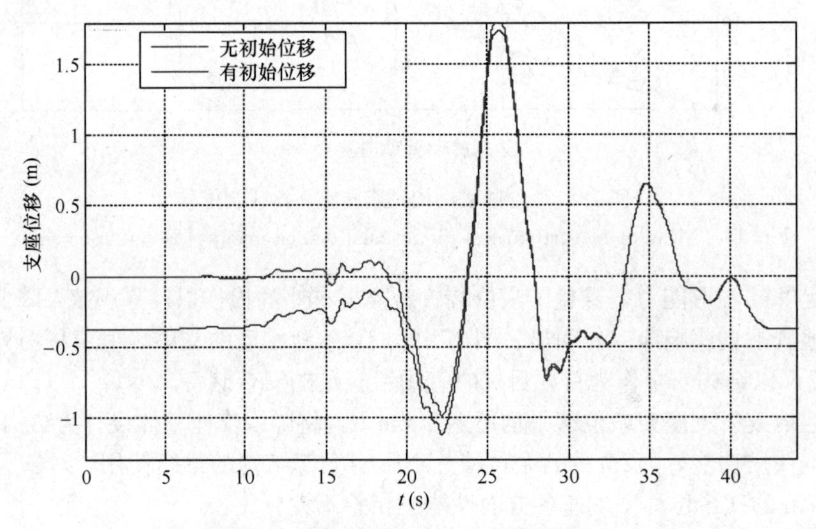

图 19 工况 8 隔震支座位移时程曲线

Fig. 19 Displacement time history curve of isolation bearing subjected to eighth earthquake wave

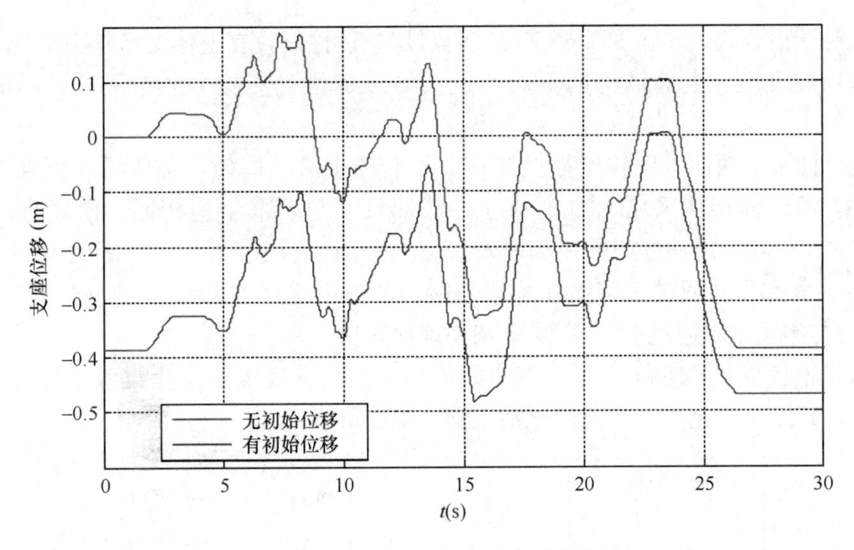

图 20　工况 3 隔震支座位移时程曲线

Fig. 20　Displacement time history curve of isolation bearing subjected to third earthquake wave

两种工况下有初位移的时程曲线与无初位移的时程曲线具有完全相同的趋势，因此对上部结构隔震效果的影响很小；尽管部分地震动中有初位移的隔震支座残余位移会增加，但增加幅度远未达到初位移的大小，整体仍明显表现为复位的趋势。可以认为地震前支座初位移并不影响隔震支座的正常使用，因此震后残余位移可以不需进一步处理。

综合考虑建筑修复费用、建筑修复时间和人员损失大位移摩擦摆隔震建筑的抗震韧性等级可达三星级。

5. 底部组合式大位移摩擦摆支座底层隔震结构韧性性能设计

本文具体以某 6 层混凝土框架结构进行底部组合式大位移摩擦摆支座隔震设计，见图 21。该建筑主体高 21.6m，长边方向宽 48.1m，短边方向宽 26.3m，共 6×3 跨。结构抗震设防烈度为八度 0.2g，场地类别为Ⅱ类，设计地震分组为第二组。结构阻尼比为 0.05。结构采用 C40 混凝土矩形柱，截面尺寸为 0.6m×0.6m，结构基本周期为 1.07s。

图 22 为该框架结构底层平面图。以图中水平方向为 X 方向，竖向为 Y 方向。结构在 X 方向上并非对称，为偏心结构，对于传统隔震支座，支反力的水平分力会带来支座初

图 21　6 层框架结构

Fig. 21　Frame structure with 6 stories

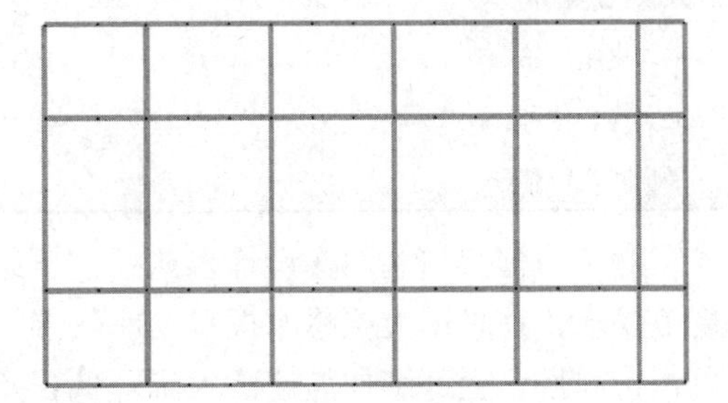

图 22　框架结构平面图

Fig. 22　Plan of frame structure with 6 stories

位移。而对于组合式大位移摩擦摆支座，隔震层整体性好，在整体变形约束下局部变形远小于普通隔震支座，上部结构的不对称仅会导致各位置支反力分布不均，不影响隔震效果。

共设计四个双曲面大位移摩擦摆支座，配套四个滑动框架，每个滑动框架 2×1 跨，每跨 10m，滑动框架柱采用 C50 混凝土，截面尺寸与上部结构相同。滑动框架边柱高 4.4m，中柱高 6m。滑动框架柱端部埋入 Q345 钢箱型滑块端头，滑块底面聚四氟乙烯厚度为 2mm。相邻滑块通过工字钢连系梁连接，连系梁规格为 Q345H 型钢 HW 号 200×204。滑块与连系梁间通过 M20 高强摩擦型螺栓连接。

考虑静止状态下竖向荷载在支座中的传递，大位移摩擦摆支座设计平面图见图 23，支座沿 X、Y 方向的允许水平位移分别达到 4m、3m。

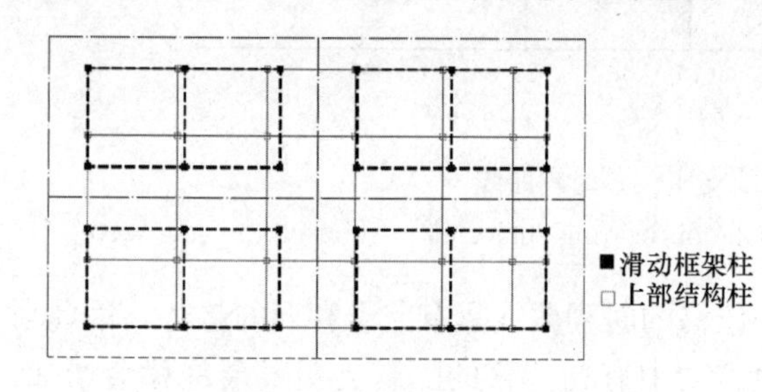

图 23　大位移摩擦摆支座平面图

Fig. 23　Plan of SLDFPB

设计滑动面半径为 63m，按滑动面边缘计算球面深度为 2.13m，无地震作用时上下滑动面边缘距离为 1.74m。根据公式 12，支座等效半径近似为 118m，经计算隔震周期近似为 20.7s。

滑动面采用 10mm 厚球面钢板，钢材等级为 Q345。上下滑动面采用 C50 混凝土，中心厚度为 1m；两侧通长布置 HPB400 钢筋，沿滑动面弯折。板厚中间部分设置直径不小于 12mm，间距不大于 300mm 的双向钢筋网。设计时分别进行支座运动过程中滑动面正截面抗弯验算、混凝土冲切验算、斜截面受剪验算及局部受压验算。

表 3 为地震下原框架结构与增加底部组合式大位移摩擦摆支座隔震措施的框架结构地震下响应比较。计算分析时按 Y 方向进行地震加载。

框架结构与隔震结构响应比较　　表 3

Comparison of response between frame structure and seismic isolation structure　　Table 3

	框架结构		隔震结构	
PGA(m/s^2)	0.7	0.7	4.0	6.2
层间位移角	1/490	1/1433	1/953	1/815
层加速度(m/s^2)	0.83	0.59	1.07	1.25
最大支座位移(m)	—	0.09	1.12	1.77
平均支座位移(m)	—	0.04	0.23	0.50
最大残余位移(m)	—	0.03	0.35	0.58
平均残余位移(m)	—	0.01	0.11	0.18

对于该框架结构，在设计多遇地震下结构已可能有部分进入塑性，抗震能力难以满足规范要求；在采用大位移摩擦摆支座底层隔震后，结构在设计罕遇地震、极罕遇地震下均可保持弹性，上部结构加速度大幅降低，支座位移远未达到设计位移，残余位移不影响上部结构正常工作，隔震效果及结构整体韧性十分理想。

6. 结论

本文介绍了大位移摩擦摆隔震结构及其多面层隔震体系的概念与构造，重点研究底部组合式大位移摩擦摆支座底层隔震结构的力学性能及隔震韧性，得到主要结论如下：

（1）大位移摩擦摆支座采用摩擦摆隔震原理，以一整层作为滑动面，一榀或几榀框架作为滑块，整体形成隔震层，具有充足的隔震层变形能力，显著提高了隔震层的安全性；更大的隔震半径也有利于提高隔震效果。多层面隔震体系振动模式清晰，抗倾覆能力好，适用于高层结构隔震。

（2）底部组合式大位移摩擦摆支座不同于传统柱下摩擦摆支座，力学性能受上部结构影响。大位移摩擦摆支座的等效半径略小于上部结构实际运动半径，等效摩擦系数略大于滑动面摩擦系数，等效地震作用略大于实际地震作用。本文得到的等效支座参数计算结构响应结果准确，能够直接用于隔震设计。

（3）底部组合式大位移摩擦摆支座结构隔震效果很好，在罕遇地震下上部结构整体保持弹性。在极罕遇地震下也能够保证主体结构不出现损伤，个别非结构构件可能出现损伤但容易修复，隔震层工作性能稳定。将隔震层设计为结构基础，不考虑其他使用功能残余位移并不影响上部结构的继续使用，有残余位移的隔震支座仍具有基本相同的隔震效果。结构整体具备最高等级韧性。

参考文献

[1] HOLLING C S. Resilience and stability of ecological systems[J]. Annual Review of Ecology and Systematics，1973，4：1-23.

[2] LANCE H G，LOWEL L P. Resilience and the behavior of large-scale systems[M]. Washington DC：Island Press，2002：3-20.

[3] 周颖，吕西林．摇摆结构及自复位结构研究综述[J]. 建筑结构学报，2011，32(9)：1-10.（ZHOU Ying，LU Xilin. State-of-the-art on rocking and selfcentering structures[J]. Journal of Building Structures，2011，32(9)：1-10.（in Chinese））

[4] WANG J，DING S，HAN B，et al. Self-healing properties of reactive powder concrete with nanofillers[J]. Smart Materials and Structures，2018，27(11)

[5] MENG J，ZHONG J，XIAO H，et al. Enhancement of strength and ductility of cement-based composites by incorporating silica nanoparticle coated polyvinylalcohol fibers[J]. AIP Advances，2019，9(6).

[6] CHANG S E，SHINOZUKA M. Measuring improvements in the disaster resilience of communities[J]. Earthquake Spectra，2004，20(3)：739-755.

[7] GIAN P C，ANDREI M R，MICHEL B. Framework for analytical quantification of disaster resilience[J]. Engineering Structures，2010，32(11) ：3639-3649.

[8] 吕西林，陈云，毛苑君．结构抗震设计的新概念——可恢复功能结构[J]. 同济大学学报(自然科学版)，2011，39(07)：941-948.（LV Xilin，CHEN Yun，MAO Yanjun. New concept of structural seismic design：earthquake resilient structures[J]. Journal of Tongji University（Natural Science），2011，39(07)：941-948.（in Chinese））

[9] FILIATRAULT A，SULLIVAN T. Performance-based seismic design of nonstructural building com-

ponents：The next frontier of earthquake engineering[J]. Earthquake Engineering and Engineering Vibration，2014，13(S1)：17-46.

[10] 卢啸．钢筋混凝土框架核心筒结构地震韧性评价[J/OL]．建筑结构学报，2020，8：1-9. https：//doi. org/10. 14006/j. jzjgxb. 2019. 0406.（Seismic resilience evaluation of a reinforced concrete frame core tube structure[J/OL] Journal of Building Structures：1-9[2020-08-14]．https：//doi. org/10. 14006/j. jzjgxb. 2019. 0406.（in Chinese））

[11] API. API 2A Ed. 21 Recommended practice for planning，designing and constructing fixed offshore platforms-working stress design[S]. USA：API，2007.

[12] MYRTLE R C，MASRI S F，NIGBOR R L，et al. Classification and prioritization of essential systems in hospitals under extreme events[J]. Earthquake Spectra，2005，21(3)：779-802.

[13] 尚庆学，李吉超，王涛．医疗系统抗震韧性评估指标体系[J]．工程力学，2019，36(S1)：106-110.（SHANG Qingxue，LI Jichao，WANG Tao. Indicators system used in seismic resilience assessment of hospital system[J]. Engineering Mechanics，2019，36(S1)：106-110.（in Chinese））

[14] 武沛松，王建，欧进萍．隔震建筑抗极罕遇地震能力与主要破坏模式分析[J]．防灾减灾工程学报，2020，40(03)：317-325.（WU Peisong，WANG Jian，OU Jinping. Research and design of main failure modes of seismically isolated structure subjected to very rare earthquakes [J]. Journal of Disaster Prevention and Mitigation Engineering，2020，40(03)：317-325.（in Chinese））

[15] 武沛松．隔震建筑抗超大震性能分析与设计[D]．哈尔滨工业大学，2015.（WU Peisong. Research and design on seismic performance of high-rise isolation structure in very-rare earthquake [D]. Harbin Institute of Technology，2015.（in Chinese））

[16] ZAYAS V，LOW S，BOZZO L，MAHIN S. Feasibility and performance studies on improving the earthquake resistance of new and existing buildings using the friction pendulum system[R]. Technical Report UBC /EERC-89 /09，University of California at Berkeley，1989.

[17] ZAYAS V，LOW S and MAHIN S. A simple pendulum technique for achieving seismic isolation [J]. Earthquake Spectra，1990，6.

[18] OU J P，WU P S. Resilient Isolation-Structure Systems with Super-Large Displacement Friction Pendulum Bearings[J].

[19] MOSQUEDA G，WHITTAKER A S，FENVES G L. Characterization and modeling of friction pendulum bearings subjected to multiple components of excitation[J]. Journal of Structural Engineering，2004，130(3)：433-442.

[20] PETTI L，POLICHETTI F，PALAZZO B. Analysis of seismic performance of FPS base isolated structures subjected to near fault events[J]. International Journal of Engineering and Technology，2014，5(6)：5233-40.

[21] XIA J，NING X，TAN P，et al. Impact of the equivalent center of mass separating from the sliding surface on the isolation performance of friction pendulum bearings[J]. Earthquake Engineering and Engineering Vibration，2015，14(4)：695.

[22] 住房和城乡建设部. 建筑抗震韧性评价标准(征求意见稿)．北京．2018.

[23] 宁晓晴．重要建筑地震安全性及韧性评价方法研究[D]．中国地震局工程力学研究所，2018.（NING Xiaoqing. Research on Evaluation Methodology of Seismic Safety and Resilience for Significant Buildings[D]. Institute of Engineering Mechanics，China Earthquake Administration，2018.（in Chinese））

3 Synthesis of Structural Vibration Control and Health Monitoring for a Novel Seismic Resistance Technology

Youlin Xu
(Faculty of Construction and Environment, The Hong Kong Polytechnic University,
Hong Kong, China)

Abstract: Strong earthquakes would result in serious damage to or entire collapse of building structures, causing huge economic loss and human death. Hence, structural vibration control (SVC) technology has been developed in the past fifty years to reduce seismic responses of building structures and prevent them from dysfunction and collapse. In recent ten years, structural health monitoring (SHM) technology has been applied to monitor seismic responses of building structures in real time, identify structural damage, assess structural performance and guide structural maintenance and repair. However, the SVC system and the SHM system are generally treated as two independent systems in research and practice. This study thus explores how to best combine the two systems together to reduce system cost on one hand and to enhance control quality by using the real time feedback information from the SHM system on the other hand. Moreover, the use of the updated structural model and the real time structural responses from the SHM system can significantly reduce uncertainties in seismic performance assessment. In this regard, this paper first addresses the two important issues in the synthesis of SVC and SHM: (1) collective placements of control devices and sensors; and (2) synthetic method. The experimental investigation is then carried out to verify the proposed synthetic method. Finally, this paper presents the SHM-based seismic performance assessment method for building structures. The results clearly show that SVC and SHM can be synthesized as a novel seismic resistance technology.

Keywords: building structures, seismic resistance, structural vibration control, structural health monitoring, synthesis

1. Introduction

Structural vibration control (SVC) technologies have been developed for building structures to reduce excessive vibrations caused by strong winds, severe earthquakes or other disturbances. Structural health monitoring (SHM) technologies have been developed for building structures to identify their dynamic characteristics and parameters (system identification) and to detect their possible damage (damage detection). Although SVC systems and SHM systems both require the use of sensors, data acquisition and signal transmission for real implementation, the areas of SVC and SHM have generally been treated separately according to their respective primary objectives. This separate approach is neither practical nor cost-effective if building structures require both a SVC system and a SHM system. This separate approach is also unsuitable for creating smart building structures with their own sensors (nervous systems), processors (brain systems), and actuators (muscular systems, thus mimicking biological systems.

Substantial efforts have been thus made for the synthesis of both SVC and SHM technologies for building structures against earthquakes. Based on the variations of natural frequencies and mode shapes, Chen and Xu[1] and Xu and Chen[2] proposed and numerically investigated an integrated SVC and SHM system using semi-active friction dampers to fulfill model updating, seismic response control and damage detection of building structures. More recently, on the basis of frequency response functions (FRFs), an integrated method for system identification and damage detection of controlled buildings equipped with semi-active friction dampers was proposed by Huang et al. [3] and experimentally validated via a complex building structure with a 12-story main building and a 3-story podium structure[4].

Compared with the frequency domain integrated approach mentioned above, more attentions have been paid to the development of time domain integrated approach mainly due to the integrated system in the time domain being able to act on-line. During an event, in an ideal time domain integrated approach, the structural parameters should be accurately estimated and immediately employed for the determination of optimal control force on one hand, and the measured control force should be used to improve the identification accuracy on the other hand. In this regard, SHM system and SVC system are tightly interconnected and interacted to make the whole system self-diagnose and self-adapt in an effective manner. Some attempts have been made to realize such kind of smart building structures. For example, on the basis of a direct adaptive control algorithm, Gattulli and Romeo[5] proposed an integrated procedure for both vibration suppression and health monitoring of multi-degree-of-freedom (MDOF) shear-type building structures. Chen et al. [6] proposed a general time domain approach to the integration of the vibration control and health monitoring of building structures accommodating various types of control devices and online damage detection. More recently, Ding and Law[7] proposed a method for integrating structural control and evaluation into large-scale structural systems. The control system was implemented with the linear quadratic Gaussian (LQG) and the pseudo negative stiffness (PNS) controls for the vibration mitigation of building structures, whereas in the structural evaluation system, a modified adaptive regularization method was developed based on model updating techniques with the aim of providing updated structural parameters for the control system. Lin et al. [8] proposed a hybrid health monitoring system linked to an adaptive structural control algorithm to improve the control performance; this was verified via a three-story steel frame structure with a damaged column and a damaged joint. Karami and Amini[9] proposed an algorithm including integrated online health monitoring and a semi-active control strategy for reducing both the damage to and seismic response of the main structure caused by strong seismic disturbance. Lei et al. [10] proposed an integrated algorithm for the decentralized structural control of shear-type tall buildings and for identification of unknown earthquake-induced ground motion. Based on the extended Kalman filter and error tracking techniques, an on-line integration technique was also proposed by Lei et al. [11] for the health monitoring and active optimal vibration control of the undamaged/damaged structures. He et al. [12] proposed a time-domain integrated vibration control and health monitoring approach for the simultaneous vibration mitigation and damage detection of building structures without a prior knowledge of external excitations. An integrated semi-active control and damage detection system was considered by Amini et al. [13] to locate and characterize the damage in base-isolated structures and to mitigate base displacements under the effects of seismic excitation. On the basis of recursive least-square estimation, Xu et al. [14] proposed a real-time integrated procedure for accurately identifying time-varying structural parameters and unknown excitations, as well as optimally mitigating excessive vibration in a building structure.

Although the effectiveness of most integrated methodologies has been examined through numerical examples, several experimental investigations exploring the possibility of establishing such a smart building structure can also be found. For example, a new real-time tuning algorithm that is able to identify the instantaneous frequency of linear time-varying systems and tune smart mass dampers for vibration attenuation were developed and experimentally investigated by Nagarajaiah[15]. By using a model reference adaptive control algorithm, a hybrid real-time health monitoring and control system for building structures during earthquakes was presented by Yang et al. [16] and experimentally validated using a three-story aluminum frame structure. Moreover, by employing magneto-rheological (MR) dampers, an experimental investigation of an integrated smart building structure subject to seismic excitations was conducted by He et al. [17].

In the framework of performance-based earthquake engineering (PBEE), the seismic performance of a (controlled) building structure after an earthquake event is often assessed in a probabilistic manner, in which the earthquake damage-to-ground motion relationship, also called fragility curve, is a key component. A fragility curve describes the probability of a structure or a structural component reaching or exceeding a specific damage state (DS) in terms of certain engineering demand parameters (EDPs) as a function of ground motion intensity measure (IM). The fragility curves are then used together with the simulated structural responses for assessing the damage states of the structure or structural components after an earthquake event[18]. With the help of SHM technology, the seismic responses of a tall building can be measured by the SHM system during an earthquake. The use of measured structural responses can help update the building model to make it being best representative of the prototype and reduce uncertainties in determining fragility curves[19]. Therefore, the SHM system installed in a tall building is supposed to be a promising tool for near real-time and automated seismic performance assessment.

In brief, this paper first discusses the two important issues in the synthesis of SVC and SHM: (1) collective placements of control devices and sensors; and (2) synthetic method. The experimental investigation is then presented to verify the proposed synthetic method. Finally, this paper presents the SHM-based seismic performance assessment method for building structures.

2. Collective Placement of Control Devices and Sensors

The optimal sensor locations are often determined for the purpose of SHM such as system identification and damage detection without consideration of control performance, whereas the optimal locations of control devices always refer to control performance by assuming that structural responses, which act as feedbacks in control algorithm and indices in performance functions, can be measured by the sensors without consideration of their availability in both numbers and locations. However, it is impractical and uneconomical to install the sensors at all the required locations when full-state feedbacks are used for the control of a building structure. It is also not economical to install two sensor systems with one for SHM and one for SVC when a civil structure needs both SHM and SVC. Therefore, it is highly desirable to develop the techniques to locate sensors and control devices collectively and cost-effectively to make the building structure of self-sensing, self-adaptive and self-diagnostic ability. In this section, in terms of the increment-based approach for control device placement and the response reconstruction-based sensor placement approach, a collective placement method for the determination of the minimal number and optimal location of both control devices and sensors is presented for vibration control of building structures under earthquake excitation. The feasibility and accuracy of the proposed method are investigated numerically through a 20-story shear building structure under the El-Centro ground excitation. The number and location of sensors determined by this collective placement method can also be used for seismic performance assessment of building structures.

2.1 Increment-Based Approach for Optimal Placement of Control Devices

Consider a seismic-excited tall building modeled by an n degree-of-freedom (DOF) system with active/passive control devices. The matrix equation of motion of the controlled building can be written as

$$\boldsymbol{M}\ddot{\boldsymbol{x}}(t)+\boldsymbol{C}\dot{\boldsymbol{x}}(t)+\boldsymbol{K}\boldsymbol{x}(t)=-\boldsymbol{M}\boldsymbol{I}\,\ddot{x}_g+\boldsymbol{D}\boldsymbol{U}(t) \tag{1}$$

in which $\boldsymbol{x}(t)$ is the n-dimensional displacement vector relative to the ground; $\boldsymbol{M}$, $\boldsymbol{K}$ and $\boldsymbol{C}$ are respectively the $n\times n$ mass, stiffness and damping matrix of the building structure; $\boldsymbol{I}$ is an n-dimensional vector with all elements being unity; $\ddot{x}_g$ is the ground acceleration; $\boldsymbol{U}(t)$ is the m-dimensional control force vector; and $\boldsymbol{D}$ is the $n\times m$ matrix denoting the location of the control force. Equation (1) can also be converted to the following continuous state-space equation,

$$\dot{\boldsymbol{Z}} = \boldsymbol{AZ} + \boldsymbol{BU} + \boldsymbol{W} \tag{2}$$

where

$$\boldsymbol{A} = \begin{bmatrix} \boldsymbol{0} & \boldsymbol{I}_{\mathrm{n}} \\ -\boldsymbol{M}^{-1}\boldsymbol{K} & -\boldsymbol{M}^{-1}\boldsymbol{C} \end{bmatrix};\quad \boldsymbol{B} = \begin{bmatrix} \boldsymbol{0} \\ \boldsymbol{M}^{-1}\boldsymbol{D} \end{bmatrix};\quad \boldsymbol{Z} = \begin{Bmatrix} \boldsymbol{x} \\ \dot{\boldsymbol{x}} \end{Bmatrix};\quad \boldsymbol{W} = \begin{Bmatrix} \boldsymbol{0} \\ -\boldsymbol{1} \end{Bmatrix} \ddot{x}_{\mathrm{g}} \tag{3}$$

The following linear quadratic performance index is often chosen for study in structural vibration control under random disturbance[20, 21].

$$J = \frac{1}{2} E\left\{\int_0^{t_{\mathrm{f}}} (\boldsymbol{Z}^{\mathrm{T}}\boldsymbol{QZ} + \boldsymbol{U}^{\mathrm{T}}\boldsymbol{RU})\,\mathrm{d}t\right\} \tag{4}$$

where $\boldsymbol{Q}$ is a $2n \times 2n$ positive semi-definite weighting matrix for the structure response; $\boldsymbol{R}$ is an $m \times m$ positive definite weighting matrix for the control force; E is the expectation operator; and t_{f} is the duration defined to be longer than that of the earthquake. For a close-loop control configuration with the ground motion being a white noise random process, minimizing Equation (4) subject to the constraint of Equation (2) results in the following optimal control force,

$$\boldsymbol{U} = -\boldsymbol{GZ} \tag{5}$$

where $\boldsymbol{G}$ is the control gain, given by

$$\boldsymbol{G} = \boldsymbol{R}^{-1}\boldsymbol{B}^{\mathrm{T}}\boldsymbol{P} \tag{6}$$

The matrix P is the solution of the classical Riccati equation,

$$\boldsymbol{A}^{\mathrm{T}}\boldsymbol{P} + \boldsymbol{PA} - \boldsymbol{PBR}^{-1}\boldsymbol{B}^{\mathrm{T}}\boldsymbol{P} + \boldsymbol{Q} = \boldsymbol{0} \tag{7}$$

The substitution of Equation (5) into Equation (2) and Equation (4) leads to

$$\dot{\boldsymbol{Z}} = (\boldsymbol{A} - \boldsymbol{BG})\boldsymbol{Z} + \boldsymbol{W} \tag{8}$$

$$J = \frac{1}{2} E\left\{\int_0^{t_{\mathrm{f}}} (\boldsymbol{Z}^{\mathrm{T}}\bar{\boldsymbol{Q}}\boldsymbol{Z})\,\mathrm{d}t\right\} \tag{9}$$

in which

$$\bar{\boldsymbol{Q}} = \boldsymbol{Q} + \boldsymbol{G}^{\mathrm{T}}\boldsymbol{RG} \tag{10}$$

The solution of Equation (8) with an initial condition $\boldsymbol{Z}(t=0) = \boldsymbol{Z}_0$ can be expressed as

$$\boldsymbol{Z}(t) = \boldsymbol{\psi}(t)\boldsymbol{Z}_0 + \int_0^t \boldsymbol{\psi}(t-\tau)\boldsymbol{W}(\tau)\,\mathrm{d}\tau \tag{11}$$

where

$$\boldsymbol{\psi}(t) = \exp\{(\boldsymbol{A} - \boldsymbol{BG})t\} = \boldsymbol{\Phi}\mathrm{diag}(e^{\lambda_1 t}, e^{\lambda_2 t}, \cdots, e^{\lambda_{2n} t})\boldsymbol{\Phi}^{-1} \tag{12}$$

in which λ_1, λ_2, $\cdots$, λ_{2n} are the eigenvalues of matrix $(\boldsymbol{A} - \boldsymbol{BG})$; and $\boldsymbol{\Phi}$ consists of all the eigenvectors.

Assume that the initial condition of the building structure and the ground motion are completely uncorrelated and note that the mean value of the ground excitation is zero.

Then, the substitution of Equation (11) into Equation (9) with some mathematical manipulation yields,

$$J = \frac{1}{2}\int_0^{t_f} Z_0^T \psi(t)^T \bar{Q} \psi(t) Z_0 \mathrm{d}t + \frac{1}{2}E\left\{\int_0^{t_f}\int_0^{t} W(\tau)^T \psi(t-\tau)^T \bar{Q} \psi(t-\tau) W(\tau) \mathrm{d}\tau \mathrm{d}t\right\}$$

$$= \frac{1}{2} Z_0^T S Z_0^T + \frac{1}{2}E\left\{\int_0^{t_f} W(\tau)^T S W(\tau) \mathrm{d}\tau\right\} \tag{13}$$

in which

$$S = \int_0^{t_f} S_0(t) \mathrm{d}t \tag{14}$$

$$S_0(t) = \psi(t)^T \bar{Q} \psi(t) \tag{15}$$

Consider that only a few control devices are moved away from a tall building having a large number of story units equipped with control devices and note that some uncertainties will be involved in the trial-and-error selection of the new weighting matrices for the controlled building with a few control devices removed. The position matrix of control devices $\boldsymbol{B}$ will then have a change $\Delta\boldsymbol{B}$, but the control gain $\boldsymbol{G}$ may be assumed to remain unchanged. As a result,

$$\Delta S_0(t) = \Delta\psi(t)^T \bar{Q} \psi(t) + \psi(t)^T \bar{Q} \Delta\psi(t) = -t[G^T \Delta B^T S_0(t) + S_0(t) \Delta B G] \tag{16}$$

$$\Delta S(t) = \int_0^{t_f} \Delta S_0(t) \mathrm{d}t = -G^T \Delta B^T S_m - S_m \Delta B G \tag{17}$$

in which

$$S_m = \int_0^{t_f} t[S_0(t)] \mathrm{d}t \tag{18}$$

The change of the position matrix of control devices $\boldsymbol{B}$ with $\Delta\boldsymbol{B}$ leads to the increment of the performance index ΔJ. Then, use of the trace theorem of a matrix yields

$$\Delta J = \frac{1}{2}\mathrm{tr}\left\{\left[Z_0 Z_0^T + E\left(\int_0^{t_f} W(\tau) W(\tau)^T \mathrm{d}\tau\right)\right]\Delta S\right\} = -\mathrm{tr}\{(R_0 + F) S_m \Delta B G\} \tag{19}$$

in which

$$R_0 = Z_0 Z_0^T \tag{20}$$

$$F = E\left\{\int_0^{t_f} W(\tau) W(\tau)^T \mathrm{d}\tau\right\} \tag{21}$$

If the i th control device is removed, the increment of the performance index is then

$$\Delta J_i = -\mathrm{tr}\{(R_0 + F) S_m \Delta B_i G\} \quad (i = 1, 2, 3, \cdots, m) \tag{22}$$

The contribution percentage (CP) index can be then defined as

$$CP_i = \frac{\Delta J_i}{\sum_{i=1}^{j} \Delta J_i} \tag{23}$$

The increment of the control performance index due to the removal of the i-th control device, ΔJ_i, calculated by Equation (22) reflects the sensitivity of the i-th control device to the performance index. Therefore, based on the calculated increment from the removal of each control device, the sequence of importance of all the control devices can be obtained. It can be seen from Equation (23) that the larger value of the CP_i indicates the more important influence of the i-th control device on the total control performance. Consequently, the CP index provides a great convenience for determining the number and location of control devices according to the predetermined control performance. Different from the classical integer heuristic programming methods such as the sequential search algorithm (SSA)[22] or the Worst-Out-Best-In (WOBI) algorithm[23], the control device placement method basing on the sequence of the calculated CP_i is relatively simple because the value of CP_i does not need to be re-calculated when other control device is removed.

After the number and location of the control devices are determined according to the sequence of CP_i, the control algorithm shall be decided and the number and location of sensors shall be selected, in which the structural responses measured by the sensors can be used as the feedbacks for vibration control. However, it is often difficult in practice to install enough sensors to obtain the required structural responses as feedbacks for vibration control of a large building structure. This is particularly true if LQR or LQG control algorithms are selected with complete structural responses required for feedbacks. In this regard, the determination of the number and location of limited sensors of the controlled building structure to fulfill the vibration control task becomes necessary.

2.2 Response Reconstruction-Based Approach for Optimal Placement of Sensors

To keep all signs in this subsection are consistent, the equation of motion of a controlled building structure of multi-degrees-of-freedom (MDOFs) under earthquake excitation is given again by

$$\boldsymbol{M}\ddot{\boldsymbol{x}}+\boldsymbol{C}\dot{\boldsymbol{x}}+\boldsymbol{K}\boldsymbol{x}=-\boldsymbol{M}\boldsymbol{I}\,\ddot{x}_{\mathrm{g}}+\boldsymbol{H}_{\mathrm{c}}\boldsymbol{U} \tag{24}$$

in which $\boldsymbol{x}$, $\dot{\boldsymbol{x}}$ and $\ddot{\boldsymbol{x}}$ are the displacement, velocity, and acceleration response vector, respectively; $\boldsymbol{M}$, $\boldsymbol{K}$ and $\boldsymbol{C}$ are the mass, stiffness and damping matrix of the building structure, respectively; $\boldsymbol{U}$ is the control force vector; $\boldsymbol{H}_{\mathrm{c}}$ is the matrix denoting the location of the control force; and $\ddot{x}_{\mathrm{g}}$ is the ground acceleration. Equation (24) can also be converted to the following continuous state-space equation,

$$\dot{\boldsymbol{X}}=\boldsymbol{A}_{\mathrm{c}}\boldsymbol{X}+\boldsymbol{B}_{\mathrm{c}}\ddot{x}_{\mathrm{g}}+\boldsymbol{D}_{\mathrm{c}}\boldsymbol{U} \tag{25}$$

where

$$\boldsymbol{X}=[\boldsymbol{x}\quad \dot{\boldsymbol{x}}]^{\mathrm{T}} \tag{26}$$

$$\boldsymbol{A}_{\mathrm{c}}=\begin{bmatrix}\boldsymbol{0} & \boldsymbol{I}\\ -\boldsymbol{M}^{-1}\boldsymbol{K} & -\boldsymbol{M}^{-1}\boldsymbol{C}\end{bmatrix};\quad \boldsymbol{B}_{\mathrm{c}}=\begin{bmatrix}\boldsymbol{0}\\ -\boldsymbol{1}\end{bmatrix};\quad \boldsymbol{D}_{\mathrm{c}}=\begin{bmatrix}\boldsymbol{0}\\ \boldsymbol{M}^{-1}\boldsymbol{H}_{\mathrm{c}}\end{bmatrix} \tag{27}$$

Since it is almost impossible to install control devices and sensors at all the possible locations of a building structure, especially for large building structures, it is necessary to find an optimal or suboptimal way to install limited control devices and sensors at their appropriate locations to achieve best/better performance in vibration control and/or health monitoring.

Since the unmeasured structural responses can be reconstructed from the limited structural response measurements, these reconstructed responses could also be employed as feedbacks for vibration control so as to improve the applicability of the control system for building structures. In this regard, the minimal number and optimal placement of sensors are determined in this section for a controlled structure with the objective that the reconstructed structural responses can be used as feedbacks for vibration control while the predetermined control performance is maintained. It is noted that although the increment-based approach is capable of optimally locating active/passive control device, only active control devices are considered in this section because the feedbacks are not required for passive control devices.

By minimizing the linear quadratic performance index, the optimal control force vector can be expressed as,

$$\boldsymbol{U}=-\boldsymbol{G}\boldsymbol{X} \tag{28}$$

where $\boldsymbol{G}$ is the control gain and expressed by Equation (6). The substitution of Equation (28) into Equation (25) yields

$$\dot{\boldsymbol{X}}=[\boldsymbol{A}_{\mathrm{c}}-\boldsymbol{D}_{\mathrm{c}}\boldsymbol{G}]\boldsymbol{X}+\boldsymbol{B}_{\mathrm{c}}\ddot{\boldsymbol{x}}_{\mathrm{g}}=\boldsymbol{A}_{1}\boldsymbol{X}+\boldsymbol{B}_{\mathrm{c}}\ddot{\boldsymbol{x}}_{\mathrm{g}} \tag{29}$$

In practice, the responses to be measured as feedbacks for vibration control of a building structure under earthquake excitation are often the absolute acceleration responses which can be directly measured by accelerometers. The corresponding observation equation of the controlled building structure can then be written as

$$\boldsymbol{Y}=\boldsymbol{C}_{\mathrm{c}}\boldsymbol{X}+\boldsymbol{F}_{\mathrm{c}}\boldsymbol{U} \tag{30}$$

where Y denotes the measured absolute acceleration responses of the building structure. $\boldsymbol{C}_{\mathrm{c}}$ and $\boldsymbol{F}_{\mathrm{c}}$ can be found as follows:

$$\boldsymbol{C}_{\mathrm{c}}=[-\boldsymbol{M}^{-1}\boldsymbol{K}\quad -\boldsymbol{M}^{-1}\boldsymbol{C}];\quad \boldsymbol{F}_{\mathrm{c}}=[\boldsymbol{M}^{-1}\boldsymbol{H}_{\mathrm{c}}] \tag{31}$$

By using Equation (28), Equation (30) can be rewritten as

$$\boldsymbol{Y}=[\boldsymbol{C}_{\mathrm{c}}-\boldsymbol{F}_{\mathrm{c}}\boldsymbol{G}]\boldsymbol{X}=\boldsymbol{C}_{1}\boldsymbol{X} \tag{32}$$

In reality, the measured responses are discretely sampled with a time interval of Δt. Moreover, the measurement noise and process noise always exist. Consequently, the state-space Equation (29) and the observation Equation (30) shall be converted to the following discrete forms,

$$\boldsymbol{X}_{k+1}=\boldsymbol{A}_{2}\boldsymbol{X}_{k}+\boldsymbol{B}_{\mathrm{d}}\ddot{\boldsymbol{x}}_{\mathrm{g},k}+\boldsymbol{w}_{k} \tag{33}$$

$$\boldsymbol{Y}_k = \boldsymbol{C}_1 \boldsymbol{X}_k + \boldsymbol{v}_k \tag{34}$$

where $\boldsymbol{X}_{k+1}$ is the discrete state vector; $\boldsymbol{w}_k$ and $\boldsymbol{v}_k$ are the process noise and measurement noise, respectively, which are assumed as zero-mean white noise processes with variance matrices equal to $\boldsymbol{Q}_1$ and $\boldsymbol{R}_1$ respectively; $\boldsymbol{C}_1$ is defined in Equation (32); $\boldsymbol{A}_2$ denotes the discrete-state control matrix; and $\boldsymbol{B}_\mathrm{d}$ denotes the discrete-state input matrix. They can be expressed as follows:

$$\boldsymbol{X}_k = \boldsymbol{X}(k \cdot \Delta t) \quad (k = 1, 2, 3, \cdots) \tag{35}$$

$$\boldsymbol{A}_2 = e^{\boldsymbol{A}_1 \Delta t}; \qquad \boldsymbol{B}_\mathrm{d} = \int_0^{\Delta t} e^{\boldsymbol{A}_1 t} \mathrm{d}t \cdot \boldsymbol{B}_\mathrm{c} \tag{36}$$

in which $\boldsymbol{A}_1$ and $\boldsymbol{B}_\mathrm{c}$ are defined in Equation (29) and Equation (27), respectively.

The Kalman filter provides an unbiased and recursive algorithm to optimally estimate the unknown state vector. It is employed here for the optimal placement of sensors as well as the response reconstruction. For the active control of a building structure, the Kalman filter algorithm involves the two sets of equations. The first set of equations is the time update equations:

$$\widehat{\boldsymbol{X}}_{k+1|k} = \boldsymbol{A}_2 \widehat{\boldsymbol{X}}_k + \boldsymbol{B}_\mathrm{d} \ddot{\boldsymbol{x}}_{\mathrm{g},k} \tag{37}$$

$$\boldsymbol{P}_{k+1|k} = \boldsymbol{A}_2 \boldsymbol{P}_k \boldsymbol{A}_2^\mathrm{T} + \boldsymbol{Q}_1 \tag{38}$$

in which $\widehat{\boldsymbol{X}}_{k+1|k}$ and $\boldsymbol{P}_{k+1|k}$ denote a priori state estimate and a priori error covariance matrix, respectively. The second set of equations is the measurement update equations:

$$\widehat{\boldsymbol{X}}_{k+1|k+1} = \widehat{\boldsymbol{X}}_{k+1|k} + \boldsymbol{K}_{k+1}^{\mathrm{KF}} \{\boldsymbol{Y}_{k+1} - \boldsymbol{C}_1 \widehat{\boldsymbol{X}}_{k+1|k}\} \tag{39}$$

$$\boldsymbol{P}_{k+1|k+1} = [\boldsymbol{I} - \boldsymbol{K}_{k+1}^{\mathrm{KF}} \boldsymbol{C}_1] \cdot \boldsymbol{P}_{k+1|k} \tag{40}$$

$$\boldsymbol{K}_{k+1}^{\mathrm{KF}} = \boldsymbol{P}_{k+1|k} \boldsymbol{C}_1^\mathrm{T} \cdot [\boldsymbol{C}_1 \boldsymbol{P}_{k+1|k} \boldsymbol{C}_1^\mathrm{T} + \boldsymbol{R}_1]^{-1} \tag{41}$$

in which $\widehat{\boldsymbol{X}}_{k+1|k+1}$ and $\boldsymbol{P}_{k+1|k+1}$ are the posterior state estimate and the posterior error covariance matrix, respectively; and $\boldsymbol{K}_{k+1}^{\mathrm{KF}}$ is the optimal Kalman gain matrix.

However, it will be computationally prohibited to directly apply the aforementioned Kalman filter algorithm to building structures with a large number of DOFs involved as described by Equation (24) or (25) . In consideration that under the earthquake excitation the structural responses are mainly denominated by the first several modes of vibration of the structure and the contribution of the modes of vibration with high frequencies could be ignored, the mode superposition method can be employed to lift computation prohibition. Letting $\boldsymbol{x} = \boldsymbol{\Phi}_\mathrm{s} \cdot \boldsymbol{q}_\mathrm{s}$, Equation (25) in the state space can then be expressed by

$$\dot{\boldsymbol{Z}} = \boldsymbol{A}'_\mathrm{c} \boldsymbol{Z} + \boldsymbol{B}'_\mathrm{c} \ddot{\boldsymbol{x}}_\mathrm{g} + \boldsymbol{D}'_\mathrm{c} \boldsymbol{U} \tag{42}$$

in which

$$\boldsymbol{Z} = [\boldsymbol{q}_\mathrm{s} \quad \dot{\boldsymbol{q}}_\mathrm{s}]^\mathrm{T}; \quad \boldsymbol{q}_\mathrm{s} = [q_1 \quad q_2 \quad \cdots \quad q_\mathrm{s}]^\mathrm{T} \tag{43}$$

$$A'_c = \begin{bmatrix} 0 & I \\ -\omega_s^2 & -2\xi_s\omega_s \end{bmatrix}; \quad B'_c = \begin{bmatrix} 0 \\ -\Phi_s^T MI \end{bmatrix}; \quad D'_c = \begin{bmatrix} 0 \\ \Phi_s^T H_c \end{bmatrix} \tag{44}$$

where q_s is the vector of modal coordinates; subscript s denotes the number of the selected modes of vibration; Φ_s is the selected mass-normalized displacement mode shape matrix; ξ_s and ω_s are the modal damping ratio matrix and modal frequency matrix with respect to the selected modes of vibration, respectively. In the modal domain, the control force shown in Equation (28) could be rearranged as

$$U = -GX = -G\begin{bmatrix} \Phi_s & 0 \\ 0 & \Phi_s \end{bmatrix} Z \tag{45}$$

Consequently, Equation (11. 20) can be rewritten as

$$\dot{Z} = \left[A'_c - D'_c G \begin{bmatrix} \Phi_s & 0 \\ 0 & \Phi_s \end{bmatrix} \right] Z + B'_c \ddot{x}_g = A'_1 Z + B'_c \ddot{x}_g \tag{46}$$

As mentioned before, the measured responses of the building structure are the absolute accelerations and Equation (42) can then be rewritten as

$$Y = C_1 \begin{bmatrix} \Phi_s & 0 \\ 0 & \Phi_s \end{bmatrix} Z = C'_1 Z \tag{47}$$

To implement the Kalman filter algorithm in the modal domain, the matrices A_1, B_c, and C_1 in Equations (33) - (41) should be respectively substituted by A'_1, B'_c, and C'_1 in Equations (46)、(47) . Furthermore, by comparing Equation (29) with Equation (46), one can find that the dimension of the state vectors in the modal domain is significantly reduced. Since the higher-frequency modes of vibration, which may be falsely excited by noise, are truncated in this procedure, it is not only computationally economic but also likely to improve the estimation accuracy to some extent. From this point of view, the modal domain provides a promising way for the use of the Kalman filter algorithm in the vibration control of large civil structures.

In this study, the observation equation is used not only to represent the measured responses but also to reconstruct structural responses at unmeasured key locations. Therefore, three types of structural responses in the discrete forms are introduced according to Equation (47) as

$$Y_{e,k} = C'_e Z_k; \quad \hat{Y}_{e,k} = C'_e \hat{Z}_k; \quad Y_{m,k} = C'_m Z_k \tag{48}$$

where Y_e, $\hat{Y}_e$ and Y_m represent the real structural responses at the interested locations, the reconstructed structural responses at the interested locations and the measured structural responses from the sensors, respectively; the matrix C'_e depends on the locations where the responses are of interest; and C'_m depends on the limited number of sensors for measurements.

The accuracy of the reconstructed responses can be measured by the reconstruction error δ_k.

$$\boldsymbol{\delta}_k = \hat{\boldsymbol{Y}}_{e,k} - \boldsymbol{Y}_{e,k} = \boldsymbol{C}'_e(\hat{\boldsymbol{Z}}_k - \boldsymbol{Z}_k) \tag{49}$$

Therefore, the asymptotic covariance matrix of the reconstruction error can be expressed as

$$\Delta = \text{cov}(\boldsymbol{\delta}) = \boldsymbol{C}'_e \boldsymbol{P} \boldsymbol{C}'^{T}_e \tag{50}$$

Notably, the output influence matrices $\boldsymbol{C}'_e$ and $\boldsymbol{C}'_m$ probably tend to be ill-conditioned or badly scaled especially when only few responses are measured because the absolute acceleration responses of the building structure at different locations may have different orders of magnitude. Without appropriate pre-treatment of the matrix, the inverse operation for the determination of optimal Kalman gain matrix may lead to inaccurate results. Consequently, the standard deviation of the corresponding sensor noise is employed to normalize the matrices $\boldsymbol{C}'_e$ and $\boldsymbol{C}'_m$ as

$$\bar{\boldsymbol{C}}'_e = \boldsymbol{R}_e^{-1/2} \boldsymbol{C}'_e; \quad \bar{\boldsymbol{C}}'_m = \boldsymbol{R}_m^{-1/2} \boldsymbol{C}'_m \tag{51}$$

where $\boldsymbol{R}_e$ and $\boldsymbol{R}_m$ are the signal noise matrices with different dimensions. For example, if the measurements are absolute acceleration responses, $\boldsymbol{R}_m$ can be expressed as

$$\boldsymbol{R}_m = E(\boldsymbol{v}\boldsymbol{v}^T) = \boldsymbol{\sigma}_a^2 \boldsymbol{I} \tag{52}$$

in which $\boldsymbol{\sigma}_a^2$ is the measurement noise variance matrix of acceleration responses. Hence, the reconstruction error and the corresponding covariance matrix in Equation (49) and Equation (50) should be normalized accordingly in consideration of the un-bias estimation.

$$\bar{\boldsymbol{\delta}}_k = \bar{\boldsymbol{C}}'_e(\hat{\boldsymbol{Z}}_k - \boldsymbol{Z}_k) \tag{53}$$

$$\bar{\boldsymbol{\Delta}} = \text{cov}(\bar{\boldsymbol{\delta}}) = \bar{\boldsymbol{C}}'_e \boldsymbol{P} \bar{\boldsymbol{C}}'^{T}_e \tag{54}$$

Moreover, since the normalized output influence matrix is used, the optimal Kalman gain shown in Equation (41) should be updated and given by

$$\boldsymbol{K}_{k+1}^{KF} = \boldsymbol{P}_{k+1|k} \bar{\boldsymbol{C}}'^{T}_m \cdot [\bar{\boldsymbol{C}}'_m \boldsymbol{P}_{k+1|k} \bar{\boldsymbol{C}}'^{T}_m + \boldsymbol{I}]^{-1} \boldsymbol{R}_m^{-1/2} \tag{55}$$

It can also be seen that each diagonal element of the $\bar{\boldsymbol{\Delta}}$ matrix in Equation (54) represents the normalized variance of the reconstruction error for the corresponding response. Therefore, the maximum diagonal element denotes the maximum reconstruction error, whereas the trace of the matrix $\bar{\boldsymbol{\Delta}}$ represents the sum of the reconstruction errors at all the locations of interest. From this point of view, the optimal sensor placement can be performed with the objective to minimize the sum of the normalized reconstruction error.

Object function: $$\min \boldsymbol{tr}(\bar{\boldsymbol{\Delta}}) \tag{56}$$

subject to $$\bar{\boldsymbol{\sigma}}_{max} \leqslant \boldsymbol{\sigma}_{max} \tag{57}$$

in which $\bar{\boldsymbol{\sigma}}_{max}$ is the maximum estimation error and defined as

$$\bar{\boldsymbol{\sigma}}_{\max} = \max(\operatorname{diag}(\bar{\boldsymbol{\Delta}})) \tag{58}$$

and $\boldsymbol{\sigma}_{\max}$ is the preset allowable error. It is understood that the maximum value of reconstruction error as well as the trace of the matrix $\bar{\boldsymbol{\Delta}}$ would be increased when the number of sensors is reduced. A simple iterative procedure can then be conducted, in which the candidate sensors are removed one by one until the target error level is reached. In each step, only one sensor location, the removal of which leads to a minimal trace of the matrix $\bar{\boldsymbol{\Delta}}$, will be deleted. Thus, the sensor with minimal contribution on the response reconstruction will be removed at each step, and this procedure for sensor location is thus suboptimal. Nevertheless, this suboptimal procedure is beneficial and applicable for large and complex civil structures, for the dimension of the state vectors in the modal domain is significantly reduced when only first several modes of vibration are used.

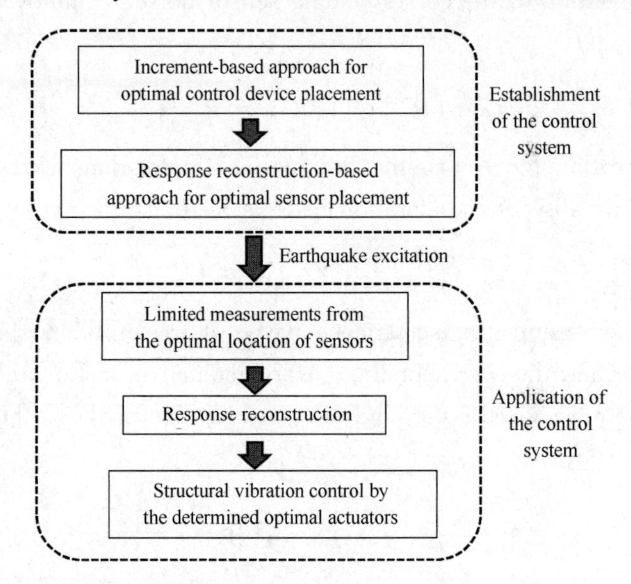

Figure 1 The schematic diagram for the establishment and application of the proposed control system

It could be seen that the controlled system with optimal locations of both actuators and sensors can be established according to the approaches presented in this section, and the schematic diagram is plotted in Figure 1 to show the establishment and application of the proposed control system. The number of the control devices and sensors in this control system would be rather small. The feasibility and accuracy of the proposed method is investigated in the following subsection by using a 20-story shear building under earthquake excitation as a numerical example.

2.3 Case Study

A 20-story shear building as shown in Figure 2 is employed to investigate the feasibility and accuracy of the presented method. The mass and stiffness coefficients of the 20-story shear building are listed in Table 1. The Rayleigh damping assumption with the proportional coefficient of 0.8 for the mass matrix and 1×10^{-5} for the stiffness matrix is used to construct the damping matrix. The increment of performance index for vibration control of the building is first computed for the optimal control device placement. The eigenvalue analysis is then performed to extract the mode shapes for the optimal sensor placement and response reconstruction. The El-Centro ground excitation with a peak acceleration of 0.34g and the Kobe ground excitation with a peak acceleration of 0.81g are finally selected as the input to assess the control performance of the building equipped with the selected control system. The comparison of the control performance subject to two more earthquakes, i.e. Northridge earthquake and Hachinohe earthquake, is also conducted. It should be noted the building model is linear, and the nonlinear effect on the structural dynamics is not considered and analyzed in this study.

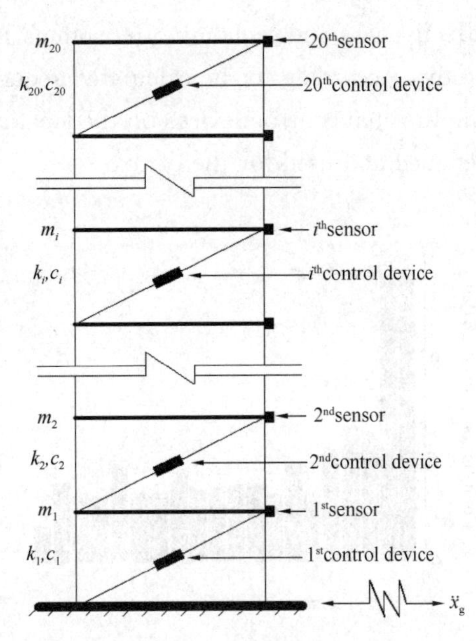

Figure 2 Structural model of a tall building with control devices and sensors

Structural parameters of the 20-story shear building **Table 1**

Floor number	Mass (kg)	Stiffness (N/m)	Floor number	Mass (kg)	Stiffness (N/m)
1~4	8000	9.2×10^7	13~16	8000	7.5×10^7
5~8	8000	8.8×10^7	17~20	8000	7.0×10^7
9~12	8000	8.0×10^7	—	—	—

2.3.1 Determination of optimal control devices

For the determination of the optimal locations of control devices and sensors, a ground excitation composed of white noise random signals is applied to the shear building. The control devices are initially installed on each floor together with the braces. By removing the control devices one by one, the increment of performance index is calculated and the contribution percentage of every control device can be then obtained from Equation (23) and shown in Figure 3. It can be easily found from Figure 3 that the sequence of the control devices' locations is 1, 2, 4, 3, 5, 7, 16, 13, 19, 9, 10, 11, 12, 15, 18, 14, 8, 6, 17, and 20. Moreover, the relationship between the summation of *CP* value and the displacement reduction of the top floor is given in Figure 4. It can be seen that with the increase of the summation of *CP* value, the structural responses are reduced accordingly. The dash line indicates the response reduction when twenty control devices are all installed. It can also be found from Figure 4 that when the summation reaches 70%, the trend for vibration attenuation is becoming slow. Therefore, for the consideration of both effectiveness and economy, the summation of the CP value is assumed to be 70% in this study. According to the calculated sequence and the desired summation of the CP value, the first ten control devices on the locations of 1, 2, 4, 3, 5, 7, 16, 13, 19, and 9 shall be retained and used for the control.

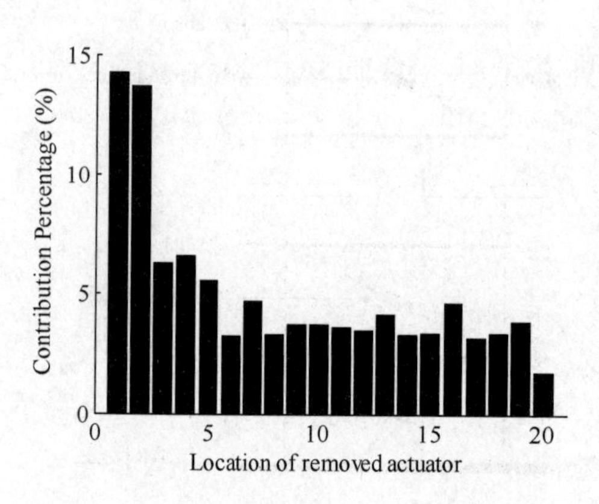

Figure 3 Optimal sequence of control devices

Figure 4 the relationship between the summation of CP value and displacement reduction

It is known that in practical situations the structural parameter uncertainties resulting from modeling errors often exist. Thus, the influence of such uncertainties on the number and location of the control devices is discussed here. Since it is relatively time-consuming for the consideration of the uncertainties existing in all the structural parameters, only two structural parameters with uncertainties (e. g. k_2 and k_6) are considered for the demonstration purpose. The values of k_2 and k_6 are approximated as a normal distribution with a mean value of 9.2×10^7 and 8.8×10^7, respectively, and the standard deviation of 5% of the corresponding mean value. Although some variations exist in the calculated CP values, for example in some cases CP_{16} becomes larger than CP_7, the aforementioned ten control devices are still retained when the two random parameters are involved. It can thus be concluded, to some extent, from these results that the increment-based approach is not much sensitive to the structural parameter variations.

Furthermore, two integer heuristic programming methods, i. e. the sequential search algorithm (SSA)[22] and the Worst-Out-Best-In (WOBI) algorithm[23], are employed for determining the optimal control device location and comparing with the results obtained from the proposed method. In these two algorithms, ten optimal locations out of twenty possible locations are selected for the placement of the control devices. In this regard, the two performance indices of the building structure are considered. The first one is the maximum peak acceleration at the top floor described by

$$PI_1 = \max\{|\ddot{x}_{top}|\} \tag{59}$$

The second one is the maximum peak inter-story drift as follows:

$$PI_2 = \max\{|\bar{x}_{pi}|\} \quad (i = 1, 2, \cdots, 20) \tag{60}$$

in which $|\ddot{x}_{top}|$ and $|\bar{x}_{pi}|$ denotes the absolute value of peak acceleration at the top floor and the absolute value of peak inter-story drift of the i-th story, respectively. The objective function to find the optimal location of the control devices is to minimize the maximum peak acceleration response at the top floor (PI_1) and the maximum peak inter-story drift (PI_2), respectively.

The control device locations determined by the SSA and WOBI methods are shown in Table 2. The result obtained from the increment-based approach is also listed in Table 2. For the sake of easy comparison, the locations are listed in the ascending order. Although there are several different control device locations, the results obtained by those three algorithms are close to each other. It can also be seen from Table 2 that even for the identical algorithm, the determined locations are not the same if different objective functions are used. It should be noted that for r control devices to be placed in n possible locations, there would be $nr-[r(r-1)/2]$ configurations for SSA and n evaluations of the objective function in each iteration for WOBI. This means that 155 combinations are required to be considered for SSA and 20 location strategies are required to be evaluated in each iteration for WOBI. Moreover, if all of the possible locations are considered, it would be more time-consuming. For example, in our case, i. e. 10 control devices being placed in 20 possible locations, the total number of possible combinations of the control device locations is $n!/[r!(n-r)!]=20!/(10!\times10!)=184,756$ [24].

The comparison of the control device locations determined by three algorithms Table 2

Algorithm	Objective function	Control device locations
SSA	*Minimize*(PI_1)	1, 2, 4, 5, 7, 9, 10, 13, 15, 19
	Minimize (PI_2)	1, 3, 5, 7, 9, 13, 14, 16, 18, 20
WOBI	*Minimize*(PI_1)	2, 3, 5, 6, 9, 10, 13, 16, 19, 20
	Minimize(PI_2)	1, 3, 5, 7, 10, 13, 15, 17, 19, 20
Increment-based approach	*Maximize* (CP)	1, 2, 3, 4, 5, 7, 9, 13, 16, 19

2.3.2 Determination of optimal sensors

To determine the optimal placement of the sensors, the approach introduced in Section 2.2 is employed. The first five modes of vibration with the corresponding natural frequencies of 1.26Hz, 3.65Hz, 6.05Hz, 8.43Hz and 10.74Hz are employed. Twenty accelerometers, which are used to measure the acceleration response of each floor, are used as initial candidate locations for sensors. For practical consideration, all the measured structural responses are simulated by the numerically computed structural responses superimposed with a white noise of a 5% noise-to-signal ratio in terms of the root mean square (RMS). The predefined threshold value of the maximum estimation error is applied to determine the number of sensors, and it is defined as the ratio of standard deviation of reconstruction error variance to that of noise, which is used to quantify the estimation accuracy. A smaller value of the ratio corresponds to higher estimation accuracy and represents a more stringent criterion on the estimation error, and of course more sensors are required. It should also be noted if the value of the ratio is too large, the allowable reconstruction error would be too large which may result in the reconstructed responses being probably incorrect or contaminated too much by the noise. In this study, the target maximum estimation error $\boldsymbol{\sigma}_{max}$ in Equation (58) is set to be 3.0, which means the standard deviation of reconstruction error variance is three times to that of noise. By following the procedure for the determination of optimal sensor placement, four accelerometers which are respectively located on the 3rd floor, 12th floor, 16th floor, and 20th floor are finally selected. It can be found the number of the sensors is rather reduced as compared with the initial candidate set. Since the locations of the control devices and sensors both are determined, the optimal control system for this shear building is established, which includes ten control devices and four accelerometers.

As mentioned before, the standard deviation of the sensor noise is employed to normalize the reconstruction error and the corresponding covariance matrix. It is thus anticipated that the number and location of the selected sensors will be altered if the measurement noise covariance matrix $\boldsymbol{R}_1$ is changed. However, in many cases, the measurement noise covariance is often evaluated prior to the actual operation of the Kalman filter. Thus, for the same sensors under the same conditions, the variation of the measurement noise covariance matrix $\boldsymbol{R}_1$ is small and the influence of $\boldsymbol{R}_1$ matrix can be controlled in a reasonable range. $\boldsymbol{Q}_1$ matrix defined in Equation (33) is the process noise covariance matrix which is mainly used for the consideration of the modeling errors. The structural parameter uncertainties mentioned before is used for investigating the effect of the modeling errors on the optimal sensor placement. Likewise, the values of k_2 and k_6 are approximated as a normal distribution with a mean value of 9.2×10^7 and 8.8×10^7, respectively, and the standard deviation of 5% of the corresponding mean value. With the consideration of these parameter uncertainties, the determined optimal location of the sensors was found to remain unchanged. It should be noted that one basic premise of the presented response reconstruction-based technique is that the structural model is relatively accurate and updated using the appropriate model updating technique. From this point of view, although the process noise covariance $\boldsymbol{Q}_1$ matrix is generally difficult to be estimated, the influence of $\boldsymbol{Q}_1$ matrix on the number and location of sensors should be relatively small. Some statements can also be found in Zhang [25].

2.3.3 Investigation of control performance and response-reconstruction accuracy

To show the efficiency of the control system with limited control devices and sensors, the El-Centro ground excitation is applied to the shear building with and without the control system. Only the acceleration responses at the 3rd floor, 12th floor, 16th floor and 20th floor of the building are assumed to be measured. The measured responses are then contaminated by 5% white noise. It is noted that although only four acceleration responses are measured, the remaining structural responses can be reconstructed and used for vibration control. Four cases are considered: Case 1: the control devices and sensors are installed on each floor of the shear building for vibration control; Case 2: the control devices are installed on the determined optimal position (the aforementioned ten locations) but the sensors are installed on each floor of the building for feedbacks; Case 3: the control devices and sensors both are installed at the determined positions (the aforementioned ten and four positions respectively) and the reconstructed responses are used as feedbacks for vibration control; and Case 4: no control devices and sensors are installed in the building (without control) .

The time histories of acceleration response of the building at the top floor are computed and plotted in Figure 5 for the four cases. For the sake of clarification, only the time segment of the acceleration responses from 8 s to 16 s are given in Figure 5 although the time duration of the displacement response is from 0 to 30 s. On one hand, it can be seen that the structural responses of the uncontrolled building are significantly reduced with the control system. On the other hand, the control performance of the optimal control system, defined as Case 3, is close to that of the fully controlled building defined as Case 1. Moreover, it can also be seen that the control performance in Case 3 is in good agreement with that in Case 2, which indicates that the reconstructed responses can be employed for vibration control with acceptable accuracy. Though only the displacement and acceleration responses of the building at the top floor are plotted in Figure 5, similar results for the remaining building floors can be obtained as well. To have a more comprehensive comparison of the control performance, the maximum displacement and acceleration responses of the building at each floor are depicted in Figure 6 for the four cases. It can be seen that the maximum responses are significantly reduced when the control devices are employed. It can also be seen that the results of Case 2 and Case 3 are close to each other, implying that the utilization of the reconstructed responses for control is reliable. Though the maximum displacement and acceleration responses of the building at several upper floors from Case 3 are relatively larger than those from Case 1, the control performance of the optimal control

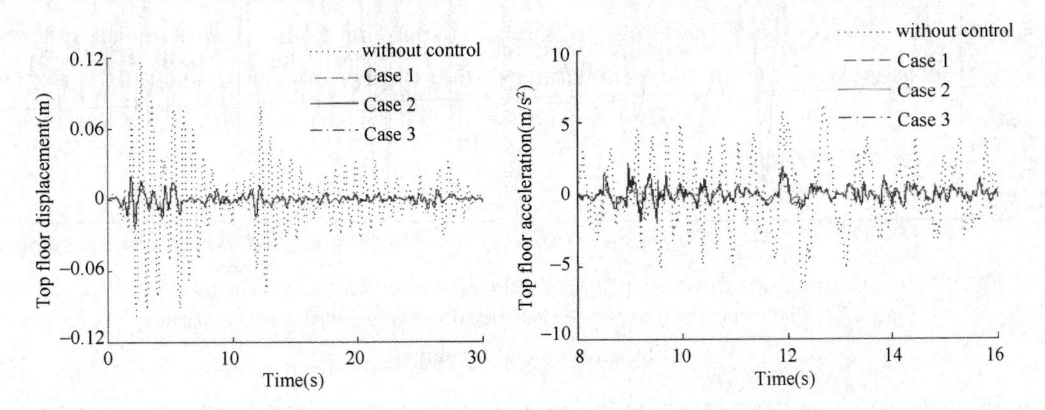

Figure 5 Time histories of displacement and acceleration responses at the top floor (El-Centro ground excitation)

system with only ten control devices and four sensors is still acceptable with consideration of both control-effectiveness and cost-effectiveness.

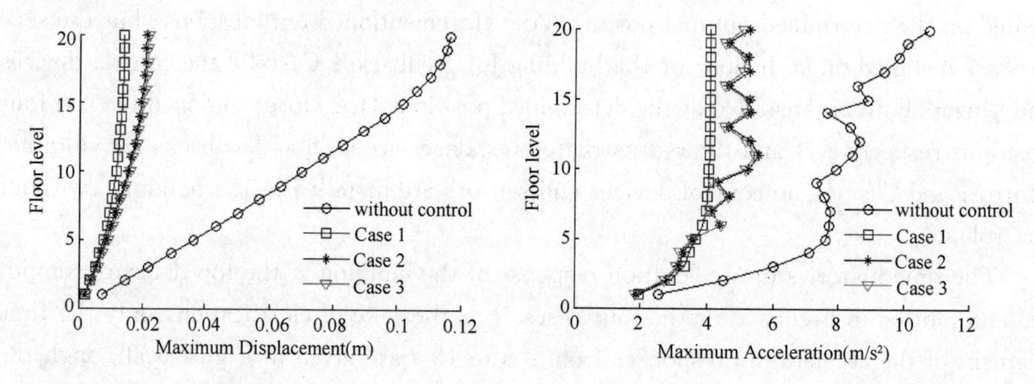

Figure 6 The comparison of the maximum displacement and acceleration responses

(El-Centro ground excitation)

Moreover, Figure 7 gives the time histories of the reconstructed displacement and acceleration responses (dashed lines) and the corresponding actual ones (solid lines). Only the displacement and acceleration responses of the building at the top floor are shown in Figure 7 as an example, and only the time segments from 8 s to 16 s are demonstrated for clarification. It is clear that the reconstructed responses are in good agreement with the corresponding actual responses, confirming that the reconstructed responses could be reliably employed as feedbacks for vibration control. Moreover, since the reconstructed responses are very close to the actual ones, it could be used for the purpose of system identification and seismic performance assessment as well. More details of the system identification and seismic performance response based on the response reconstruction technique will be given in Section 3 and Section 4 respectively.

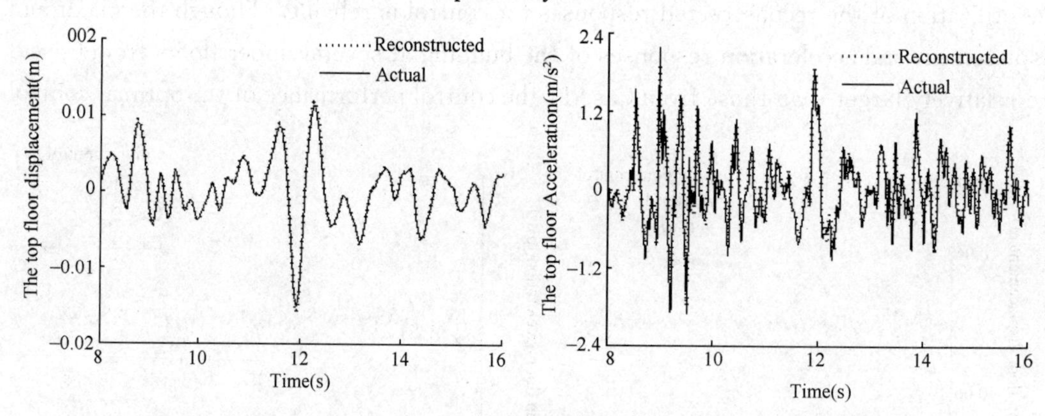

Figure 7 The comparison of the reconstructed responses with the actual ones

(El-Centro ground excitation)

The control performances of the control devices determined by SSA and WOBI as shown in Table 2 are also computed and compared with those from the increment-based approach. The maximum displacement and acceleration responses of the building at each floor are shown in Figure 8. The cases of the SSA with the objective function of minimizing PI_1 and PI_2 are respectively denoted as SSA1 and SSA2 in Figure 8. Similar definition of WOBI1 and WOBI2 can also be found in Figure 8. It can be seen that although several control device locations are different according to these algorithms, the control performance is still close to each other. It should also be noted that the complete structural responses, such as the displacement and velocity responses at all floor are assumed to be known for control in SSA and WOBI algorithm whereas only four accelerometers are required in the proposed control system. Moreover, basing on the peak value of the control force, one more performance index is considered and shown as follows [26],

$$PI_3 = \left\{ \frac{\max |\{U(t)\}|}{W} \right\} \tag{61}$$

where $U(t)$ is the control force; and W is the total weight of the building. The values of PI_3 for the aforementioned cases of the full control (i. e. the control devices being installed on each floor of the building), increment-based approach, SSA1, SSA2, WOBI1, and WOBI2 are 0.018, 0.042, 0.044, 0.033, 0.037, and 0.032, respectively. The total energy consumed in the full control, increment-based approach, SSA1, SSA2, WOBI1, and WOBI2 are 1.313×10^4, 1.288×10^4, 1.291×10^4, 1.297×10^4, 1.296×10^4, 1.295×10^4 kN · m, respectively. By comparing with the case of the full control, it can be found that with the reduction of the number of control devices, the control forces of the retained control devices are increased in order to achieve an acceptable control performance. However, from the viewpoint of the energy consumption, with the identical input energy (e. g. the identical earthquakes applied to the system), the larger the system vibration reduction is achieved, the more the energy is consumed by the control devices. Therefore, although the control force for each control device in the case of full control is the smallest, it can be seen that with twenty control devices in this case the total energy consumption is the largest.

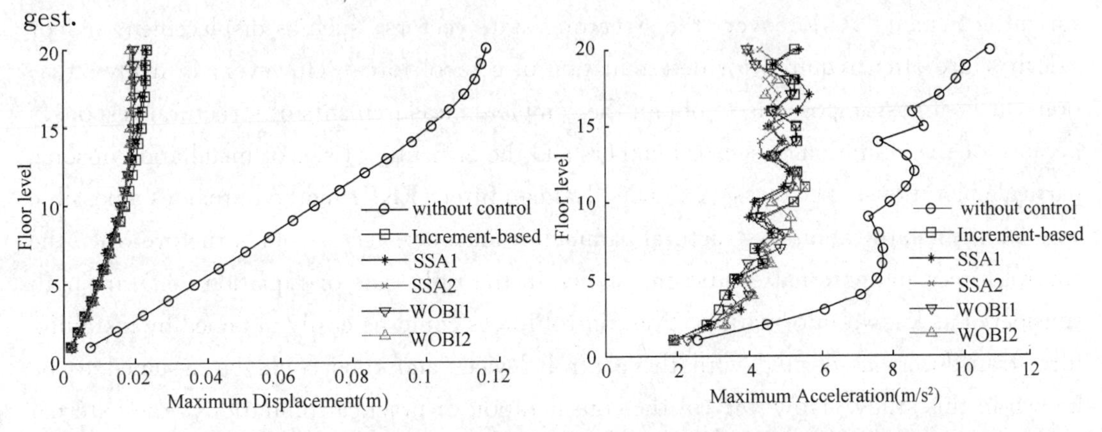

Figure 8 The comparison of the control performance with different algorithms (El-Centro ground excitation)

3. Synthesis of SVC and SHM Systems

This section first presents an integrated health monitoring and vibration control system in the time domain in terms of the projection matrix and the extended Kalman filter (EKF) for simultaneous consideration of vibration mitigation and time-invariant parameter identification of building structures without the knowledge of the external excitations[12]. The efficiency and accuracy of the proposed approach is then experimentally validated via a five-story building structure equipped with magneto-rheological (MR) dampers.

3.1 Equation of Motion of Integrated System

The second-order differential equation of motion of the controlled building structure with n degrees-of-freedom (DOFs) can be given as follows:

$$\boldsymbol{M}\ddot{\boldsymbol{x}}(t)+\boldsymbol{C}\dot{\boldsymbol{x}}(t)+\boldsymbol{K}\boldsymbol{x}(t)=\boldsymbol{\varphi}^{*}\boldsymbol{f}^{*}(t)+\boldsymbol{\varphi}\boldsymbol{f}(t) \tag{62}$$

in which $\boldsymbol{M}$, $\boldsymbol{C}$, and $\boldsymbol{K}$ represent the mass, damping and stiffness matrices of the building structure, respectively; $\ddot{\boldsymbol{x}}(t)$, $\dot{\boldsymbol{x}}(t)$, and $\boldsymbol{x}(t)$ denote structural acceleration, velocity and displacement response vectors, respectively; $\boldsymbol{f}^{*}(t)$ is the control force vector; $\boldsymbol{f}(t)$ is the external excitation vector; $\boldsymbol{\varphi}^{*}$ and $\boldsymbol{\varphi}$ denote the influence matrices associated with $\boldsymbol{f}^{*}(t)$ and $\boldsymbol{f}(t)$, respectively.

Although vibration control system is used for mitigating structural vibration and trying to keep the structure in a healthy state, it is still inevitable that damage may occur to some structural members under severe earthquake. If damage does occur to some structural members, the control forces should be adjusted accordingly because the structural parameters of the damaged structural members shall be employed for the determination of control force so as to maintain control efficiency in active, semi-active or hybrid control. The structural parameters of the damaged structural members are assumed to be time-invariant (constant) . Moreover, the structural state vectors, such as displacement and/or velocity, are often required for determination of control force. However, in many situations, it is almost impossible to obtain the complete measurements of structural responses because of the limitation of sensor numbers and the difficulty of sensor installation in some particular locations. The classic extended Kalman filter (EKF) method provides a possible way for the identification of structural parameters and state vectors of a structure when the control forces and external excitations shown in the right side of Equation (62) are both considered as known information. The control forces could be easily obtained by installing force transducers associated with the control devices and thus could be assumed to be known in this study. However, in the consideration of practical limitations, the external excitations applied in the controlled building structure, such as earthquakes, are assumed to be unknown in this section. Since the external excitations are not available, the classic

EKF method cannot be directly employed. An integrated vibration control and health monitoring approach, based on the projection matrix and the EKF method, is proposed in this section for constant parameter and excitation identification as well as vibration mitigation. The formula of the proposed approach is presented in the following subsections.

3.2 Identification of Structural Parameters and Excitations

Consider an extended state vector consisting of structural displacement, velocity, and structural parameters to be identified,

$$\boldsymbol{Z}(t)=[\boldsymbol{x}(t)^{\mathrm{T}} \quad \dot{\boldsymbol{x}}(t)^{\mathrm{T}} \quad \boldsymbol{\theta}^{\mathrm{T}}]^{\mathrm{T}} \tag{63}$$

in which $\boldsymbol{\theta}$ is an m-unknown parametric vector composed of structural stiffness and damping coefficients. As the unknown parameters are assumed to be constant during the event, i. e. $\dot{\boldsymbol{\theta}}=\mathbf{0}$, one can obtain the following nonlinear state equation,

$$\frac{\mathrm{d}\boldsymbol{Z}(t)}{\mathrm{d}t}=\begin{bmatrix}\dot{\boldsymbol{x}}(t)\\ \boldsymbol{M}^{-1}(-\boldsymbol{C}\dot{\boldsymbol{x}}(t)-\boldsymbol{K}\boldsymbol{x}(t)+\boldsymbol{\varphi}^{*}\boldsymbol{f}^{*}(t)+\boldsymbol{\varphi}\boldsymbol{f}(t))\\ \mathbf{0}\end{bmatrix}=\boldsymbol{u}(\boldsymbol{Z}(t),\boldsymbol{f}(t),\boldsymbol{f}^{*}(t),t) \tag{64}$$

Let $\hat{\boldsymbol{Z}}_{k|k}$ and $\hat{\boldsymbol{f}}_{k|k}$ respectively be the estimates of $\boldsymbol{Z}_k$ and $\boldsymbol{f}_k$ at time $t=k\cdot\Delta t$ with Δt being the sampling interval. Equation (64) can be linearized with respect to $\hat{\boldsymbol{Z}}_{k|k}$ and $\hat{\boldsymbol{f}}_{k|k}$ as follows,

$$\boldsymbol{u}(\boldsymbol{Z}_k,\boldsymbol{f}_k,\boldsymbol{f}_k^{*},k\Delta t)=\boldsymbol{u}(\hat{\boldsymbol{Z}}_{k|k},\hat{\boldsymbol{f}}_{k|k},\boldsymbol{f}_k^{*},k\Delta t)+\boldsymbol{U}_{k|k}(\boldsymbol{Z}_k-\hat{\boldsymbol{Z}}_{k|k})+\boldsymbol{B}_{k|k}(\boldsymbol{f}_k-\hat{\boldsymbol{f}}_{k|k}) \tag{65}$$

in which

$$\boldsymbol{U}_{k|k}=\left.\frac{\partial\boldsymbol{u}(\boldsymbol{Z}_k,\boldsymbol{f}_k,\boldsymbol{f}_k^{*},k\Delta t)}{\partial\boldsymbol{Z}_k}\right|_{\boldsymbol{Z}=\hat{\boldsymbol{Z}}_{k|k},\,\boldsymbol{f}_k=\hat{\boldsymbol{f}}_{k|k}} \tag{66}$$

$$\boldsymbol{B}_{k|k}=\left.\frac{\partial\boldsymbol{u}(\boldsymbol{Z}_k,\boldsymbol{f}_k,\boldsymbol{f}_k^{*},k\Delta t)}{\partial\boldsymbol{f}_k}\right|_{\boldsymbol{Z}=\hat{\boldsymbol{Z}}_{k|k},\,\boldsymbol{f}_k=\hat{\boldsymbol{f}}_{k|k}}=\begin{bmatrix}\mathbf{0}\\ \boldsymbol{M}^{-1}\boldsymbol{\varphi}\\ \mathbf{0}\end{bmatrix} \tag{67}$$

Substituting $\frac{\boldsymbol{Z}_{k+1}-\boldsymbol{Z}_k}{\Delta t}=\frac{\mathrm{d}\boldsymbol{Z}_k}{\mathrm{d}t}=\boldsymbol{u}(\boldsymbol{Z}_k,\boldsymbol{f}_k,\boldsymbol{f}_k^{*},k\Delta t)$ into the left side of Equation (65), one obtains,

$$\boldsymbol{Z}_{k+1}=\boldsymbol{Z}_k+\Delta t[\boldsymbol{u}(\hat{\boldsymbol{Z}}_{k|k},\hat{\boldsymbol{f}}_{k|k},\boldsymbol{f}_k^{*},k\Delta t)+\boldsymbol{U}_{k|k}(\boldsymbol{Z}_k-\hat{\boldsymbol{Z}}_{k|k})+\boldsymbol{B}_{k|k}(\boldsymbol{f}_k-\hat{\boldsymbol{f}}_{k|k})] \tag{68}$$

The discretized observation equation associated with Equation (63) at time $t=k\cdot\Delta t$ can be described as

$$\boldsymbol{y}_k=\boldsymbol{K}\boldsymbol{x}_k+\boldsymbol{C}\dot{\boldsymbol{x}}_k-\boldsymbol{\varphi}\boldsymbol{f}_k+\boldsymbol{v}_k=\boldsymbol{h}(\boldsymbol{Z}_k)-\boldsymbol{\varphi}\boldsymbol{f}_k+\boldsymbol{v}_k \tag{69}$$

in which $\boldsymbol{y}_k=-\boldsymbol{M}\ddot{\boldsymbol{x}}_k+\boldsymbol{\varphi}^{*}\boldsymbol{f}_k^{*}$ is the measurement vector which can be obtained from the measured structural acceleration responses $\ddot{\boldsymbol{x}}_k$ and the measured control force $\boldsymbol{f}_k^{*}$; $\boldsymbol{h}(\boldsymbol{Z}_k)$ is a combination of state vector and equal to $\boldsymbol{K}\boldsymbol{x}_k+\boldsymbol{C}\dot{\boldsymbol{x}}_k$; $\boldsymbol{v}_k$ is the measurement noise vector assumed to be a Gaussian white noise vector with zero mean and a covariance matrix $E[\boldsymbol{v}_k\boldsymbol{v}_j^{\mathrm{T}}]=\boldsymbol{R}_k\delta_{kj}$, where δ_{kj} is the Kroneker delta. It can be found from Equation (69) that there are two variable vectors $\boldsymbol{Z}_k$ and $\boldsymbol{f}_k$ to be predicted and estimated. Consequently, Equation (69) is a multiple regression equation which means the classic EKF method cannot be directly applied. To transform the multiple regression problem described by Equation (69) to a single regression problem, a straightforward way in terms of the projection matrix is proposed here.

From Equation (69), one can obtain

$$\boldsymbol{\varphi f}_k=-\boldsymbol{y}_k+\boldsymbol{h}(\boldsymbol{Z}_k)+\boldsymbol{v}_k \tag{70}$$

By assuming that the number of observed DOFs is larger than the number of excitations, the closest solution of $\boldsymbol{f}_k$ in Equation (70) can then be given by the following equation through least-square estimation (LSE),

$$\boldsymbol{f}_{k,\mathrm{LS}}=(\boldsymbol{\varphi}^{\mathrm{T}}\boldsymbol{\varphi})^{-1}\boldsymbol{\varphi}^{\mathrm{T}}[-\boldsymbol{y}_k+\boldsymbol{h}(\boldsymbol{Z}_k)+\boldsymbol{v}_k] \tag{71}$$

The error of the solution from Equation (71) is given by

$$\begin{aligned}\boldsymbol{err}&=\boldsymbol{\varphi f}_k-\boldsymbol{\varphi f}_{k,LS}\\&=[-\boldsymbol{y}_k+\boldsymbol{h}(\boldsymbol{Z}_k)+\boldsymbol{v}_k]-\boldsymbol{\varphi}(\boldsymbol{\varphi}^{\mathrm{T}}\boldsymbol{\varphi})^{-1}\boldsymbol{\varphi}^{\mathrm{T}}[-\boldsymbol{y}_k+\boldsymbol{h}(\boldsymbol{Z}_k)+\boldsymbol{v}_k]\\&=(\boldsymbol{I}_n-\boldsymbol{\varphi}(\boldsymbol{\varphi}^{\mathrm{T}}\boldsymbol{\varphi})^{-1}\boldsymbol{\varphi}^{\mathrm{T}})[-\boldsymbol{y}_k+\boldsymbol{h}(\boldsymbol{Z}_k)+\boldsymbol{v}_k]\end{aligned} \tag{72}$$

in which $\boldsymbol{I}_n$ is the $n\times n$ identity matrix; the matrix $\boldsymbol{\varphi}(\boldsymbol{\varphi}^{\mathrm{T}}\boldsymbol{\varphi})^{-1}\boldsymbol{\varphi}^{\mathrm{T}}$ is the projection matrix that projects the vector $[-\boldsymbol{y}_k+\boldsymbol{h}(\boldsymbol{Z}_k)+\boldsymbol{v}_k]$ on to the space spanned by the columns of $\boldsymbol{\varphi}$. As a limit, the error in Equation (72) tends to be zero, leading to

$$\boldsymbol{\Phi y}_k=\boldsymbol{\Phi h}(\boldsymbol{Z}_k)+\boldsymbol{\Phi v}_k \tag{73}$$

where $\boldsymbol{\Phi}=\boldsymbol{I}_n-\boldsymbol{\varphi}(\boldsymbol{\varphi}^{\mathrm{T}}\boldsymbol{\varphi})^{-1}\boldsymbol{\varphi}^{\mathrm{T}}$ for simplicity of presentation. It is noted that the projection matrix and naturally the matrix $\boldsymbol{\Phi}$ has two properties: (1) $\boldsymbol{\Phi}$ is a symmetric matrix, i. e. $\boldsymbol{\Phi}^{\mathrm{T}}=\boldsymbol{\Phi}$; and (2) $\boldsymbol{\Phi}^2=\boldsymbol{\Phi}$. It can be found from Equation (73) the multiple regression equation expressed by Equation (69) is transformed into a simple regression equation.

It is obvious from Equation (71) that if the estimate vector $\hat{\boldsymbol{Z}}_{k|k}$ of $\boldsymbol{Z}_k$ at the k-th time step is obtained, the unknown excitation could be accordingly determined as

$$\hat{\boldsymbol{f}}_{k|k}=(\boldsymbol{\varphi}^{\mathrm{T}}\boldsymbol{\varphi})^{-1}\boldsymbol{\varphi}^{\mathrm{T}}[-\boldsymbol{y}_k+\boldsymbol{h}(\hat{\boldsymbol{Z}}_{k|k})] \tag{74}$$

Let $\hat{\boldsymbol{Z}}_{k+1|k}$ be the priori estimate of state $\boldsymbol{Z}_{k+1}$, and linearize $\boldsymbol{h}(\boldsymbol{Z}_{k+1})$ with respect to $\hat{\boldsymbol{Z}}_{k+1|k}$:

$$\boldsymbol{h}(\boldsymbol{Z}_{k+1})=\boldsymbol{h}(\hat{\boldsymbol{Z}}_{k+1|k})+\boldsymbol{H}_{k+1|k}(\boldsymbol{Z}_{k+1}-\hat{\boldsymbol{Z}}_{k+1|k}) \tag{75}$$

in which

$$\boldsymbol{H}_{k+1|k}=\left.\frac{\partial\boldsymbol{h}(\boldsymbol{Z}_{k+1})}{\partial\boldsymbol{Z}_{k+1}}\right|_{\boldsymbol{Z}_{k+1}=\hat{\boldsymbol{Z}}_{k+1|k}} \tag{76}$$

The priori estimation state $\widehat{\boldsymbol{Z}}_{k+1|k}$ could be given according to the first-order Taylor expansions,

$$\widehat{\boldsymbol{Z}}_{k+1|k} = \widehat{\boldsymbol{Z}}_{k|k} + \Delta t[\boldsymbol{u}(\widehat{\boldsymbol{Z}}_{k|k}, \widehat{\boldsymbol{f}}_{k|k}, \boldsymbol{f}_k^*, k\Delta t)] \tag{77}$$

It should be noted that integrating the actual nonlinear equation, i. e. Equation (64), forward at each sampling interval is usually implemented in the EKF approach to determine $\widehat{\boldsymbol{Z}}_{k+1|k}$ for the purpose of improvement of the accuracy of the estimate. The priori estimation error $\boldsymbol{\varepsilon}_{k+1|k}$ of the unknown state vector $\boldsymbol{Z}_{k+1}$ at time $t=(k+1)\Delta t$ can then be obtained by the combination of Equation (68),

$$\boldsymbol{\varepsilon}_{k+1|k} = \boldsymbol{Z}_{k+1} - \widehat{\boldsymbol{Z}}_{k+1|k} = (\boldsymbol{I} + \Delta t\boldsymbol{U}_{k|k})(\boldsymbol{Z}_k - \widehat{\boldsymbol{Z}}_{k|k}) + \Delta t\boldsymbol{B}_{k|k}(\boldsymbol{f}_k - \widehat{\boldsymbol{f}}_{k|k}) \tag{78}$$

in which $\boldsymbol{U}_{k|k}$ and $\boldsymbol{B}_{k|k}$ are defined in Equations (67-68), respectively. By the combination of Equations (67, 71 and 74), the second term in right side of Equation (78) can be transformed as

$$\Delta t\boldsymbol{B}_{k|k}(\boldsymbol{f}_k - \widehat{\boldsymbol{f}}_{k|k}) = \Delta t\begin{bmatrix} \boldsymbol{0} \\ \boldsymbol{M}^{-1}\boldsymbol{\varphi}\Big((\boldsymbol{\varphi}^{\mathrm{T}}\boldsymbol{\varphi})^{-1}\boldsymbol{\varphi}^{\mathrm{T}}[-\boldsymbol{y}_k + \boldsymbol{h}(\boldsymbol{Z}_k) + \boldsymbol{v}_k] - (\boldsymbol{\varphi}^{\mathrm{T}}\boldsymbol{\varphi})^{-1}\boldsymbol{\varphi}^{\mathrm{T}}[-\boldsymbol{y}_k + \boldsymbol{h}(\widehat{\boldsymbol{Z}}_{k|k})]\Big) \\ \boldsymbol{0} \end{bmatrix}$$

$$= \Delta t\begin{bmatrix} \boldsymbol{0} \\ \boldsymbol{M}^{-1}(\boldsymbol{I} - \boldsymbol{\Phi})[\boldsymbol{H}_{k|k}(\boldsymbol{Z}_k - \widehat{\boldsymbol{Z}}_{k|k}) + \boldsymbol{v}_k] \\ \boldsymbol{0} \end{bmatrix} \tag{79}$$

in which $\boldsymbol{H}_{k|k}$ can be calculated according to Equation (76) when $\boldsymbol{Z}_k = \widehat{\boldsymbol{Z}}_{k|k}$. Substituting Equation (79) into Equation (78), one obtains

$$\boldsymbol{\varepsilon}_{k+1|k} = \left(\boldsymbol{I} + \Delta t\boldsymbol{U}_{k|k} + \Delta t\begin{bmatrix} \boldsymbol{0} \\ \boldsymbol{M}^{-1}(\boldsymbol{I} - \boldsymbol{\Phi})\boldsymbol{H}_{k|k} \\ \boldsymbol{0} \end{bmatrix}\right)(\boldsymbol{Z}_k - \widehat{\boldsymbol{Z}}_{k|k}) + \begin{bmatrix} \boldsymbol{0} \\ \Delta t\boldsymbol{M}^{-1}(\boldsymbol{I} - \boldsymbol{\Phi}) \\ \boldsymbol{0} \end{bmatrix}\boldsymbol{v}_k$$

$$= \left(\boldsymbol{I} + \Delta t\boldsymbol{U}_{k|k} + \Delta t\begin{bmatrix} \boldsymbol{0} \\ \boldsymbol{M}^{-1}(\boldsymbol{I} - \boldsymbol{\Phi})\boldsymbol{H}_{k|k} \\ \boldsymbol{0} \end{bmatrix}\right)\boldsymbol{\varepsilon}_{k|k} + \begin{bmatrix} \boldsymbol{0} \\ \Delta t\boldsymbol{M}^{-1}(\boldsymbol{I} - \boldsymbol{\Phi}) \\ \boldsymbol{0} \end{bmatrix}\boldsymbol{v}_k$$

$$= \boldsymbol{A}_1\boldsymbol{\varepsilon}_{k|k} + \boldsymbol{A}_2\boldsymbol{v}_k \tag{80}$$

where

$$\boldsymbol{A}_1 = \left(\boldsymbol{I} + \Delta t\boldsymbol{U}_{k|k} + \Delta t\begin{bmatrix} \boldsymbol{0} \\ \boldsymbol{M}^{-1}(\boldsymbol{I} - \boldsymbol{\Phi})\boldsymbol{H}_{k|k} \\ \boldsymbol{0} \end{bmatrix}\right); \quad \boldsymbol{A}_2 = \begin{bmatrix} \boldsymbol{0} \\ \Delta t\boldsymbol{M}^{-1}(\boldsymbol{I} - \boldsymbol{\Phi}) \\ \boldsymbol{0} \end{bmatrix} \tag{81}$$

The priori estimation error covariance can then be calculated as

$$\boldsymbol{P}_{k+1|k}=E(\boldsymbol{\varepsilon}_{k+1|k}\boldsymbol{\varepsilon}_{k+1|k}^{\mathrm{T}})=\boldsymbol{A}_1\boldsymbol{P}_{k|k}\boldsymbol{A}_1^{\mathrm{T}}+\boldsymbol{A}_2\boldsymbol{R}_k\boldsymbol{A}_2^{\mathrm{T}} \tag{82}$$

Based on Equation (73), the recursive solution for the posteriori estimation state can be given as

$$\hat{\boldsymbol{Z}}_{k+1|k+1}=\hat{\boldsymbol{Z}}_{k+1|k}+\boldsymbol{G}_{k+1}[\boldsymbol{\Phi}\boldsymbol{y}_{k+1}-\boldsymbol{\Phi}\boldsymbol{h}(\hat{\boldsymbol{Z}}_{k+1|k})] \tag{83}$$

in which $\boldsymbol{G}_{k+1}$ denotes the EKF gain matrix at the $(k+1)-th$ time step. The posteriori estimation error $\boldsymbol{\varepsilon}_{k+1|k+1}$ of the unknown state vector $\boldsymbol{Z}_{k+1}$ at time $t=(k+1)\Delta t$ can then be calculated as

$$\begin{aligned}
\boldsymbol{\varepsilon}_{k+1|k+1}&=\boldsymbol{Z}_{k+1}-\hat{\boldsymbol{Z}}_{k+1|k+1}\\
&=\boldsymbol{Z}_{k+1}-\hat{\boldsymbol{Z}}_{k+1|k}-\boldsymbol{G}_{k+1}[\boldsymbol{\Phi}\boldsymbol{y}_{k+1}-\boldsymbol{\Phi}\boldsymbol{h}(\hat{\boldsymbol{Z}}_{k+1|k})]\\
&=\boldsymbol{Z}_{k+1}-\hat{\boldsymbol{Z}}_{k+1|k}-\boldsymbol{G}_{k+1}[\boldsymbol{\Phi}\boldsymbol{h}(\boldsymbol{Z}_{k+1})+\boldsymbol{\Phi}\boldsymbol{v}_{k+1}-\boldsymbol{\Phi}\boldsymbol{h}(\hat{\boldsymbol{Z}}_{k+1|k})]\\
&=\boldsymbol{Z}_{k+1}-\hat{\boldsymbol{Z}}_{k+1|k}-\boldsymbol{G}_{k+1}[\boldsymbol{\Phi}(\boldsymbol{h}(\hat{\boldsymbol{Z}}_{k+1|k})+\boldsymbol{H}_{k+1|k}(\boldsymbol{Z}_{k+1}-\hat{\boldsymbol{Z}}_{k+1|k}))+\boldsymbol{\Phi}\boldsymbol{v}_{k+1}-\boldsymbol{\Phi}\boldsymbol{h}(\hat{\boldsymbol{Z}}_{k+1|k})]\\
&=(\boldsymbol{I}-\boldsymbol{G}_{k+1}\boldsymbol{\Phi}\boldsymbol{H}_{k+1|k})(\boldsymbol{Z}_{k+1}-\hat{\boldsymbol{Z}}_{k+1|k})-\boldsymbol{G}_{k+1}\boldsymbol{\Phi}\boldsymbol{v}_{k+1}\\
&=(\boldsymbol{I}-\boldsymbol{G}_{k+1}\boldsymbol{\Phi}\boldsymbol{H}_{k+1|k})\boldsymbol{\varepsilon}_{k+1|k}-\boldsymbol{G}_{k+1}\boldsymbol{\Phi}\boldsymbol{v}_{k+1}
\end{aligned} \tag{84}$$

Similarly, the posteriori estimation error covariance can then be found as

$$\begin{aligned}
\boldsymbol{P}_{k+1|k+1}&=E(\boldsymbol{\varepsilon}_{k+1|k+1}\boldsymbol{\varepsilon}_{k+1|k+1}^{\mathrm{T}})\\
&=(\boldsymbol{I}-\boldsymbol{G}_{k+1}\boldsymbol{\Phi}\boldsymbol{H}_{k+1|k})\boldsymbol{P}_{k+1|k}(\boldsymbol{I}-\boldsymbol{G}_{k+1}\boldsymbol{\Phi}\boldsymbol{H}_{k+1|k})^{\mathrm{T}}+\boldsymbol{G}_{k+1}\boldsymbol{\Phi}\boldsymbol{R}_{k+1}\boldsymbol{\Phi}\boldsymbol{G}_{k+1}^{\mathrm{T}}
\end{aligned} \tag{85}$$

To obtain the optimal value of the gain matrix $\boldsymbol{G}_{k+1}$ that can minimize the estimation error covariance $\boldsymbol{P}_{k+1|k+1}$ at time $t=(k+1)\Delta t$, the differentiation of $\boldsymbol{P}_{k+1|k+1}$ in Equation (85) with respect to $\boldsymbol{G}_{k+1}$ produces

$$\partial\boldsymbol{P}_{k+1|k+1}/\partial\boldsymbol{G}_{k+1}=2\boldsymbol{G}_{k+1}\boldsymbol{\Phi}(\boldsymbol{H}_{k+1|k}\boldsymbol{P}_{k+1|k}\boldsymbol{H}_{k+1|k}^{\mathrm{T}}+\boldsymbol{R}_{k+1})\boldsymbol{\Phi}-2\boldsymbol{P}_{k+1|k}\boldsymbol{H}_{k+1|k}^{\mathrm{T}}\boldsymbol{\Phi} \tag{86}$$

By setting the value of the partial derivative to zero, one can obtain

$$\boldsymbol{G}_{k+1}=\boldsymbol{P}_{k+1|k}\boldsymbol{H}_{k+1|k}^{\mathrm{T}}\boldsymbol{\Phi}[\boldsymbol{\Phi}(\boldsymbol{H}_{k+1|k}\boldsymbol{P}_{k+1|k}\boldsymbol{H}_{k+1|k}^{\mathrm{T}}+\boldsymbol{R}_{k+1})\boldsymbol{\Phi}]^{-1} \tag{87}$$

It can be found that the corresponding two sets of equations for EKF with unknown excitations, i. e. time update equations including Equation (77) and Equation (82) as well as measurement update equations including Equation (83), Equation (85), and Equation (87), are established. The unknown loadings at the current time step can then be identified according to Equation (74) by using the state vectors estimated through these two sets of equation. Notably, if the excitations are all assumed to be known, the matrix $\boldsymbol{\Phi}$ would be equal to an identify matrix $\boldsymbol{I}$ and the aforementioned two sets of equations reduce to the classical EKF method.

The presented algorithm can identify the unknown external excitations and time-invariant structural parameters simultaneously, and the flowchart of the proposed algorithm for the identification is shown in Figure 9.

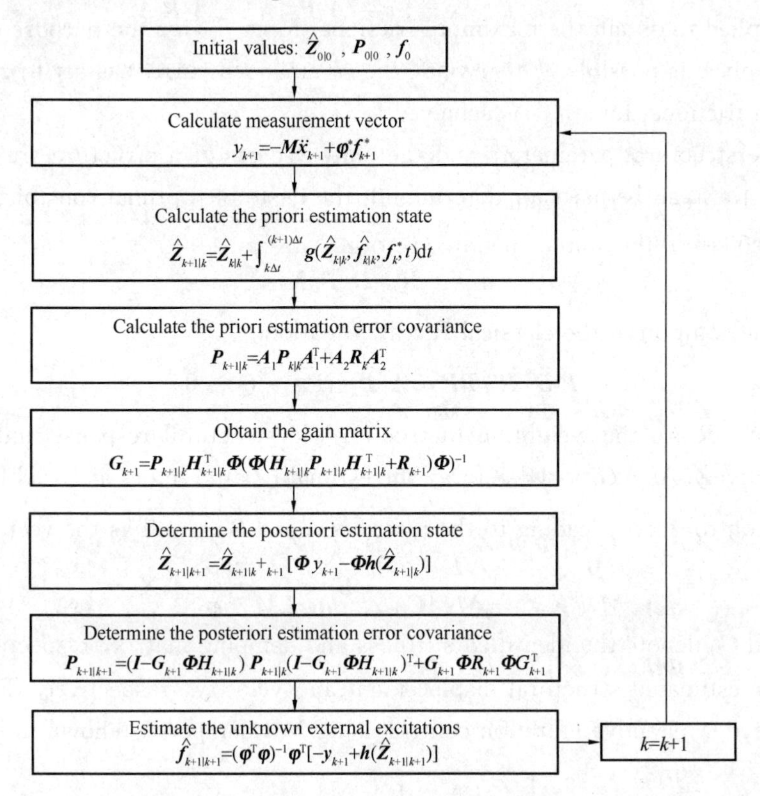

Figure 9 Flowchart of the proposed method for identifying time-invariant structural parameters and unknown excitations

3.3 Structural Vibration Control with Identified Time-Invariant Parameters

In this study, vibration control is also simultaneously considered for the building structure. Magneto-rheological (MR) dampers rather than semi-active friction dampers are used here because MR dampers are applicable for large civil structures. A variety of semi-active control strategies for the adjustment of the properties of MR dampers for the purpose of seismic control have been proposed and investigated[27-31]. Here, a switching control algorithm is considered and can be expressed as follows[31],

$$u_{\mathrm{d}}=\begin{cases} c_{\mathrm{d}}\dot{e}+f_{\mathrm{d}}^{\max}\operatorname{sgn}(\dot{e}) & (u\dot{e}<0) \\ c_{\mathrm{d}}\dot{e}+f_{\mathrm{d}}^{\min}\operatorname{sgn}(\dot{e}) & (u\dot{e}\geqslant 0) \end{cases} \tag{88}$$

where u_{d} is the MR damper force; c_{d} is the damping coefficient determined by the viscosity of MR fluid; $f_{\mathrm{d}}^{\max}$ and $f_{\mathrm{d}}^{\min}$ respectively denote the maximum value and minimum value of the frictional forces which are related to the yielding shear stress and can be achieved by adjusting the applied field; $\dot{e}$ is the velocity of the MR damper; and u is the optimal control force determined by the linear quadratic Gaussian (LQG) control algorithm. In case the optimal control force and the structural motion are in opposite direction, the maximum level of the current is applied to obtain the maximum frictional force $f_{\mathrm{d}}^{\max}$ for the purpose of mitigating vibration as much as possible. Otherwise, the commanded current is set to zero and thus the minimum frictional force $f_{\mathrm{d}}^{\min}$ is achieved.

Since the structural parameters and the state vectors of a structure are estimated in Section 3.2, they can be used for determining the desirable optimal control force $\boldsymbol{u}$ in the LQG control algorithm.

$$\boldsymbol{u}=-\bar{\boldsymbol{R}}^{-1}\bar{\boldsymbol{D}}^{\mathrm{T}}\bar{\boldsymbol{P}}\{\bar{\boldsymbol{X}}\} \tag{89}$$

where $\bar{\boldsymbol{P}}$ is the solution of the classical Riccati equation.

$$\bar{\boldsymbol{P}}\bar{\boldsymbol{B}}^{\mathrm{T}}\bar{\boldsymbol{R}}^{-1}\bar{\boldsymbol{B}}\bar{\boldsymbol{P}}-\bar{\boldsymbol{A}}^{\mathrm{T}}\bar{\boldsymbol{P}}-\bar{\boldsymbol{P}}\bar{\boldsymbol{A}}-\bar{\boldsymbol{Q}}=\boldsymbol{0} \tag{90}$$

in which $\bar{\boldsymbol{Q}}$ and $\bar{\boldsymbol{R}}$ are the weighting matrix for the structural response and the control force, respectively. As mentioned before, the estimate state vectors are used for the determination of control force, leading to the matrices $\bar{\boldsymbol{A}}$ and $\bar{\boldsymbol{B}}$ as well as the vector $\bar{\boldsymbol{X}}$.

$$\bar{\boldsymbol{A}}=\begin{bmatrix}\boldsymbol{0} & \boldsymbol{I}\\ -\boldsymbol{M}^{-1}\boldsymbol{K}_{id} & -\boldsymbol{M}^{-1}\boldsymbol{C}_{id}\end{bmatrix};\ \bar{\boldsymbol{B}}=\begin{bmatrix}\boldsymbol{0}\\ \boldsymbol{M}^{-1}\boldsymbol{\varphi}^{*}\end{bmatrix};\bar{\boldsymbol{X}}=\begin{Bmatrix}\boldsymbol{x}_{id}\\ \dot{\boldsymbol{x}}_{id}\end{Bmatrix} \tag{91}$$

where $\boldsymbol{K}_{id}$ and $\boldsymbol{C}_{id}$ denote the identified stiffness and damping matrix, respectively; $\boldsymbol{x}_{id}$ and $\dot{\boldsymbol{x}}_{id}$ denote the estimated structural displacement and velocity, respectively. The flowchart of the proposed semi-active vibration control using MR dampers is shown in Figure 10.

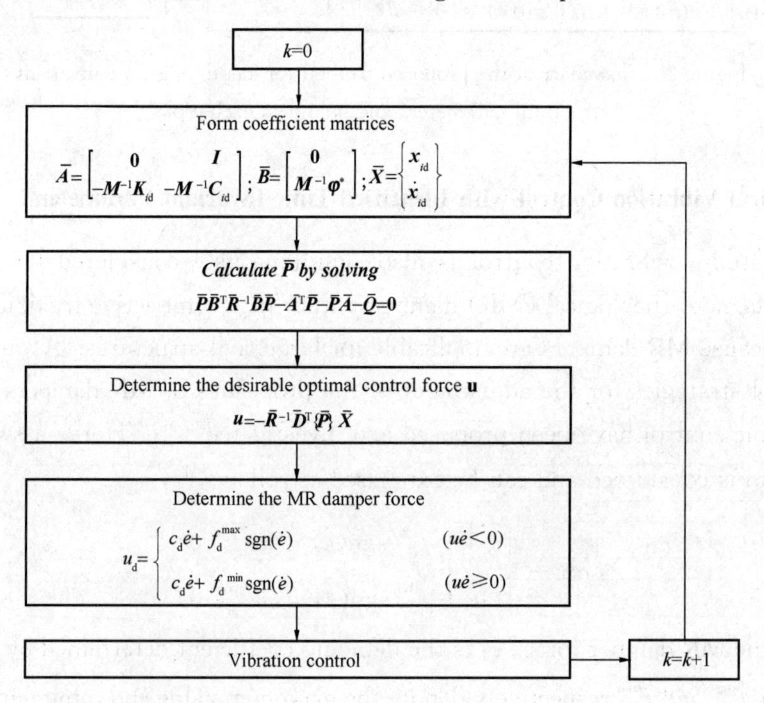

Figure 10 Flowchart of semi-active control with MR damper

3.4 Implementation Procedure of Integrated System with Time-Invariant Parameters

Based on the equations presented in the previous two subsections, the integrated procedure can be implemented for time-invariant structural parameter identification, excitation identification, and vibration control of a building structure step by step as follows:

Step 1: Calculate the observation equation based on the measured acceleration responses and MR damper forces;

Step 2: Determine the priori estimation state $\hat{\boldsymbol{Z}}_{k+1|k}$ according to Equation (77);

Step 3: Determine the priori estimation error covariance $\boldsymbol{P}_{k+1|k}$ *using Equation* (82);

Step 4: *Obtain the gain matrix* $\boldsymbol{G}_{k+1}$ according to Equation (87);

Step 5: Obtain the posteriori estimation state $\hat{\boldsymbol{Z}}_{k+1|k+1}$ according to Equation (83);

Step 6: Update the posteriori estimation error covariance $\boldsymbol{P}_{k+1|k+1}$ *by Equation* (85);

Step 7: *Estimate the unknown excitations* $\hat{\boldsymbol{f}}_{k+1|k+1}$ according to Equation (74);

Step 8: Form the coefficient matrices according to Equation (91) with the aid of the estimated state $\hat{\boldsymbol{Z}}_{k+1|k+1}$ determined in Step 5;

Step 9: Calculate $\bar{\boldsymbol{P}}$ by solving Equation (90) and determine the optimal control force according to Equation (89); and

Step 10: Determine the MR damper force according to Equation (88).

It is clear that in the proposed integrated approach, the identified structural parameters and responses are employed in the control devices and control algorithm on one hand, and the measured control forces are used for parameter and excitation identification on the other hand. A schematic diagram for the implementation of such integrated health monitoring and vibration control system for the building structure is shown in Figure 11.

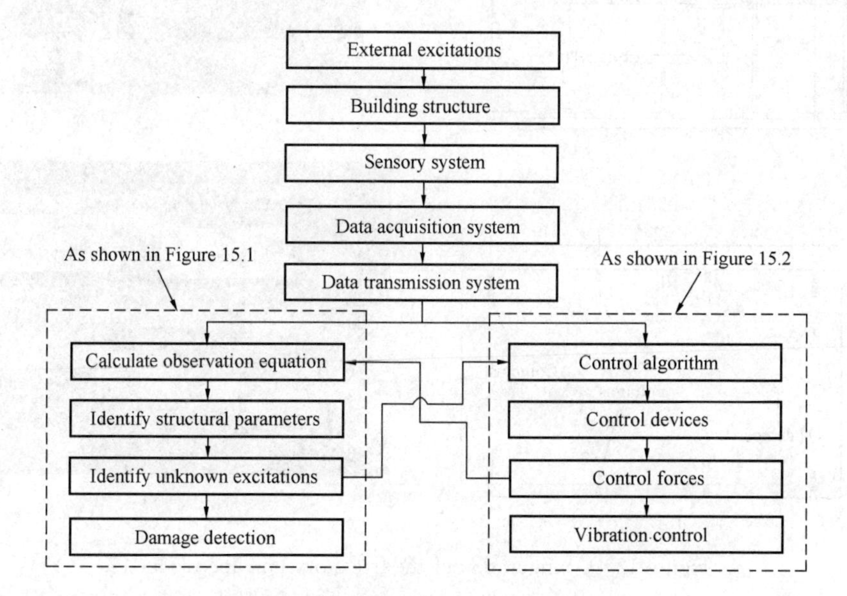

Figure 11 A schematic diagram of the implementation of the integrated system with time-invariant parameters

4 Experimental Investigation of Integrated System

4.1 Experimental Setup

To experimentally investigate the performance of the proposed integrated system for damage detection and vibration control of a building structure, a five-story building model was designed and built in the Structural Dynamics Laboratory of The Hong Kong Polytechnic University, as shown in Figure 12.

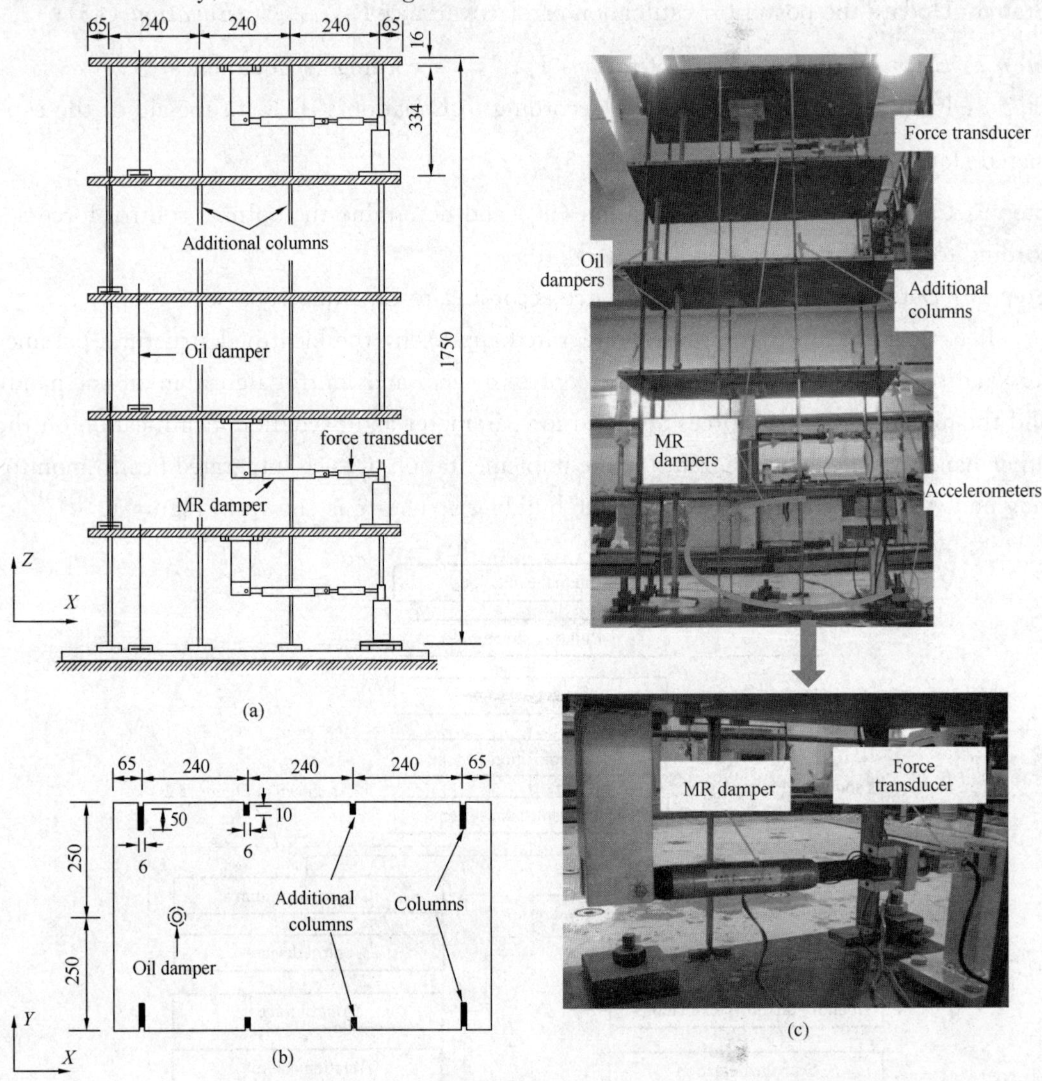

Figure 12 Configuration of the five-story building model

(a) Elevation view (unit: mm); (b) Plane view (unit: mm); (c) Photograph of the model

The building model consisted of five rigid plates of 850 mm × 500 mm × 16 mm and four equal sized rectangular columns of the cross section of 50 mm × 6 mm. The total height of this building model was 1750 mm, with the height of each story being identical. The cross section of the column was arranged in such a way that the stiffness of the building model in the y-direction was much larger than that in the x-direction. Each steel floor plate was highly rigid in the horizontal direction compared with the columns, which led to a shear type deformation. Four additional columns of the cross section of 10 mm × 6 mm were added in the building model to simulate damage when they were symmetrically cut off in some stories. Each column was embedded into and welded to the plates to ensure the rigid joints formed at the connections. All of the columns at the first floor were eventually welded to a thick steel plate that was in turn bolted firmly to a shaking table. A series of silicon oil dampers were designed and installed between each of the adjacent floor plates to increase the structural damping. As can be seen from Figure 12, the lumped mass for each floor was composed of the masses of the plate, columns, oil damper, MR damper, and the auxiliary for connecting the dampers to the plate. With the measurement of each component, the mass matrix for the building model can be determined as $\boldsymbol{M} = \mathrm{diag}[67.955, 61.918, 56.837, 63.079, 59.922]$ (unit: kg) in which diag [·] denotes a diagonal matrix. The Rayleigh damping assumption was used to construct the damping matrix in this study.

The building model was fixed on the shaking table of 3m × 3m, which was built by MTS Corporation. During the tests, the structural acceleration response of each floor was measured by a piezoelectric accelerometer with a sampling frequency of 1000 Hz. The ground acceleration induced by the shaking table was also measured by the accelerometer and used for the comparison with the identified one. Moreover, the force transducers were placed in series with the MR dampers to measure the control forces which were then used for the parameter and excitation identification. The displacement response of each floor was also obtained but solely utilized for the assessment of the control performance.

Three MR dampers (RD-1097-01) manufactured by the Lord Corporation, USA, were respectively installed on the 1st, 2nd, and 5th floor, and used to provide the control forces to the building model. To achieve the best performance of the MR damper, the Rheonetic Wonder Box device controller kit (RD-3002-03) designed by the Lord Corporation was used along with the MR damper. The output current supplied to the MR damper by the Rheonetic Wonder Box device depended on the command voltage (the input voltage) from the computer. Through the calibration of the Wonder Box device (see Figure 13), it was found that the output current of such device could reach its saturation value when the input voltages were larger than a certain value (about 1.8 V). Therefore, the maximum and minimum input voltages were set to be 1.8 V and 0 V, respectively. A Simulink model was built, and a dSPACE real-time simulator and control system was employed to process and analyze the aforementioned measurements and provide the desired control signals according to Equation (88).

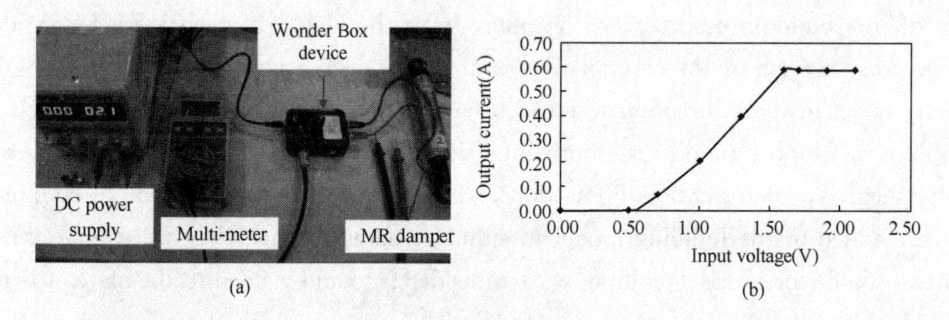

Figure 13 Calibration of Wonder Box device

(a) Calibration setup of Wonder Box; (b) Output voltages versus input voltages

4.2 Damage Scenarios

The static tests of the building model were first conducted for determining the structural stiffness in the cases without and with additional columns. As the number of the calibrated mass block increased, the increased deformation of the building model in the x-direction was recorded by dial gauges and used for the determination of the structural stiffness. The measured results are given in Table 3.

The measured stiffness of the building model **Table 3**

	k_1 (N/m)	k_2 (N/m)	k_3 (N/m)	k_4 (N/m)	k_5 (N/m)
Without additional columns	2.181×10^5	2.085×10^5	2.107×10^5	2.166×10^5	2.213×10^5
With two additional columns	2.418×10^5	2.323×10^5	2.363×10^5	2.412×10^5	2.483×10^5
With four additional columns	2.657×10^5	2.576×10^5	2.619×10^5	2.639×10^5	2.713×10^5

Since the structural damage usually results in the reduction of structural stiffness, the damage was simulated in this study by symmetrically cutting off the additional columns in certain stories as shown in Figure 14. One advantage of such damage pattern is that the mass remains unchanged while the structural stiffness is reduced. In this experimental study, four damage scenarios were considered. In Case 1, two additional columns in the symmetrical position of the first floor were cut off; in Case 2, the remaining two additional columns (i. e. all of the four additional columns) in the first floor were cut off; in Case 3, two additional columns in the third floor were cut off; in Case 4: four additional columns in the third floor all were cut off.

For each of the aforementioned four damage scenarios, hammer tests were carried out to capture the properties of the building model when the MR dampers were removed as shown in Figure 15. The impact force was applied on the top floor of the building model and the acceleration response of each floor was measured. The modal parameters of the building, including its natural frequencies, damping ratios and modal shapes, were identified basing on the analysis of the corresponding frequency response functions. Since the mass of an MR damper was 0.525 kg only, the loss of the total mass of each floor was rel-

Figure 14　Damage pattern in the experiment

Figure 15　The hammer tests of the building model without MR dampers

atively small as compared with the original building model. Moreover, the stiffness provided by the MR damper was quite limited. From these points of view, the variations of the natural frequencies caused by the removal of the MR dampers could be small. The first two natural frequencies and damping ratios of the building model in the four damage cases are shown in Table 4.

The properties of the model determined by the hammer test　　　**Table 4**

	Natural frequencies		Damping ratios	
	f_1 (Hz)	f_2 (Hz)	ζ_1 (%)	ζ_2 (%)
Case 1	2.907	8.533	0.86	1.03
Case 2	2.845	8.408	0.89	1.08
Case 3	2.813	8.346	1.01	1.09
Case 4	2.782	8.283	1.09	1.15

4.3 Implementation of Identification and Control Algorithms

After the MR dampers were installed to the building model, the integrated system was established and the shaking table tests could be conducted for the aforementioned four damage cases. The dSPACE real-time simulator and control system was employed to realize the integration of damage detection and vibration control of the building model as schematically shown in Figure 16. It can be seen that the dSPACE real-time simulator and control system consists of three processing blocks: (1) DS2003 (A/D) block that is used for receiving analog signals from sensors and transforming them to digital signals; (2) DS2102 (D/A) block that is used for transforming digital signals to analog signals and then sending them to the control devices; and (3) the algorithms written using the MATLAB/Simulink block program, which are used to identify the structural parameters and unknown excitations, as well as to create the corresponding control forces.

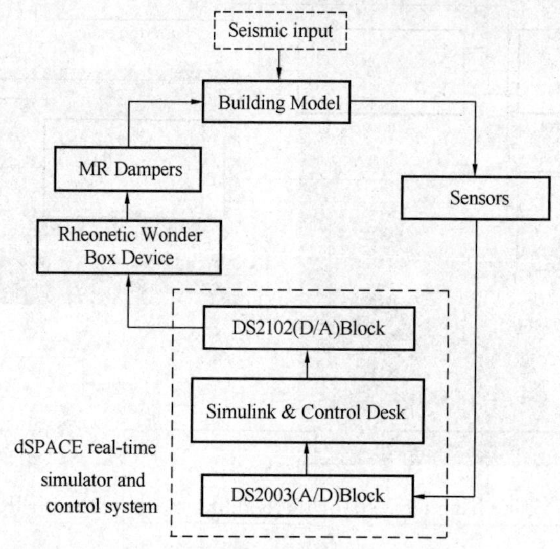

Figure 16 Schematic diagram of the integrated system

The usage of thedSPACE central processing unit (CPU) and the access to its memory can be realized by means of the main program ControlDesk of dSPACE, which offers an automatic implementation of the MATLAB/Simulink block program on the host computer via Real Time Interface (RTI) and provides a real-time interactive data display and visualization. Some details on how to build the Simulink model on the basis of the associated block programs are shown in Figure 17. It should be pointed out that each block marked by the bold line stands for the corresponding subsystem as shown in Figures (17b, c) and it is built according to the equations given in Sections 2.2 and 2.3.

After the implementation of the proposed identification and control algorithms, the shaking table tests were conducted. Kobe earthquake was considered in this experimental investigation. The structural parameter and earthquake identification results and the control performance under these two earthquakes are given in the following subsections.

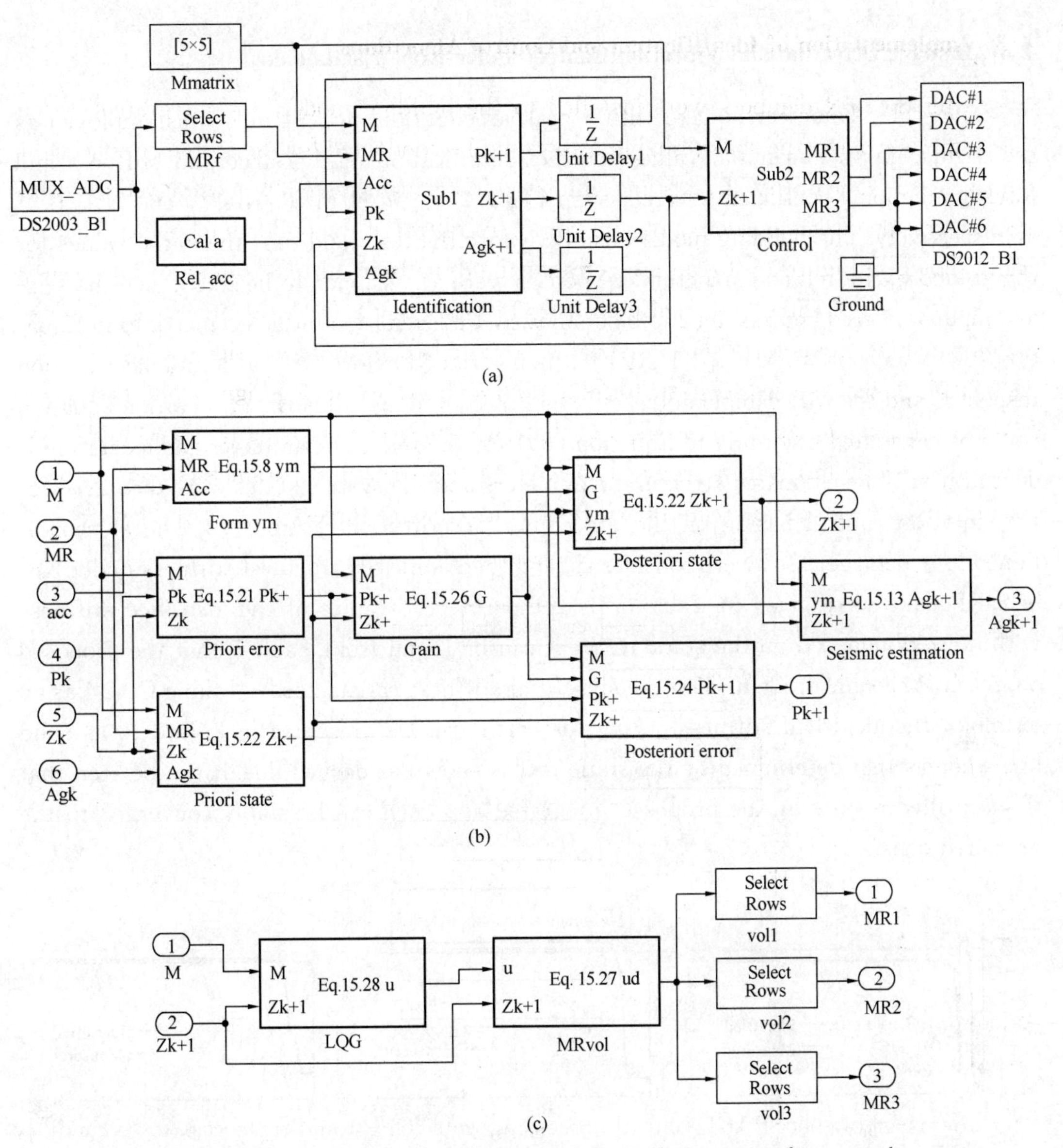

Figure 17 MATLAB/Simulink block diagrams of the implementation of integrated system
(a) Block diagram of identification and control integration; (b) Block diagram for identification in Sub1; (c) Block diagram for control in Sub2

The identified structural stiffness (under Kobe earthquake) **Table 5**

		k_1	k_2	k_3	k_4	k_5
Case 1	Identified (N/m)	2.389×10^5	2.601×10^5	2.533×10^5	2.604×10^5	2.742×10^5
	Relative error (%)	1.20	0.97	3.28	1.32	1.06
Case 2	Identified (N/m)	2.147×10^5	2.619×10^5	2.535×10^5	2.608×10^5	2.684×10^5
	Relative error (%)	1.51	1.67	3.21	1.17	1.07
Case 3	Identified (N/m)	2.111×10^5	2.638×10^5	2.388×10^5	2.662×10^5	2.765×10^5
	Relative error (%)	3.16	2.41	1.06	0.87	1.92
Case 4	Identified (N/m)	2.086×10^5	2.675×10^5	2.142×10^5	2.663×10^5	2.783×10^5
	Relative error (%)	4.31	3.84	1.64	0.91	2.58

4.4 Damage Detection and Vibration Control under Kobe Earthquake

The scaled Kobe earthquake with a peak acceleration of 0.84 m/s^2 was employed as the seismic input. The initial values for the identification of the structural stiffness and Rayleigh damping coefficients were assigned as 1.2×10^5 N/m, 0.01 s^{-1}, and 1×10^{-4} s, respectively. The building model is static before the test, and thus the initial values for the ground excitation and structural responses were all assumed to be zero. Since the seismic input was used and assumed to be unknown, the excitation influence matrix $\boldsymbol{\varphi}$ in Equation (62) was equal to-$\boldsymbol{M} \cdot [1\ 1\ 1\ 1\ 1]^{\mathrm{T}}$. As mentioned before, the structural acceleration responses and the MR damper forces all were measured and thus the observation equation could be calculated according to Equation (69) for the real-line parameter and excitation identification. The estimated structural parameters and the system states were then used for the vibration control basing on the procedures described in Section 3.4. The aforementioned four damage cases were considered, and the identified structural stiffness under Kobe earthquake were given in Table 5. By comparing the values of the identified stiffness with those obtained from the static tests, it can be found from Table 5 that the proposed approach is capable of identifying the structural stiffness satisfactorily. Taking Case 4 as an example, the identified stiffness during the earthquake is presented in Figure 18 as solid line whereas that determined by the static test is shown as dashed line. It can be seen that the identified results by the proposed integrated approach can be stably converged to the measured ones.

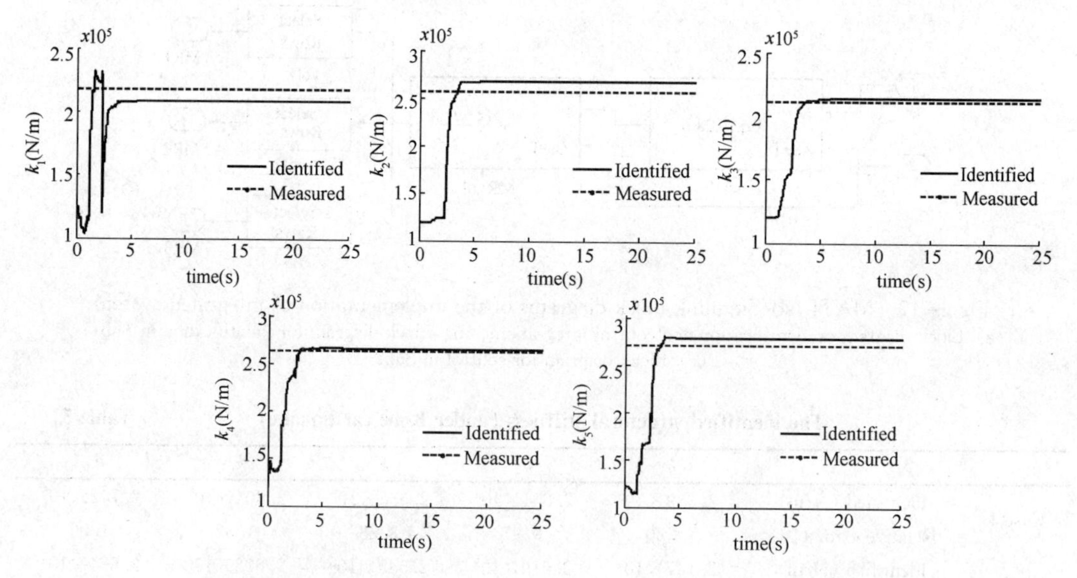

Figure 18 The identified structural stiffness under Kobe earthquake

By using the proposed approach, the Rayleigh damping coefficients can also be estimated. However, the values of Rayleigh damping coefficients are usually small and not easy for the assessment. Thus, the first two damping ratios, which are determined basing on the identified Rayleigh damping coefficients and the first two natural frequencies as shown in Table 4, are used for ease of comparison. The identified results are given in Table 6. It can be seen that, as compared with the results obtained from the hammer tests (see Table 4), the identified results by means of the proposed approach are still acceptable.

The identified damping ratios (under Kobe earthquake) **Table 6**

	Case 1	Case 2	Case 3	Case 4
ζ_1 (%)	0.67	0.86	0.82	1.06
ζ_2 (%)	0.28	0.45	0.88	0.89

Besides the identification of structural parameters, the unknown ground motion can also be simultaneously identified by the proposed approach. Also, taking Case 4 as an example, the time history of the identified ground excitation is plotted in Figure 19 as dashed curve, whereas the solid curve is the corresponding measured ground excitation obtained directly from the accelerometer. Only the segment from 4 s to 8 s is shown in Figure 19 for clarity of comparison. It can be found from Figure 19 that the identified ground excitation has a good agreement with the measured one.

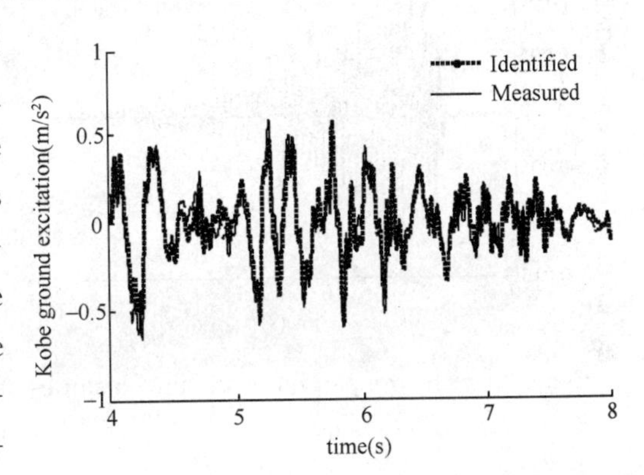

Figure 19 The comparison of the identified Kobe ground excitation

Peak responses of the building model under Kobe earthquake **Table 7**

		$\bar{s}_i$ (mm)					$\bar{a}_i$ (m/s²)				
		$\bar{s}_1$	$\bar{s}_2$	$\bar{s}_3$	$\bar{s}_4$	$\bar{s}_5$	$\bar{a}_1$	$\bar{a}_2$	$\bar{a}_3$	$\bar{a}_4$	$\bar{a}_5$
Case 1	Without control	5.83	9.22	12.58	14.74	14.65	1.95	3.09	4.12	4.66	4.80
	With control	3.45	4.19	5.17	5.69	5.79	1.11	1.37	1.30	1.45	1.61
Case 2	Without control	6.22	9.40	12.45	14.31	15.14	1.97	2.93	3.67	4.17	4.35
	With control	3.53	4.24	5.19	5.70	5.77	1.12	1.35	1.31	1.46	1.67
Case 3	Without control	5.58	8.36	11.24	12.81	13.47	1.82	2.59	3.22	3.72	3.95
	With control	3.41	3.86	4.81	5.39	5.43	1.31	1.42	1.33	1.73	1.70
Case 4	Without control	5.19	7.86	10.88	12.38	13.04	1.59	2.28	2.97	3.53	4.03
	With control	3.36	3.81	4.78	5.33	5.36	1.15	1.29	1.21	1.56	1.53

Furthermore, based on the proposed integrated approach, the structural vibration can be significantly reduced with the aid of the MR dampers. Two sets of shaking table tests, saying the building model with and without MR dampers, were conducted. The aforementioned four damage scenarios were considered. The switching control law introduced in Section 3.3 was employed for adjusting the properties of the MR dampers. The maximum acceleration and displacement responses of the building model under Kobe earthquake are given in Table 7 for comparison. $\bar{s}_i$ and $\bar{a}_i$ ($i=1, 2, \cdots, 5$) are the peak displacement and acceleration of the i-th floor obtained from the sensors, respectively. It can be seen from Table 7 that the structural vibration is significantly reduced. Moreover, the comparison of time histories of the measured displacement and acceleration of the top floor are shown in Figure 20. Only the building model in Case 4 is plotted as an example. Similar results can be obtained from the other cases.

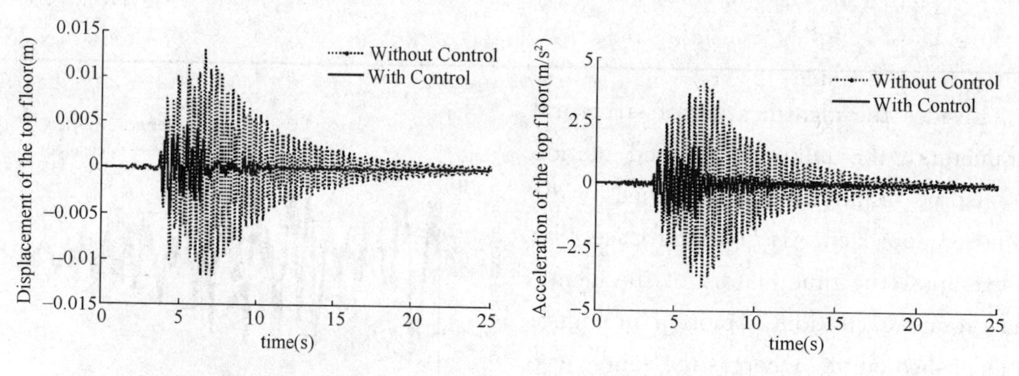

Figure 20 The comparison of the time histories of structural responses with and without control (Kobe earthquake)

5. SHM-Based Seismic Performance Assessment

A detailed seismic performance assessment for super high-rise buildings subject to long-period ground motions are essential for decision-making on post-earthquake repair, maintenance, and reoccupation. This section proposes a probabilistic seismic performance assessment framework for instrumented super high-rise buildings under bi-directional long-period ground motions, in which the probabilities of key structural components experiencing different damage levels are assessed. The fragility curves of the key structural components are obtained by performing the incremental dynamic analysis on the updated nonlinear finite element model of the super high-rise building. The estimation of the structural responses is based on the integrated optimal sensor placement and response reconstruction scheme. This scheme can also reconstruct the bi-directional long-period ground motions and can significantly reduce uncertainties in the currently-used response estimation process. The evolving mean values and variances of the structural responses are determined by using the Kalman smoothing algorithm-based Bayesian inference process, and the extreme value distribution of the structural responses is obtained in terms of Vanmarcke approximation. The generated fragility curves are then incorporated with the extreme response distributions to yield an estimation of the probabilistic damage states of the key structural components. The proposed framework is finally applied to a real super high-rise building and the numerical results manifest that the proposed framework provides a reliable way of estimating the safety and operability levels of the instrumented super high-rise building after the earthquake event.

5.1 Seismic Behavior of Super High-Rise Buildings with Hybrid Structural System

5.1.1 Super high-rise buildings with hybrid structural system

Structural systems for high-rise buildings have evolved from the relatively simple systems, such as frame system, shear-wall system, wall-frame system and core-tube system, towards more complex hybrid systems for super high-rise buildings[32]. The complex hybrid structural system for super high-rise buildings often consist of core-tubes, mega columns, and outrigger trusses. This kind of hybrid structural system has been applied to many super high-rise buildings, such as the Shanghai Tower[33], the Shanghai Jinmao Tower, and the Tianjin 117 Tower [34] among others.

Figure 21 shows schematically the major structural components and their connectivity in a hybrid structural system. In the middle, there is a core-tube consisting of shear walls and coupling beams. Mega columns are located outside of the core-tube, and the mega-trusses and outrigger trusses are used to connect the mega columns to the core-tube. The mega-columns and core-tube shear walls are supposed to withstand the axial forces and bending moments caused by the seismic loading. The coupling beams are supposed to bear mainly shear forces during earthquake excitation and the outrigger trusses/mega trusses between the core-tube and the mega-columns bears mainly shear forces and bending moments.

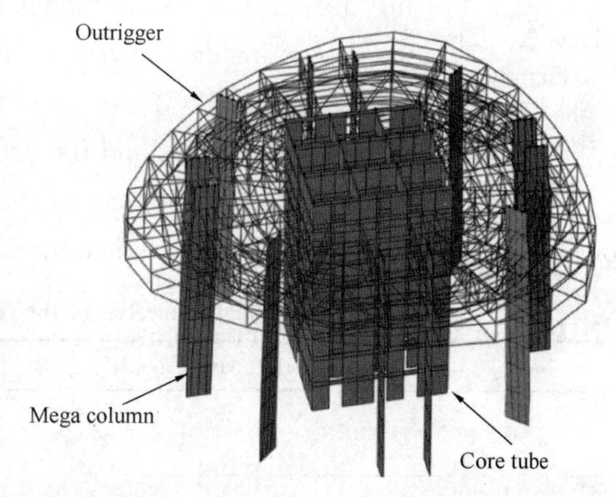

Figure 21 Schematic diagram of the hybrid structural system: a perspective view

In this study, the Shanghai Tower, a super high-rise building in Shanghai, will be taken as an example for feasibility study. The Shanghai Tower is currently the tallest building in China. The Shanghai Tower has a total architecture height of 632 m and a structural height of 580 m with 124 stories above the ground level, as shown in Figure 22. Structurally, the building is divided into nine zones along its height, separated by eight strengthening floors. A hybrid structural system is adopted, which comprises an inner core wall tube, an outer mega-frame, and a total of six levels of outriggers between the tube and the frame. The inner core wall tube is square-shaped with dimensions of 30m x 30m and divided into the nine cells at the bottom of the building. The core wall tube changes along the height of the building. The four corners of the square tube are gradually removed until zone 7 before forming a rectangular at the top of the tube. The thickness of the core wall varies from 1.2m at the bottom of the building to 0.5 m at the top. The steel plates

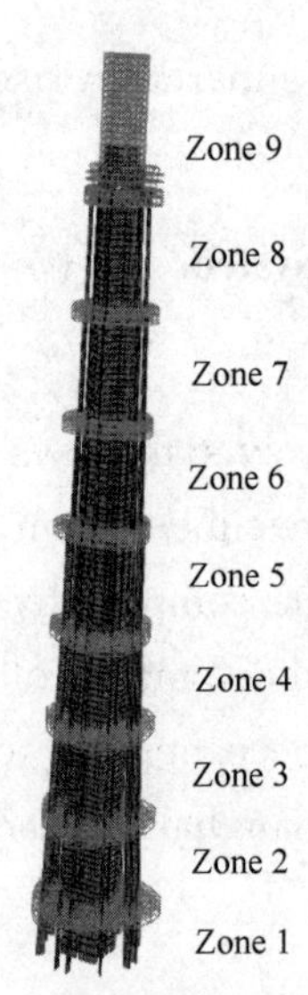

Figure 22 The configuration of the FE model of Shanghai Tower

are embedded in the walls of the tube to reduce wall thickness and improve ductility. The mega-frame consists of eight super columns, four corner columns, radial trusses, and high-box belt trusses. The outriggers are set along the height of the building at zones 2, 4, 5, 6, 7, and 8. Post-grouting bored piles are employed for the foundation of the structure[33]. The 3D nonlinear finite element (FE) model of the Shanghai Tower with the hybrid structural systems is built on the ABAQUS platform, as shown in Figure 22. The outrigger trusses are modeled using B31 beam elements. The concrete part of the mega-column is modeled by S4R shell elements with the embedded shaped steel modeled by B31 beam elements, and the two different elements are then coupled at the contact points. The shear-walls in the core-tube and the coupling beams connecting the shear-walls are modeled by S4R shell elements, and the reinforcement bars inside are generated using the "Rebar" function. Multi-points constraints (MPC) are used to model the rigid floor effect where the translational DOFs and the rotational DOFs in the horizontal plane on the same floor are coupled together. The modal information of the first 9 modes of vibration of the building are summarized in Table 8.

The modal properties of the first 9 modes of vibration **Table 8**

Mode	Type of modal response	Natural Frequencies (Hz)	Direction
1	Global	0.10942	Lateral
2	Global	0.11053	Lateral
3	Global	0.22204	Torsional
4	Global	0.30591	Lateral
5	Global	0.31386	Lateral
6	Global	0.44923	Torsional
7	Global	0.63332	Lateral
8	Global	0.65141	Lateral
9	Global	0.68741	Torsional

5.1.2 Nonlinear analysis with material and geometrical nonlinearity

To conduct the nonlinear time history analysis (NTHA), two types of nonlinearities, namely the geometric nonlinearity and the material nonlinearity, are considered in the FE modeling of the building. The geometric nonlinearities due to *P*-Δ effect and large deflection are considered by activating the "NLGEOM" module in ABAQUS, which forces the analysis to perform an iterative solution and update the stiffness matrix based on the incremental nodal displacements at each equilibrium iteration. The material nonlinearities are accounted for by adopting the nonlinear stress-strain relationships for the steel and concrete material used in the FE model of the building[35]. The concrete damaged plasticity (CDP) model is used to simulate the damage evolution of concrete. In order to simulate the inelastic behavior of concrete, the isotropic damage index considering the isotropic elastic stretching and compression plasticity is used in this model. To simulate the mechanical properties of the concrete under the monotonic, cycle, and dynamic loads, the tensile crack and compression fracture are selected as two main failure mechanisms of the concrete material. The constitutive relations of concrete (C60), as shown in Figure 23, illustrate the concrete stiffness changes over a tension-compression-tension loading cycle. In Figure 23, E_0 refers to the initial modulus of the undamaged concrete, d_t and d_c represent the tension stiffness reduction factor and compression stiffness reduction factor, respectively.

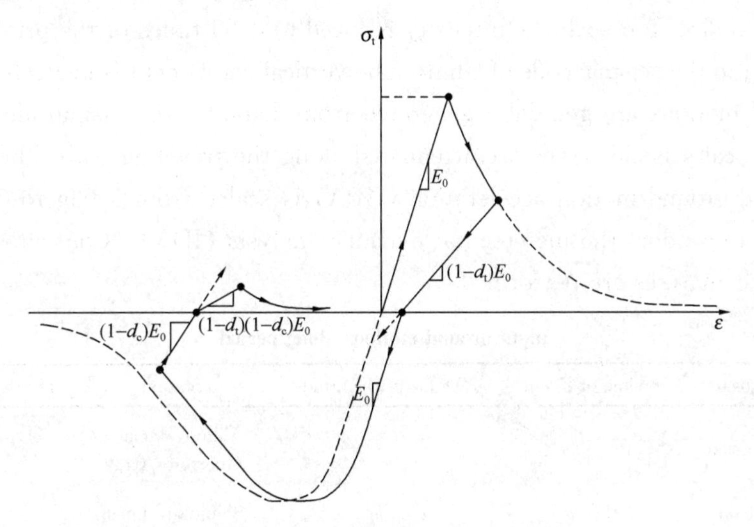

Figure 23 The stress-strain relationship in concrete damaged plasticity model

The compressive stiffness of the concrete is recovered upon crack closure as the load changes from tension to compression. On the other hand, the tensile stiffness is not recovered as the load changes from compression to tension once crushing micro-cracks are developed. A more detailed description for the CDP model can found in the ABAQUS User's Manual[36]. The bilinear stochastic hardening model[37] is employed to model the steel material, as shown in Figure 24. In Figure 24, f_y refers to the yielding strength. The model considers the Bauschinger effect and assumes no stiffness degradation during the loading cycles. The yield strength ratio for the steel is set as 1.2 and the ultimate strain is set as 0.025.

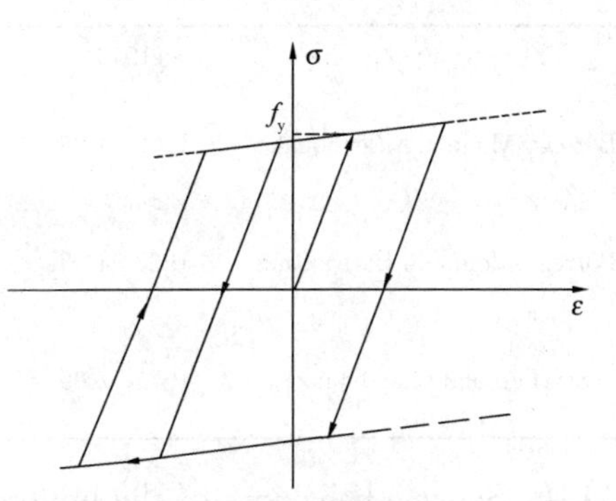

Figure 24 Bilinear hardening model of steel

5.1.3 Long-period and short-period ground motions

Two groups of ground motions (short-period and long-period) are selected as input for conducting

NTHA. For super high-rise buildings, their first natural period can be as long as 6s-10s, which is far beyond the specification in the current seismic design codes. The structural responses of the building under the two different groups of ground motions could be quite different, and the fragility curves for the structural components could be very different accordingly. A group of six long-period ground motions including two recorded and four simulated ones are employed as input for the NTHA and listed in Table 9. For comparison, a group of four short-period ground motions, listed in Table 10, are also employed in the NTHA.

The principal motion of the bi-directional ground motions is applied along the horizontal axis in the east-west direction, and the other component is applied orthogonally along the north-south direction with the intensity reduced to 0. 85 times of the principal component according to the seismic code of China. The vertical component is neglected since long-period ground motions are generally generated from distant large magnitude earthquakes where the vertical seismic waves are attenuated along the traveling path. The intensity of the 10 selected ground motion acceleration (GPA) is scaled from 0. 08g to 0. 22g with a step of 0. 02g to conduct the incremental dynamic analysis (IDA). Consequently, 80 dynamic response analyses are performed.

Input ground motions: long period **Table 9**

Name of Earthquake	Time of Event	Moment Magnitude	Station	Hypocentral Distance
Michoacan Earthquake	1985-09-19	8	Villita, Mexico Guerrero Array	47. 8km
Tokohu Earthquake	2011-03-11	9	Shinjuku, Japan	375km
Simulated Earthquake		7		300km
Simulated Earthquake		7. 5		400km
Simulated Earthquake		8		500km
Simulated Earthquake		8. 5		600km

Input ground motions: short period **Table 10**

Name of Earthquake	Time of Event	Moment Magnitude	Station	Hypocentral Distance
El Centro earthquake	1940	6. 9	El Centro, CA	8km
Borrego Mountain Earthquake	1968-04-08	6. 5	Los Angeles, CA Hollywood Storage Bldg	208km
Borrego Mountain Earthquake	1968-04-08	6. 5	Orange County, CA Eng. Bldg.	
San Fernando Earthquake	1971-02-09	6. 6	Vernon, CA CMD Bldg	49. 9km

5. 1. 4 Seismic behaviors of the building under long and short period ground motions

According to the study by Lu et al. [38], the stiffness reduction of the mega column, core tube and outrigger of the 3D FE model of the Shanghai Tower could cause a significant change in the natural periods of the first nine modes of vibration of the building. These three components are the key structural components contributing most to the total lateral stiffness of the super high-rise building, and the seismic performance assessment will be focused on mainly these key structural components in the following sections.

Vulnerabilities and seismic performance of different structural components　　Table 11

Structural Component	Seismic performance under moderate earthquake (0.1g)	Seismic performance under major earthquake (0.2g)	Damage Index
Mega-column	Elastic and intact	Elastic with local minor crack	IDR (Zone by zone)
Core-tube Shear wall	Elastic with minor crack	Local cracks	IDR (story by story)
Box belt Truss	Elastic and intact	Elastic and intact	Internal forces
Radial Truss	Elastic and intact	Partially yield	Shear forces
Coupling Beam	Moderate cracks	Cracks of concrete and yield of steel	Chord rotation

Based on a series of NTHA under seismic loadings with incremented ground motion intensities, the vulnerabilities and performance of different structural components in the high-rise building can be obtained and summarized in Table 11. The results manifest that the principal load-resisting structural components, including the mega-columns and mega trusses, remain elastic under moderate earthquakes while the core-tube shear walls experience minor cracks and the coupling beams experience moderate cracks. The performance of shear walls can be evaluated by the effective inter-story drift ratio (IDR) between two adjacent stories while the performance of coupling beams can be evaluated by its chord rotation angle that can be deduced from the inclination of its adjunct shear wall [38].

5.1.5 Component-level assessment and system-level assessment

Since the capacities of different structural components vary from each other, the seismic performance of the high-rise building should be assessed in a hierarchy way that can be categorized as:

(1) System-level performance assessment, including

- Capacity of the mega-columns
- Capacity of the radial trusses
- Overall IDR threshold

(2) Component-level performance assessment, including

- Fragility curves for coupling beams conditioned on chord rotation
- Fragility curves for shear walls conditioned on effective IDR

The system-level assessment controls the collapse risk and safety of the super high-rise building when catastrophic earthquake attacks and they can be assessed in a data-driven manner where the measured responses are used directly for the assessment. The component-level assessment is in charge of the in-service operation of the building and quantifies the damage states of the structural components undergoing minor or moderate earthquakes. The system-level assessment in terms of capacity is relatively straight forward and will not be discussed in this study. This study focuses only on the component-level assessment, which is performed through the data-model interaction where the measured data from multi-type sensors are employed in conjunction with the linear FE model for the complete response reconstruction.

5.2 Generation of Analytical Component-Level Fragility Curves

Damage fragility curves are key components of conducting probabilistic seismic performance assessment for the high-rise building. These curves relate the likelihood/probability of a structural component reaching or exceeding a specific damage state to a single EDP or multiple EDPs.

5.2.1 Damage states and damage indexes

Generally, damage states are characterized by different stages sustained by a building structure when subjected to external loading, as listed in Table 12. The indicators (descriptions) of different damage states are, but not limited to, concrete crack initiation and growth, concrete crushing, reinforcement yielding, buckling and fracture among others[39]. The most commonly-used damage index for structures and structural components is the Park-Ang damage model[40], which accounts for the cumulative damage of reinforced-concrete components under cyclic loading. To determine the damage state of an RC member based on the Park-Ang damage model, one needs to obtain the maximum deformation, equivalent yield force and hysteretic energy absorbed by the component during the earthquake. This model is difficult to be applied to a super high-rise building of a large number of structural components. An alternative solution is to use the equivalent strain of the element as the damage index to indicate various damage states of the target structural components[37, 41]. This is because (1) the equivalent strain value of the element reflects the stiffness degradation of the RC member due to concrete cracking or fracturing during the

Damage states, descriptions and indexes **Table 12**

Label	Damage state	Description	Compressive strain of concrete	Tensile strain of reinforcement
DS1	Slight Damage	Minor cracks; partial crushing of concrete. No structural repairs are necessary.	0.002~0.0035	0.0017~0.01
DS2	Moderate Damage	Diagonal crack occurrence; spalling of cover concrete	0.0035~0.005	0.01~0.03
DS3	Severe Damage	Extensive crashing of concrete; disclosure of reinforcement	0.005~0.008	0.03~0.05
DS4	Complete Damage	Damage is so extensive that repair of most structural elements is not feasible.	>0.008	>0.05

excessive cyclic loading; (2) the strain responses are more sensitive to local or minor damages; and (3) the dynamic responses of the RC components in the super high-rise building are greatly influenced by the mechanical properties of the materials. The damage states, the descriptions, and the corresponding damage indexes in terms of the equivalent strains of the material are summarized in Table 4, referring to the findings in references[37, 41].

5.2.2 Engineering demand parameters

To correlate the damage states of structural components with the measurable structural responses, damage-sensitive engineering demand parameters should be selected for different types of structural components. The engineering demand parameters (EDPs) refer to the structural responses that are most relevant to the damage experienced by the structural components. In the super high-rise building, the EDPs for the coupling beams and shear walls can be selected as the chord rotation angle and the effective inter-story drift ratio, respectively.

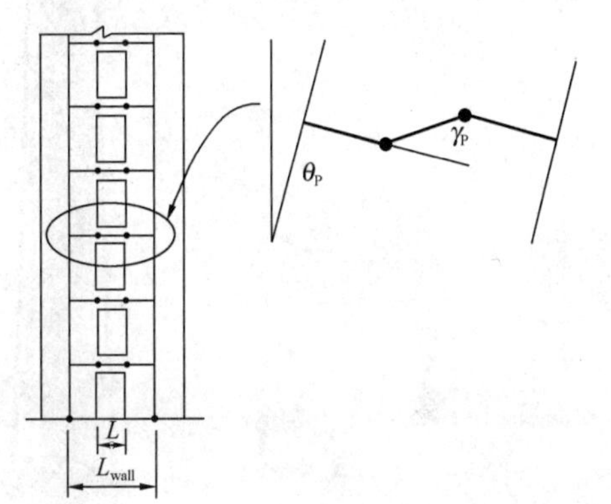

Figure 25 Calculation of chord rotation angle of the coupling beam

The performance of a coupling beam is controlled by the chord rotation angle γ_P (see Figure 25), which can be calculated as

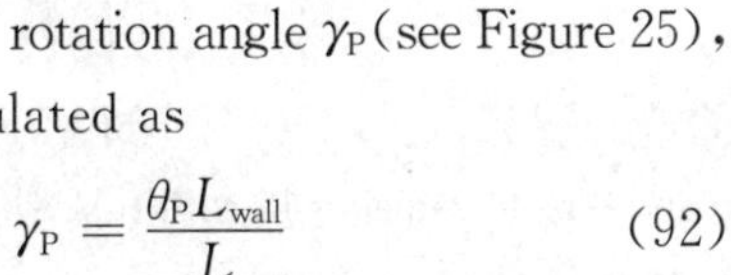

$$\gamma_P = \frac{\theta_P L_{wall}}{L} \tag{92}$$

where θ_P is the inclination of the connected shear wall; L_{wall} is the distance between centroid axes of the shear wall; and L is the effective distance of the coupling beam.

Since the super high-rise building with the hybrid structural system is predominated by flexural-shear deformation rather than pure flexural or shear deformation, the fragility curve developed for the shear wall is the function of the IDR of the wall panel, which is different from the IDR estimated for a story of a building[42]. Figure 26 shows the locations of compressive concrete damage and tensile steel damage of the core-tube shear walls obtained from the NTHA of the building under the simulated bi-directional long-period ground motions The PGA of the input ground motion along the principle direction (east-west) is scaled to 0.16g and the PGA of the other component (north-south) is scaled to 0.136g. Figure 27 displays the maximum overall IDR of the building as well as the maximum effective IDR of the shear walls along the height of the building. It can be observed that the damage locations of the shear walls coincide with the maximum effective IDR responses of the high-rise building rather than the maximum overall IDR. Thus, the effective IDR is a better EDP for indicating the damages of the shear walls in the super high-rise building. It can also be seen from Figures 26 and 27 that the damage of the shear walls occurs mainly in the floors adjacent to the strengthened floors by the mega trusses in Zones 7 and 8.

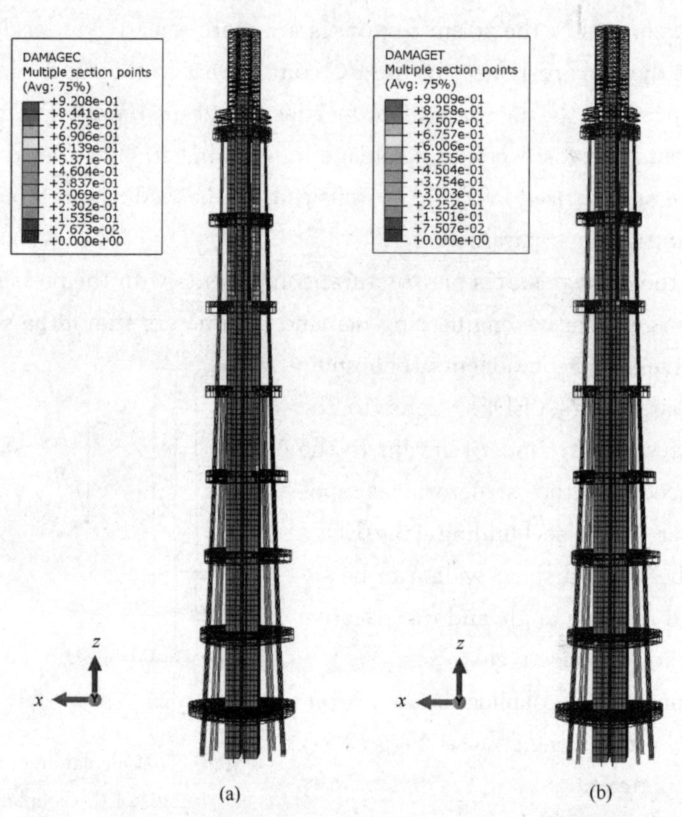

Figure 26 Locations of compressive and tensile damages in core-tube shear walls:

(a) compressive concrete damage; (b) tensile steel damage

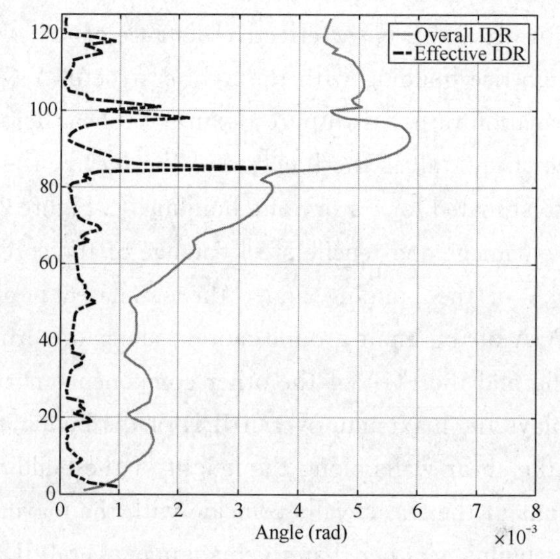

Figure 27 Maximum overall IDR vs maximum effective

IDR (scaled for illustration) of the shear wall along the height of the building

5.2.3 Fragility curve fitting based on simulation results

Based on the computed damage state (damage index) and corresponding EDP of the high rise building, the cumulative distribution function (CDF) of the EDP, defined as the probability of exceeding each damage state for a specific magnitude of EDP[39], can be calculated as

$$F_{\mathrm{dm}}(EDP)=P(DS\geqslant ds_i\mid EDP)=\Phi\left(\frac{\ln(EDP/\hat{\mu})}{\hat{\beta}}\right) \tag{93}$$

where Φ is the standard cumulative normal distribution function, and the mean and logarithmic standard deviation are calculated as $\hat{\mu}=E(EDP)$ and $\hat{\beta}=\sigma_{\ln EDP}$; DS refers to the damage experienced by the structural component; and ds_i is the threshold value of various levels of damage states.

To determine the parameters in the CDF of the lognormal distribution expressed in Equation (93), the maximum likelihood method (MLE) is adopted in this study for the parameter inference[43]. Let us assume m out of n components are damaged, and the likelihood of the given dataset being observed is the product of the likelihoods for each component. The likelihood function can be formulated as

$$Likelihood=\left[\prod_{i=1}^{m}\Phi\left(\frac{\ln(EDP_i/\mu)}{\beta}\right)\right]\left[\prod_{j=1}^{n-m}\left(1-\Phi\left(\frac{\ln(EDP_i/\mu)}{\beta}\right)\right)\right] \tag{94}$$

The concept of the MLE method is to estimate the parameters μ and β of the distribution that maximizes the likelihood function. It is equal to maximize the logarithmic likelihood function which is easier to implement.

$$\{\hat{\mu},\hat{\beta}\}=\underset{\mu,\beta}{\operatorname{argmax}}\left\{\sum_{i=1}^{m}\ln\Phi\left(\frac{\ln(EDP_i/\mu)}{\beta}\right)+\sum_{j=1}^{n-m}\ln\left[1-\Phi\left(\frac{\ln(EDP_i/\mu)}{\beta}\right)\right]\right\} \tag{95}$$

After running a series of NTHA on the 3D FE model of the super high-rise building under bi-directional ground motions, the damage states of the key structural components in terms of strain responses and corresponding EDPs such as effective inter-story drift or chord rotation, can be exported from ABAQUS using PYTHON and post-processed in MATLAB to obtain the fragility curves through the MLE method. The framework used in this study for generating the fragility curves of structural components in the target building is plotted in Figure 28.

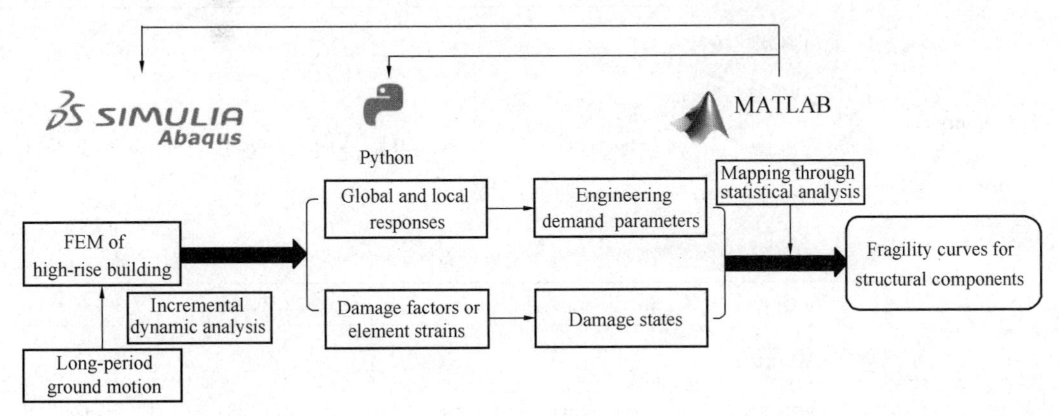

Figure 28 Flowchart of developing fragility curves for structural components

5.2.4 Fragility curves for coupling beams in Zones 7 and 8

The layout of the coupling beams in Zones 7 and 8 are shown in Figure 29 (a) .

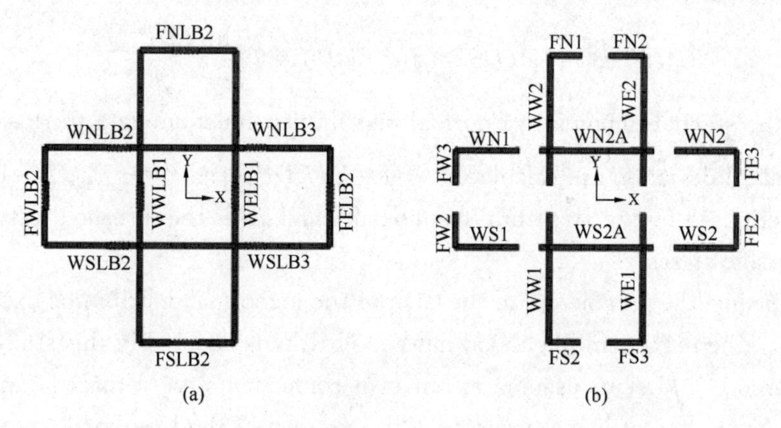

Figure 29 Structural layout of the core-tube in Zones 7 and 8:
(a) coupling beams; (b) shear walls

The coupling beams are divided into two categories (A and B) according to their sizes and the reinforcement ratios. The fragility curve parameters for the two categories of the coupling beams in Zone 8 under short-period and long-period ground motions are summarized in Table 13 for various damage states. Figure 30 shows the fragility curves for the coupling beams in Zone 8 with various damage states, which will be used to compute the damage probabilities of the coupling beams in the subsequent section.

Fragility curve parameters for the two categories of coupling beams in Zone 8 under short-period and long-period ground motions **Table 13**

Category groups	Coupling beam labels	Ground motions	DS1 $\hat{\mu}$ (e−2), $\hat{\beta}$	DS2 $\hat{\mu}$ (e−2), $\hat{\beta}$	DS3 $\hat{\mu}$ (e−2), $\hat{\beta}$
Category A	FNLB2 FSLB2 FWLB2 FELB2 WWLB1 WELB1	Short-period	(0.30, 0.26)	(0.52, 0.32)	(0.78, 0.40)
		Long-period	(0.29, 0.25)	(0.49, 0.33)	(0.76, 0.39)
Category B	WNLB2 WSLB2 WNLB3 WSLB3	Short-period	(0.31, 0.28)	(0.52, 0.35)	(0.80, 0.38)
		Long-period	(0.30, 0.27)	(0.52, 0.33)	(0.78, 0.37)

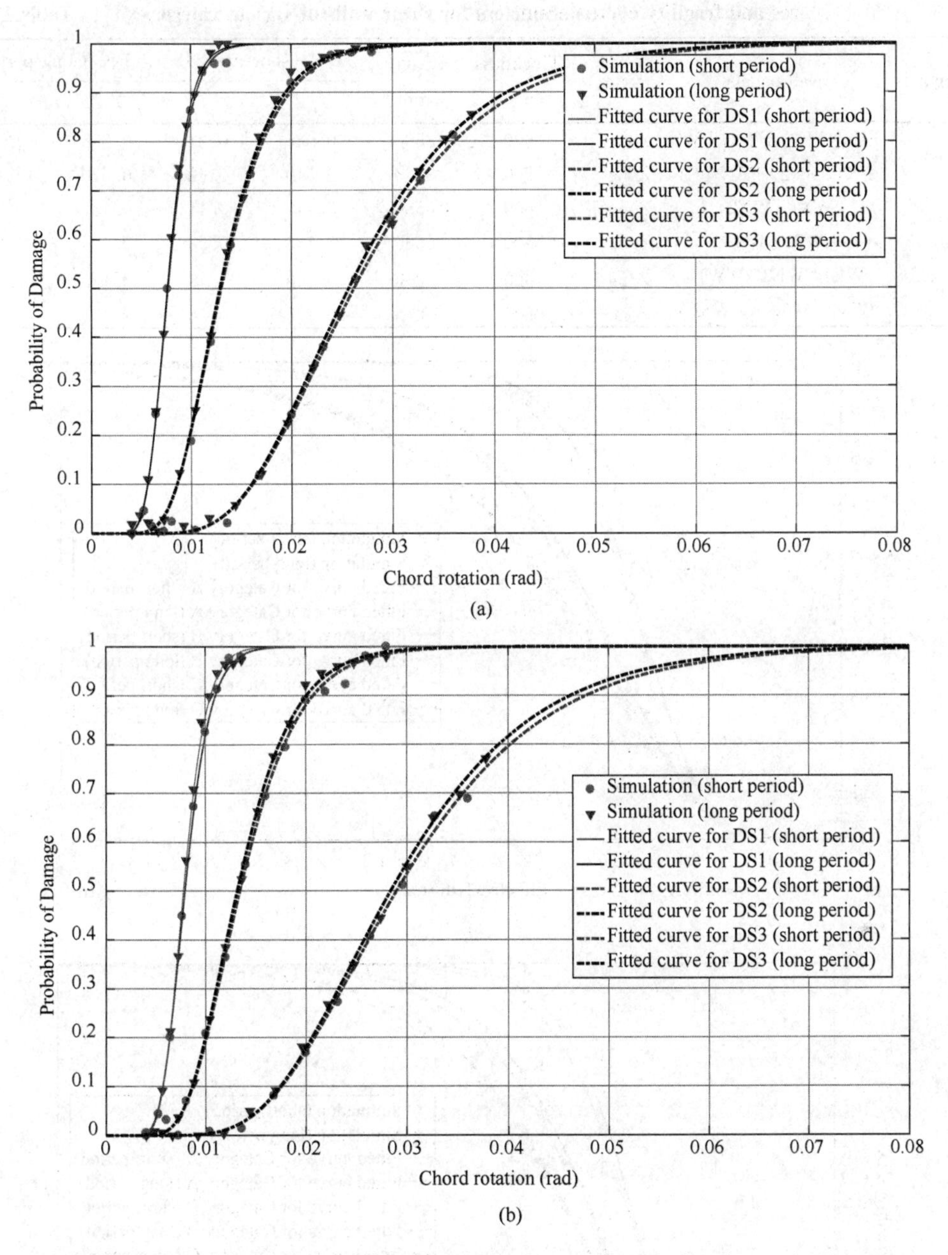

Figure 30 Fragility curves of different damage states for coupling beams in Zone 8:
(a) Category A; (b) Category B

5.2.5 Fragility curves for shear walls in Zones 7 and 8

The layout of the shear walls in Zones 7 and 8 are shown in Figure 29 (b). The sizes and fragility curve parameters for different categories of the shear walls in Zones 7 and 8 under short-period and long-period ground motions are summarized in Table 14 for damage state DS1. The fragility curves for the shear walls in Zones 7 and 8 with different categories are plotted in Figure 31 for DS1.

Sizes and fragility curve parameters for shear walls of various categories　　Table 14

Categories	wall panels	zones	Thickness (mm)	Length (mm)	DS1 (short-period)		DS1 (long-period)	
					$\hat{\mu}$ (e−3)	$\hat{\beta}$	$\hat{\mu}$ (e−3)	$\hat{\beta}$
A	FE2, FE3, FW2, FW3	Zone 7	600	3500	0.045	0.31	0.038	0.28
	FN1, FN2, FS2, FS3	Zone 8	500	3500	0.040	0.28	0.030	0.26
B	WN1, WN2,	Zone 7	600	6580	0.140	0.41	0.130	0.38
	WS1, WS2	Zone 8	500	6580	0.135	0.33	0.125	0.31
C	WE1, WE2, WW1,	Zone 7	600	13230	0.190	0.45	0.180	0.48
	WW2, WN2A, WS2A	Zone 8	500	13230	0.175	0.38	0.170	0.36

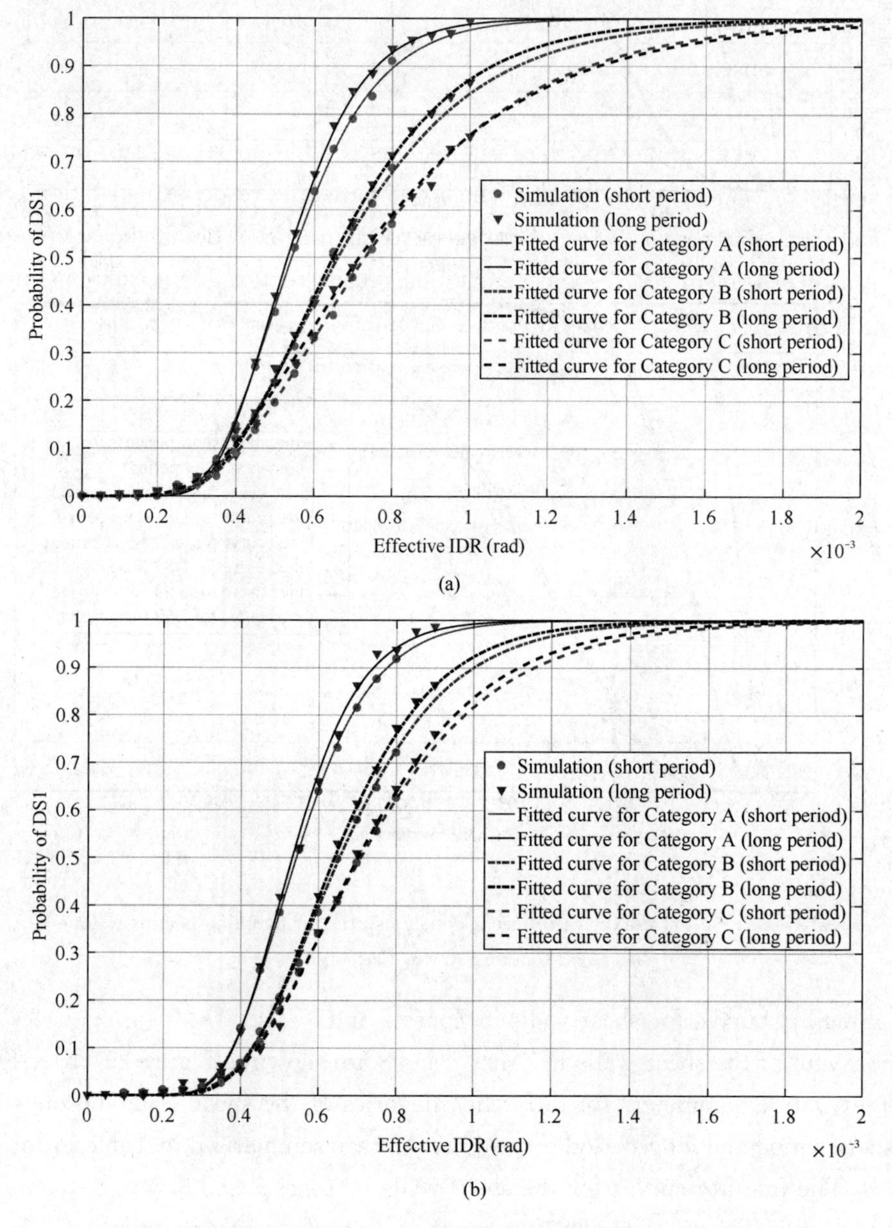

Figure 31　Fragility curves for different categories of shear walls with DS1:

(a) Zone 7; (b) Zone 8

5.3 SHM-Based Seismic Response and Ground Motion Estimation

5.3.1 Optimal sensor placement for best response and ground motion reconstruction

The optimal sensor placement (OSP) scheme introduced in Sections 3 and 4 for the best response and ground motion reconstruction are used, and the sensor candidates are selected for capturing the EDPs of the damage-sensitive structural components. The OSP scheme is not only capable of capturing the coupled bending-torsional deformation of the building but also providing an unbiased estimation of the unmeasured responses of the structural components and the bi-directional ground motions so that the uncertainties involved in the response and ground motion estimations in the previous studies can be significantly reduced.

With the target of achieving the best response-excitation reconstruction results, an optimal sensor placement configuration which minimizes the response estimation error can be obtained through a sequential sensor placement algorithm. A detailed description of the optimal sensor placement for response-excitation reconstruction for seismic monitoring of the super high-rise buildings can be found in reference[44]. The optimized sensor placement configuration obtained for the Shanghai Tower is shown in Figure 32.

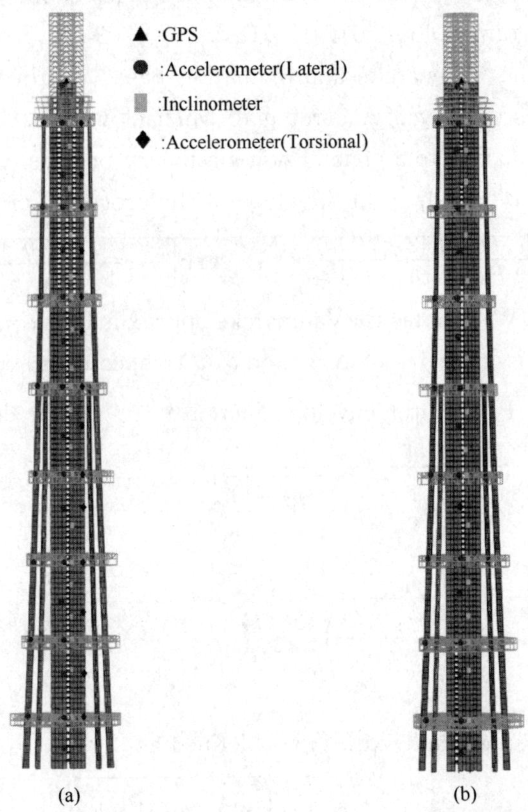

Figure 32 The configuration of the optimal sensor placement:
(a) the east-west direction (principal component); (b) the north-south direction

5.3.2 Analytical solution for EVD prediction of reconstructed structural responses

Let us consider the probability of a non-stationary process $Z(t)$ exceeding a given threshold α during an interval $(0, T)$. $Z(t)$ can be seen as a linear function of the system state $\hat{z}$ defined previously, such as the effective inter-story drift ratio. Define $Z_{max} = \max[|Z(t)|]$ and let $\eta_Z(\alpha, t)$ denote the transient up-crossing rate of $Z(t)$ above the threshold α within the time interval $(t, t+dt)$. The cumulative distribution function (CDF) of the extreme value of the non-stationary process $Z(t)$ is given by

$$P(Z_{max} > \alpha) = 1 - P(Z_{max} \leqslant \alpha) \cong 1 - \exp\left[-\int_0^T \eta_Z(\alpha, t)\,dt\right] \tag{96}$$

In general, $Z(t)$ is not a zero-mean process, and both the mean value and variance of $Z(t)$ vary with time. It thus becomes an issue of calculating the CDF of a non-stationary process with a time-varying mean to up-cross a fixed threshold. To facilitate the finding of such a CDF, the following transformation is performed.

$$X(t) = Z(t) - \mu_Z(t) \tag{97}$$

$$\beta(t) = \alpha - \mu_Z(t) \tag{98}$$

where $\mu_Z(t)$ is the time-varying mean of the non-stationary process $Z(t)$.

Therefore, the goal of determining the probability that the random variable $Z(t)$ up-crosses a threshold α is equivalent to the determination of the probability that the random variable $X(t)$ up-crosses a threshold $\beta(t)$ with $\eta_Z(\alpha, t) = \eta_X(\beta, t)$. To determine the transient up-crossing rate, the Vanmarcke approximation based on the two-state Markov assumption of crossings is employed. According to Michaelov et al. [45] and Hu and Xu[46], the Vanmarcke approximation to a general non-stationary process can be expressed as the up-crossing rate of $V(t)$, which is the envelope of the process $X(t)$ as

$$\eta_X^V(\beta,t) = \frac{[1-F_V(\beta,t)]2\eta_X(0,t)}{F_V(\beta,t)} \cdot \left\{1-\exp\left[-\frac{\eta_V(\beta,t)}{[1-F_V(\beta,t)]2\eta_X(0,t)}\right]\right\} \tag{99}$$

where the superscript "V" denotes the Vanmarcke approximation; η_X and η_V are the unconditional transient up-crossing rates of $X(t)$ and $V(t)$ respectively; and F_V is the transient CDF of $V(t)$. The joint PDF of the envelop function $V(t)$ and its derivative $\dot{V}(t)$ is:

$$f_{XX}(\beta,\dot{x},t) = \frac{1}{2\pi\sigma_X(t)\sigma_{\dot{X}}(t)q_X(t)\sqrt{1-\frac{\rho_{X\dot{X}}^2(t)}{q_X^2(t)}}} \cdot \exp\left[-\frac{1}{2\left(1-\frac{\rho_{X\dot{X}}^2(t)}{q_X^2(t)}\right)}\left(\frac{\beta^2}{\sigma_X^2(t)} - 2\frac{\beta\dot{x}\rho_{X\dot{X}}(t)}{\sigma_X(t)\sigma_{\dot{X}}(t)q_X(t)} + \frac{\dot{x}^2}{\sigma_{\dot{X}}^2(t)q_X^2(t)}\right)\right] \tag{100}$$

where $q_X(t)$ is the transient bandwidth factor defined as:

$$q_X(t) = \sqrt{1-\left\{\frac{\int_0^\infty \omega A(\omega,t)^2\,d\omega}{\sigma_X(t)\sigma_{\dot{X}}(t)}\right\}} \tag{101}$$

By using Equations (100) and (101), the formula for theVanmarke approximation originally developed by Michaelov et al. [45] can be extended for the crossing rate of $V(t)$ as:

$$\eta_X^V(\beta,t)=\frac{1}{\pi}\sqrt{1-\rho_{X\dot{X}}^2(t)}\frac{\sigma_{\dot{X}}(t)}{\sigma_X(t)}$$

$$\frac{1-\exp\left[-\sqrt{\frac{\pi}{2}}\frac{\beta}{\sigma_X(t)}\sqrt{\frac{q_X^2(t)-\rho_{X\dot{X}}^2(t)}{1-\rho_{X\dot{X}}^2(t)}}\Psi\left(\frac{\rho_{X\dot{X}}(t)\beta}{\sqrt{q_X^2(t)-\rho_{X\dot{X}}^2(t)}\sigma_X(t)}\right)\right]}{\exp\left[\frac{\beta^2}{2\sigma_X^2(t)}\right]-1} \tag{102}$$

where $\Psi(\xi)=\exp\left(-\frac{\xi^2}{2}\right)+\sqrt{2\pi}\xi\Phi(\xi)$ with $\Phi(\xi)$ denoting the cumulative normal distribution function.

Substituting the obtained up-crossing rate defined in Equation (102) into Equation (96), the CDF of the extreme value of theVanmarcke approximation of a random process can be determined.

5.3.3 Numerical results of response reconstruction and EVD prediction

A long-period ground motion accelerogram for a $M=8$, $R=400$ km earthquake scenario is generated based on the ground motion simulation scheme proposed by the authors in reference[47], where M is the earthquake magnitude and R is the distance from the earthquake source to the site of the building. The PGA of the synthetic ground motion is scaled to 0.1g as the principal component of the bi-directional ground motions, which coincides with the moderate earthquake design level in the Chinese Seismic Design Code for Shanghai. The PGA of the input is then scaled to 0.85g as the ground motion component in the north-south direction. To evaluate the accuracy of the proposed framework and the optimal multi-type sensor placement obtained, the responses of the 3D FE model of the building under the bi-directional ground motions are first computed. The computed responses at the optimal sensor locations are then contaminated with 5% root means square (RMS) noises and taken as the measured responses (acceleration, inclination and displacement responses). These measured responses are finally used to estimate (reconstruct) the responses of the 3D FE model of the building at all the response reconstruction locations as well as the ground motions. These reconstructed responses and ground motions could then be compared with the actually computed responses and input ground motions to assess the accuracy of the proposed framework. The comparative results between the actual input and reconstructed ground motions are shown in Figure 33(a) and Figure 33(b). The comparative results show a good agreement between the reconstructed ground motions and the actual input ground motions.

The structural responses of the 3D model under the bi-directional long-period ground motions are then reconstructed through the Kalman smoother-based state estimation algorithm based on the limited measured responses. Figure 34 shows the good comparison between the simulated and reconstructed results of the maximum overall IDR responses of

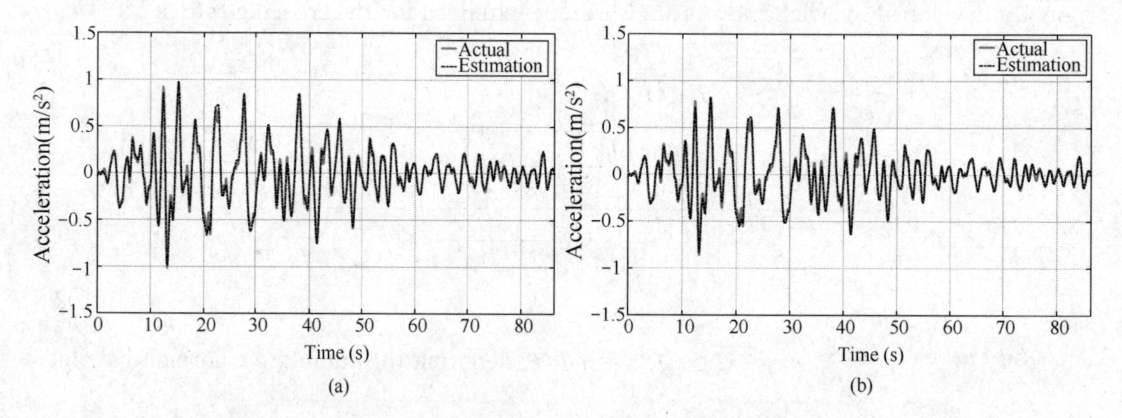

Figure 33 Comparison of actual and reconstructed bi-directional ground motions:
(a) east-west direction; (b) north-south direction

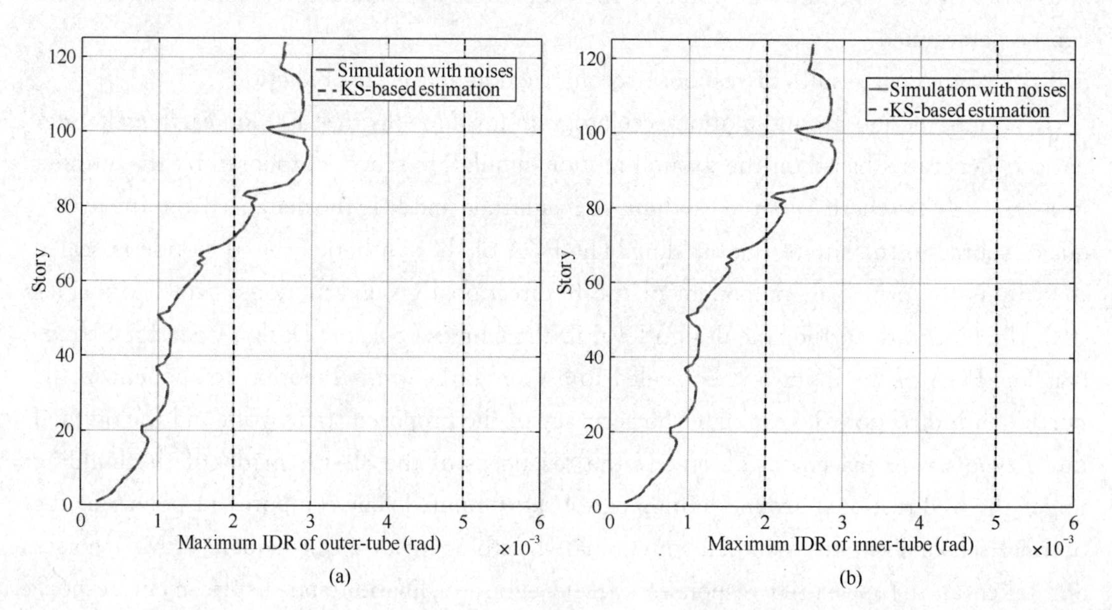

Figure 34 Comparison of the maximum IDR responses of the shear walls between the simulated and estimated results:
(a) shear wall FN1; (b) shear wall WN2A

the shear walls of the high-rise building. It can be seen that the maximum overall IDR response occurs in the 95th story of the super high-rise building.

The chord rotational angles of the coupling beams can be deduced from the inclination of the adjacent shear-walls. Figure 35 shows the chord rotation response of the beam FN-LB2 in the 95th story. The close-up view of the peak values of the chord rotation response of the beam together with the Monte Carlo (MC) simulation is shown in Figure 36. The extreme value distribution results based on the analytical prediction function and the MC simulation are shown in Figure 37. All the results show the high-quality response reconstruction.

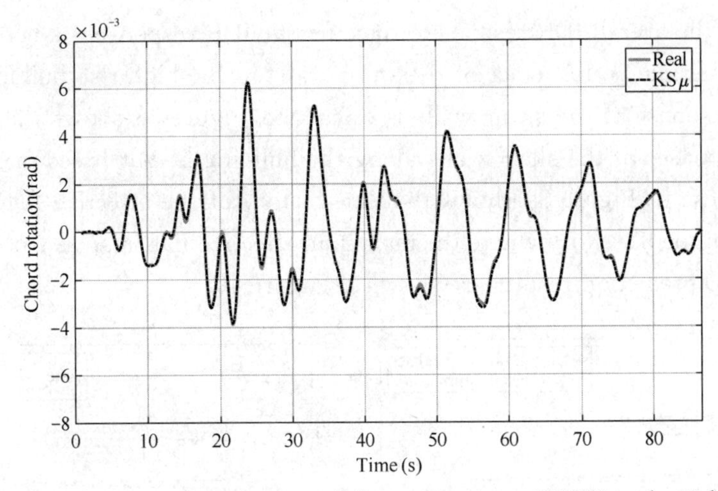

Figure 35　Chord rotation responses of the coupling beam FNLB2 in the 95th story

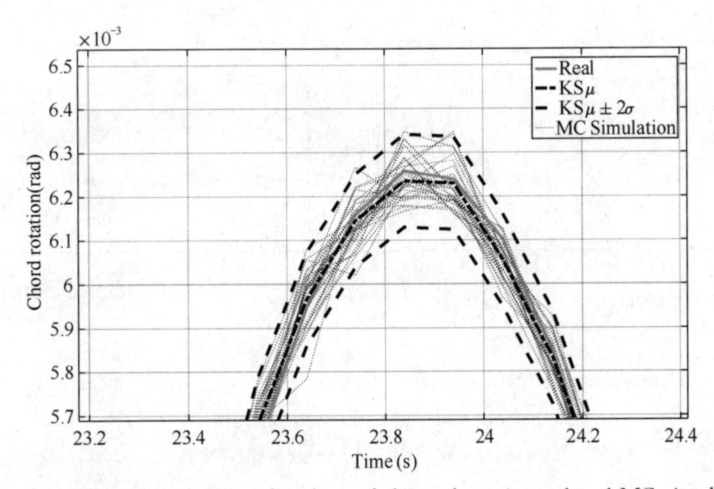

Figure 36　Close-up view of the peak values of the real, estimated and MC simulated chord rotation responses of the coupling beam FNLB2

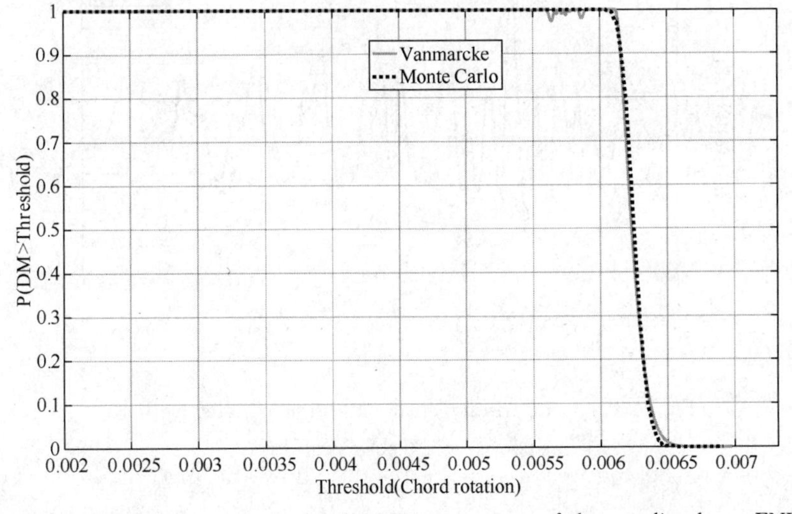

Figure 37　CDF of the extreme chord rotation responses of the coupling beam FNLB2

Compared with the overall IDR response, the effective IDR response is a better damage indicator for the seismic performance of the shear walls in the high-rise building. Thus, the effective IDR response of the shear walls is calculated. Figure 38 shows the maximum effective IDR responses of the shear walls along the building height based on the simulated and estimated results. Figure 39 shows the time history of the effective IDR responses of the shear wall in the 85th story where the maximum effective IDR response occurs. The corresponding EVD prediction results are shown in Figure 40.

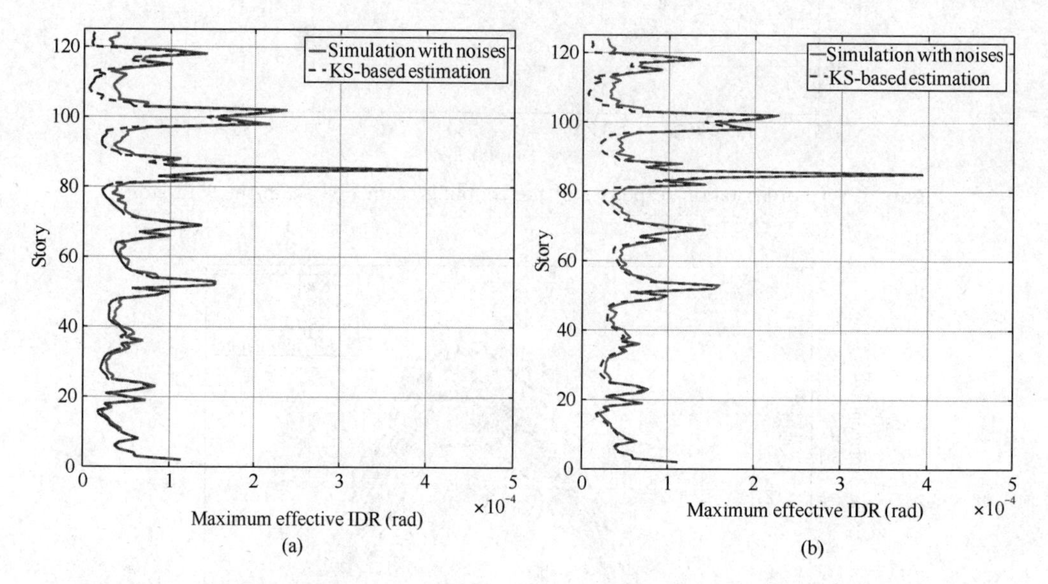

Figure 38 Comparison of the maximum effective IDR responses of shear walls between simulated and estimated results:

(a) shear wall FN1; (b) shear wall WN2A

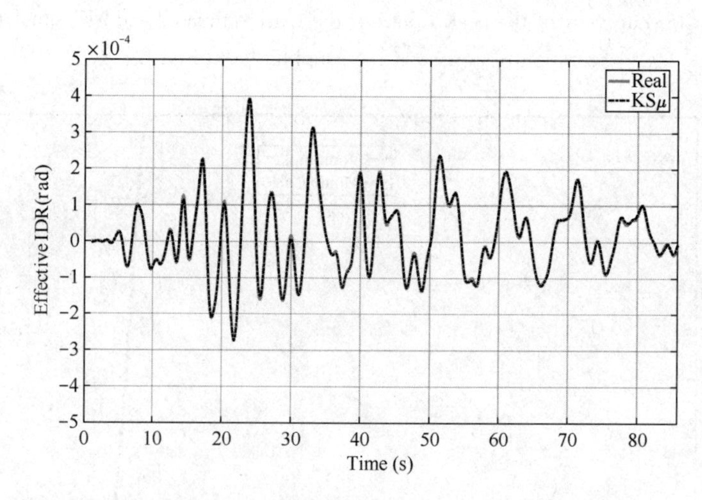

Figure 39 Comparison of real and estimated effective IDR responses of the shear wall FN1

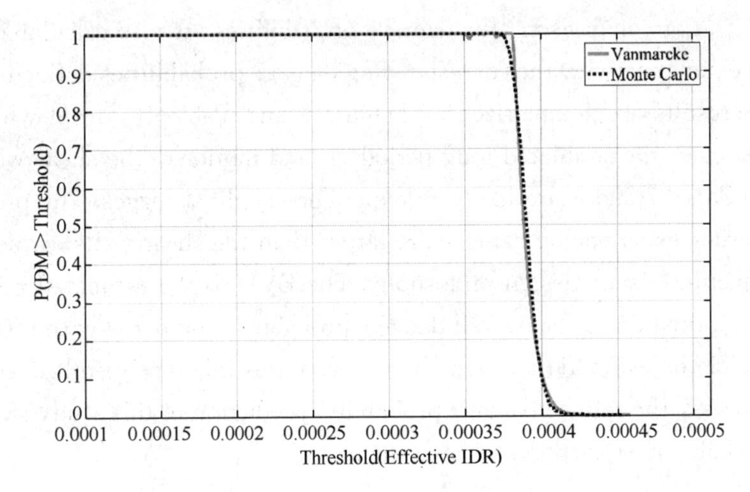

Figure 40 CDF of the extreme effective IDR response of the shear wall FN1

5.4 Component-Level Performance Assessment

By using the reconstructed responses and ground motions stemming from the measured responses, the uncertainties in the seismic performance assessment of the super high-rise building are reduced. Given that all the structural responses of the key components can be reconstructed, the component-level seismic performance assessment of the super high-rise building can be carried out. The component-level performance assessment is performed by using (a) fragility curves for the coupling beams conditioned on chord rotation and (b) fragility curves for the shear walls conditioned on effective IDR.

The estimated extreme value distribution of the effective IDR is now incorporated with the fragility curves of the shear walls to estimate the damage probabilities of the shear walls. The estimated extreme value distribution of the chord rotations is now incorporated with the fragility curves of the coupling beams to estimate the damage probabilities of the coupling beams. The joint probability of the damage state could be rewritten as

$$F_{\mathrm{dm}} = P(DS \geqslant ds_i \mid EDP = \alpha) = \int_0^{\infty} p(DS \geqslant ds_i \mid \alpha) \cdot P(Z_{\max} > \alpha)\,\mathrm{d}\alpha \qquad (103)$$

where $p(DS \geqslant ds_i \mid \alpha)$ refers to the probability density function (PDF) of damage state given the EDP α which can be obtained from the CDF formula of fragility curve; and $P(Z_{\max} > \alpha)$ is the CDF of the estimated responses up-crossing the threshold α.

The fragility curves of the structural components in the super high-rise building obtained under long-period ground motions are quite different from those obtained under short-period ground motions. Thus, the selected fragility curves should be compatible with the type of input excitation. The damage probabilities based on the EVD of the estimated responses for each type of shear walls in the 85th story and each type of coupling beams in the 95th story are calculated. In practice, the mean value of the estimated responses are used to calculate the damage probabilities of the structural components in terms of the fragility curves[48-51]. For comparison, the damage probabilities are also calculated by multiplying the mean estimation with the corresponding damage probabilities defined in the fragility curves. The results are summarized in Tables 15 and Table 16, from which it can be concluded that under the simulated long-period ground motions, the shear walls and coupling beams in Zones 7 and 8 are likely to experience slight damages. The probabilities of the coupling beams experiencing damage are larger than the shear walls, which is consistent with the ductile seismic design philosophy. The EVD of the estimated responses contributes more information to the overall damage probability compared with just considering the mean value of the estimation. Given the uncertainties inherently embedded in the currently-used process, the refined damage probability prediction in this study would yield safer and more reliable post-earthquake assessment.

Damage probability of coupling beams in the 95th story under simulated long-period ground motion **Table 15**

Category	Beam label	Damage states	Damage Probability (%)	
			Mean estimation	EVD estimation
Category A	FNLB2	DS1	17.3	28.2
	FSLB2	DS2	1.25	2.64
		DS3	<1	<1
Category B	WNLB2	DS1	16.2	26.4
	WSLB2	DS2	<1	1.75
	WNLB3	DS3	<1	<1
	WSLB3			

Damage probability of shear walls in the 85th story under simulated long-period ground motion **Table 16**

Component Type	Wall panel	Damage states	Damage Probability	
			Mean estimation	EVD estimation
Category A	FN1, FN2, FS2, FS3	DS1	11.8	15.3
Category B	WN1, WN2, WS1, WS2	DS1	7.1	8.5
Category C	WN2A, WS2A	DS1	6.9	8.2

It should be noted that the structural responses and the corresponding EDPs of the key structural components under the short-period ground motions are smaller than those under the long-period ground motions of the same PGA=0.1g. According to the simulation results in reference (Hu and Xu 2019), the maximum IDR responses under short-period ground motions is only about 70% of those under long-period ground motions, which means that the EDPs are too small to cause considerable damages to the structural components. Thus, the damage probabilities of the structural components of the building under the short-period ground motions are not given in this study.

6. Conclusions

The method for the determination of the minimal number and optimal location of both control devices and sensors for vibration control of building structures under earthquake excitation has been presented. The number and location of control devices are first determined in terms of the sequence of increments of performance index and the predetermined control performance. The response reconstruction method is then extended to the controlled building structure for the determination of the minimal number and optimal placement of sensors with the objective that the reconstructed structural responses can be used as feedbacks for the vibration control while the predetermined control performance can be maintained. The feasibility and accuracy of the proposed method are finally confirmed through the numerical case study of a 20-story shear building structure under the El-Centro ground excitation. The number and location of sensors determined by this collective placement method can be used for integrated structural vibration control and structural health monitoring of building structures.

The method for integrated structural vibration control and health monitoring of building structures has been developed in the time domain in terms of the advanced collective placement method with the projection matrix and the extended Kalman filter (EKF) for simultaneous consideration of vibration mitigation and time-invariant parameter identification of building structures without the knowledge of the external excitations. The efficiency and accuracy of the proposed method have been experimentally validated via a five-story building structure equipped with magneto-rheological (MR) dampers.

A probabilistic seismic performance assessment framework for instrumented super high-rise buildings under bi-directional long-period ground motions has been proposed and applied to a real super high-rise building. A large number of fragility curves of the key structural components in the super high-rise building of a hybrid structural system are produced. The accurate estimations of the structural responses at all the key locations as well as the unknown bi-directional ground motions are achieved using the Kalman filter/smoother-based response reconstruction scheme. The extreme value distributions (EVD) of the estimated responses are obtained, and the damage probabilities for the structural components are finally determined by incorporating the EVD values with the fragility curves for the key structural components. The results show that the results based on the EVD values could provide safer and more reliable post-earthquake assessment for the structural components and that the damage probabilities of the structural components under the short-period ground motions are much smaller than under the long-period ground motions for the super high-rise building.

7. Acknowledgements

The research works described in this paper are financially supported by the Hong Kong Research Grants Council and the Hong Kong Polytechnic University through several competitive research grants and collaborative research grants. The author would like to thank Prof. Jun Teng, Dr Jia He, Dr Qin Huang, Dr Rongpan Hu and Mr Sheng Zhan for their significant contributions to the research works described in this paper.

References

[1] Chen, B. and Xu, Y. L. (2008). Integrated vibration control and health monitoring of building structures using semi-active friction dampers: Part II-Numerical investigation, *Engineering Structures*, 30 (3), 573-587.

[2] Xu, Y. L. and Chen, B. (2008). Integrated vibration control and health monitoring of building structures using semi-active friction dampers: Part I-methodology, *Engineering Structures*, 30 (7), 1789-1801.

[3] Huang, Q. , Xu, Y. L. , Li, J. C. , Su, Z. Q. and Liu, H. J. (2012). Structural damage detection of controlled building structures using frequency response functions, *Journal of Sound and Vibration*, 331(15): 3476-3492.

[4] Xu, Y. L. , Huang, Q. , Zhan, S. , Su, Z. Q. and Liu, H. J. (2014). FRF-based structural damage detection of controlled buildings with podium structures: Experimental investigation, *Journal of Sound and Vibration*, 333(13), 2762-2775.

[5] Gattulli, V. and Romeo, F. (2000). Integrated procedure for identification and control of MDOF structures, *Journal of Engineering Mechanics*, 126(7), 730-737.

[6] Chen, B. , Xu, Y. L. and Zhao, X. (2008). Integrated vibration control and health monitoring of building structures: a time-domain approach, *Smart Structures and Systems*, 6(7), 811-833.

[7] Ding, Y. , and S. S. Law. 2011. Integration of structural control and structural evaluation for large scale structural system. In: Proceedings of SPIE 7977, Active and Passive Smart Structures and Integrated Systems 2011, San Diego, California, USA. ed. M. N. Ghasemi-Nejhad, 797724, doi: 10. 1117/12. 880690.

[8] Lin, C. H. , Sebastijanovic, N. , Yang, H. T. Y. , He, Q. and Han, X. Y. (2012). Adaptive structural control using global vibration sensing and model updating based on local infrared imaging, *Structural Control and Health Monitoring*, 19(6), 609-626.

[9] Karami, K. and Amini, F. (2012). Decreasing the damage in smart structures using integrated online DDA/ISMP and semi-active control, *Smart Materials and Structures*, 21(10), 105017.

[10] Lei, Y. , Wu, D. T. and Lin, S. Z. (2013). Integration of decentralized structural control and the identification of unknown inputs for tall shear building models under unknown earthquake excitation, *Engineering Structures*, 52, 306-316.

[11] Lei, Y. , Zhou, H. and Liu, L. J. (2014). An on-line integration technique for structural damage detection and active optimal vibration control, *International Journal of Structural Stability and Dynamics*, 14(5), 1440003, DOI: 10. 1142/S0219455414400033.

[12] He, J. , Huang, Q. and Xu, Y. L. (2014). Synthesis of vibration control and health monitoring of

building structures under unknown excitation. Smart Materials and Structures, 23: 105025. doi: 10.1088/0964-1726/23/10/105025.

[13] Amini, F., Mohajeri, S. A. and Javanbakht, M. (2015). Semi-active control of isolated and damaged structures using online damage detection. Smart Materials and Structures, 24: 105002. doi: 10.1088/0964-1726/24/10/105002.

[14] Xu, Y. L., Huang, Q., Xia, Y. and Liu, H. J. (2015). Integration of health monitoring and vibration control for smart building structures with time-varying structural parameters and unknown excitations. Smart Structures and Systems, 15(3): 807-830.

[15] Nagarajaiah, S. (2009). Adaptive passive, semiactive, smart tuned mass dampers: Identification and control using empirical mode decomposition, Hilbert transform, and short-term Fourier transform, *Structural Control and Health Monitoring*, 16(7-8), 800-841.

[16] Yang, H. T. Y., Shan, J. Z., Randall, C. J., Hansma, P. K. and Shi, W. X. (2014). Integration of health monitoring and control of building structures during earthquakes, *Journal of Engineering Mechanics*, 140(5), 04014013.

[17] He, J., Xu, Y. L., Zhan, S. and Huang, Q. (2016). Structural control and health monitoring of building structures with unknown ground excitations: Experimental investigation, Journal of Sound and Vibration.

[18] Moehle, J., and Deierlein, G. G. (2004). A framework methodology for performance-based earthquake engineering, *Proc.*, 13*th World Conf. on Earthquake Engineering*. Vancouver, BC: WCEE.

[19] Cremen, G. and Baker, J. W. (2018). Quantifying the benefits of building instruments to FEMA P-58 rapid post-earthquake damage and loss predictions, *Engineering Structures*, 176, 243-253.

[20] Housner, G. W., Bergman, L. A., Caughey, T. K., Chassiakos, A. G., Claus, R. O., Masri, S. F., Skelton, R. E., Soong, T. T., Spencer, B. F. and Yao, J. T. P. (1997). Structural Control: Past, Present and Future. *Journal of Engineering Mechanics*, 123(9), 897 - 971.

[21] Soong, T. T. (1990). Active structural control: theory and practice. Longman; London, and Wiley: New York.

[22] Zhang, R. H. and Soong, T. T. (1992). Seismic design of viscoelastic dampers for structural applications, *Journal of Structural Engineering*, 118(5), 1375-1392.

[23] Haftka, R. T. and Adelman, H. M. (1985). Selection of actuator locations for static shape control of large space structures by heuristic integer programming, *Computers and Structures*, 20(1-3), 578-582.

[24] Agrawal, A. K. and Yang, J. N. (1999). Optimal placement of passive dampers on seismic and wind-excited buildings using combinatorial optimization, *Journal of Intelligent Material Systems and Structures*, 10(12), 997-1014.

[25] Zhang, X. H. (2012). Multi-sensing and multi-scale monitoring of long-span suspension bridges. PhD thesis, Department of Civil and Environmental Engineering, The Hong Kong Polytechnic University, Hong Kong.

[26] Spencer Jr., B. F., Dyke, S. J. and Deoskar, H. S. (1998). Benchmark problems in structural control: part II-active tendon system, Earthquake Engineering and Structural Dynamics, 27(11), 1141-1147.

[27] Dyke, S. J., Spencer Jr, B. F., Sain, M. K. and Carlson, J. D. (1996). Modeling and control of magnetorheological dampers for seismic response reduction, *Smart Materials and Structures*, 5(5), 565-575.

[28] Dyke, S. J. and Spencer Jr., B. F. (1997). A comparison of semi-active control strategies for the MR damper. In: Proceedings of Intelligent Information Systems, IEEE, Grand Bahama Island, 580-584.

[29] Jansen, L. M. and Dyke, S. J. (2000). Semiactivecontrol strategies for MR dampers: comparative study, *Journal of Engineering Mechanics*, 126(8), 795-803.

[30] Xu, Y. L., Qu W. L. and Ko J. M. (2000). Seismic response control of frame structures using magnetorheological/electrorheological dampers, *Earthquake Engineering and Structural Dynamics*, 29 (5), 557-575.

[31] Ou, J. P. (2003). Structural vibration control: active, semi-active, and intelligent control. Beijing: Science Press.

[32] Ali, M. and Moon, K. (2018). Advances in Structural Systems for Tall Buildings: Emerging Developments for Contemporary Urban Giants, *Buildings*, 8(8), 104.

[33] Su, J. Z., Xia, Y., Chen, L., Zhao, X., Zhang, Q. L., Xu, Y. L., Ding, J. M., Xiong, H. B., Ma, R. J., Lv, X. L. and Chen, A. R. (2013). Long-term structural performance monitoring system for the Shanghai Tower, *Journal of Civil Structural Health Monitoring*, 3(1), 49-61.

[34] Liu, T., Yang, B. and Zhang, Q. (2017). Health Monitoring System Developed for Tianjin 117 High-Rise Building, *Journal of Aerospace Engineering*, 30(2), B4016004.

[35] Mander, J. B., Priestley, M. J. and Park, R. (1988). Theoretical stress-strain model for confined concrete, *Journal of. Structural Engineering*, 114(8), 1804-1826.

[36] Smith, M. 2013. *ABAQUS/Standard User's Manual*, *Version* 6.13, Simulia, Providence, RI.

[37] Zhou, C., Tian, M. and Guo, K. (2019). Seismic partitioned fragility analysis for high - rise RC chimney considering multidimensional ground motion, *Structural Design of Tall Special Buildings*, 28(1), e1568.

[38] Lu, X., Lu, X., Sezen, H. and Ye, L. (2014). Development of a simplified model and seismic energy dissipation in a super-tall building, *Engineering Structures*, 67, 109-122.

[39] Sengupta, P. and Li, B. (2014). Seismic fragility evaluation of lightly reinforced concrete beam-column joints, *Journal of. Earthquake Engineering*, 18(7), 1102-1128.

[40] Park, Y. J., Ang, A. H. S. and Wen, Y. K. (1985). Seismic damage analysis of reinforced concrete buildings, *Journal of Structural Engineering*, 111(4), 740-757.

[41] GB. (2009). *GB/T* 24335-2009: *Classification of earthquake damage to buildings and special structures*. [In Chinese]China Ministry of Construction, Beijing, China: Standards Press of China.

[42] Aslani, H., and Miranda, E. (2005). Probability-based seismic response analysis, *Engineering Structures*, 27(8), 1151-1163.

[43] Lallemant, D., Kiremidjian, A. and Burton, H. (2015). Statistical procedures for developing earthquake damage fragility curves, *Earthquake Engineering Structural Dynamics*, 44(9), 1373-1389.

[44] Hu, R., Xu, Y. L., Lu, X., Zhang, C., Zhang, Q. and Ding, J. (2018). Integrated multi - type sensor placement and response reconstruction method for high - rise buildings under unknown seismic loading, *Structural Design of Tall Special Buildings*, 27(6), e1453.

[45] Michaelov, G., Lutes, L. D. and Sarkani, S. (2001). Extreme value of response to nonstationary excitation, *Journal of Engineering Mechanics*, 127(4), 352-363.

[46] Hu, L., and Xu, Y. L. (2014). Extreme value of typhoon-induced non-stationary buffeting response of long-span bridges, *Probabilistic Engineering Mechanics*, 36, 19-27.

[47] Hu, R. P., Xu, Y. L. and X. Zhao, X. (2017). Long-Period Ground Motion Simulation and its Impact on Seismic Response of High-Rise Buildings, *Journal of Earthquake Engineering*, 1-31.

[48] Naeim, F. , Hagie, S. , Alimoradi, A. and Miranda, E. (2006). Automated post-earthquake damage assessment of instrumented buildings, *Proc.* , *Advances in Earthquake Engineering for Urban Risk Reduction*, edited by S. T. Wasti and G. Ozcebe. Dordrecht, Netherlands: Springer.

[49] Naeim, F. (2013). Real-Time Damage Detection and Performance Evaluation for Buildings, *Earthquakes and Health Monitoring of Civil Structures*, 167-196. Dordrecht, Netherland: Springer.

[50] Miranda, E. (2006). Use of probability-based measures for automated damage assessment, *Structural Design of Tall Special Buildings*, 15(1), 35-50.

[51] Porter, K. , Mitrani-Reiser, J. and Beck, J. L. (2006). Near-real-time loss estimation for instrumented buildings, *Structural Design of Tall Special Building*, 15(1), 3-20.

4 Mesoscopic Mechanics Based Design and Performance Evaluation of Metallic Dampers

Hanbin Ge, Liangjiu Jia, Tianyu Gu
(Meijo University 日本名古屋)

Abstract: Metallic dampers are susceptible to ductile fracture under seismic loading. Through studies towards ductile fracture of metals at the mesoscopic level, analytical models of crack initiation, crack propagation and final rupture are proposed. This paper presents a series of plasticity models and mesoscopic mechanics based ductile fracture models for simulating mechanical behaviors of metallic dampers. By combining the plasticity and fracture models, evaluation of failure mode and damage states for metallic dampers under earthquake excitation can be achieved. Topology optimization of conventional metallic dampers can also be realized to achieve expected design objectives such as high ductility and energy dissipation capacity, and a series of new metallic dampers developed based on this concept are summarized in this paper.
Keywords: Plasticity models, Ductile fracture models, Metallic dampers

1. Introduction

Metallic dampers are widely employed in earthquake prone regions for their simple fabrication, low-cost, and easy replacement. To evaluate mechanical properties and simulate hysteretic behaviors of metallic dampers, plasticity models such as the Prager model[1], the Chaboche model[2] and the Yoshida-Uemori model[3,4] are proposed. The Chaboche model possesses a combination of independent backstresses with different formations, which can well describe the nonlinear plasticity of different materials in different strain ranges[5].

For metallic dampers, ductile fracture may happen under cyclic loading with large displacement amplitudes. However, plasticity models cannot predict fracture behaviors, leading to inaccuracy in evaluation and simulation of damping devices[6]. Founded on mesoscopic analysis, ductile fracture models considering crack initiation are proposed based on void growth and void coalescence (the McClintock model[7] and the Rice-Tracey model[8]), porous plasticity (the Gurson-Tvergaard-Needleman model[9]) and continuum damage mechanics[10-12], respectively, among which the Rice-Tracey mode describes the relationship between the growth rate of micro-voids and stress triaxiality with only one model parameter. Besides, a crack propagation rule[13] is proposed for problems involving non-uniform distribution of strain.

In this paper, several plasticity models and ductile fracture models expressed by true stress and true strain are introduced. By the combined plasticity models and fracture models, hysteretic performance of metallic dampers can be generally evaluated and numerical simulation could be more accurate. A series of new damping devices based on the concept of topology optimization are proposed.

2. Plasticity models

2.1 True stress-true strain after necking initiation

The uniaxial true stress-true strain data is required for elasto-plastic numerical simulation, which is commonly obtained from a tension coupon test. For a bar with a uniform cross section, the relationships between true stress σ, true strain ε, engineering stress s and engineering strain e before the occurrence of necking are respectively:

$$\varepsilon = \ln(1+e) \tag{1}$$

$$\sigma = s \cdot (1+e) \tag{2}$$

Instance when necking initiates can be expressed as:

$$\sigma_{\text{neck}} = \frac{\mathrm{d}\sigma}{\mathrm{d}\varepsilon} \tag{3}$$

However, stress state within the neck is triaxial, and true stress obtained by Eqs. (1) to (2) has to be modified after necking initiation. The Modified Weight Average Method[14,15] can estimate true stress-true strain relationship after necking with accuracy:

$$\sigma = \sigma_{\text{neck}} + \omega \cdot \sigma_{\text{neck}}(\varepsilon - \varepsilon_{\text{neck}}) \tag{4}$$

where ω is the optimal weight average factor, which can be obtained from coupon test. σ and ε are true stress and true strain after necking initiation, respectively, and σ_{neck} and $\varepsilon_{\text{neck}}$ are true stress and true strain when necking initiates, respectively,

2.2 Chaboche model with or without isotropic hardening

The Chaboche model is one of the most popular kinetic hardening (KH) models, which can be categorized into the Chaboche model without isotropic hardening (IH) and the Chaboche model with IH. For the two models, the KH is simulated by a combination of backstresses $\boldsymbol{\alpha}$. The evolution rules of the backstresses are defined as a combination of the Prager model and a relaxation term:

$$\mathrm{d}\boldsymbol{\alpha}_i = C_i \frac{1}{\sigma_{y0}}(\boldsymbol{S} - \boldsymbol{\alpha}_i)\mathrm{d}\varepsilon_{\text{eq}}^{\text{p}} - \gamma_i \cdot \boldsymbol{\alpha}_i \cdot \mathrm{d}\varepsilon_{\text{eq}}^{\text{p}} \tag{5}$$

where C_i and γ_i are model parameters, and the overall backstress can be calculated as:

$$\boldsymbol{\alpha} = \sum_{i=1}^{n} \boldsymbol{\alpha}_i \tag{6}$$

The Chaboche model with IH is a combined hardening model with both KH and IH components, and the formation of the KH component is the same as that of the Chaboche model without IH. The IH rule of the yield surface is given as[16]:

$$dR = k(Q_{\infty} - R)\mathrm{d}\varepsilon_{eq}^{p} \tag{7}$$

where k is a parameter to describe the IH rate, R and $\mathrm{d}R$ are the change and incremental change of yield surface, respectively. Q_{∞} is the maximum change of yield surface. For a uniaxial state, the evolution rule of IH of yield surface is expressed as:

$$\sigma = \sigma_{y0} + R = \sigma_{y0} + Q_{\infty}(1 - e^{-k \cdot \mathrm{d}\varepsilon_{eq}^{p}}) \tag{8}$$

However, it is also pointed out that the model parameters are too many and a series of cyclic tests are necessary to calibrate the model parameters.

3. Mesoscopic mechanics based ductile fracture models

3.1 Ductile crack initiation model

Rice and Tracey formulated the relationship between the radius of a void and stress triaxiality based on the analysis of a spherical void in a remote simple tension strain rate field. Assuming that the stress triaxiality, T, is constant during the void growth, the Rice-Tracey model for Mises materials is expressed as:

$$\varepsilon_{eq} = \frac{\ln \dfrac{R}{R_0}}{0.283 \cdot e^{\frac{3}{2}T}} \tag{9}$$

where ε_{eq} is equivalent strain, and R_0 and R are the initial and current radii of the void, respectively. Stress triaxiality, T, is the ratio of hydrostatic stress and equivalent stress.

The void growth rule has been given by the Rice-Tracey model, while a void coalescence rule has not been included[17]. Herein, the void coalescence is assumed to occur when R/R_0 reaches a critical value R_f/R_0:

$$\varepsilon_f = \frac{\ln \dfrac{R_f}{R_0}}{0.283 \cdot e^{\frac{3}{2}T}} = \chi \cdot e^{-\frac{3}{2}T} \tag{10}$$

To extend Eq. (11) to the case of general loading associating with non-constant T, a damage index, D_{ini}, based on the Miner's rule[18] is proposed. Assuming that damage is only due to plastic incremental deformation ε_{eq}^{p}:

$$\mathrm{d}D_{ini} \approx \frac{\mathrm{d}\varepsilon_{eq}^{p}}{\chi \cdot e^{-\frac{3}{2}T}} \tag{11}$$

For cyclic loading, it is found that negative stress triaxiality has a threshold of $-1/3$, below which no fracture will occur[19]. Thus, the model can be modified[20]:

$$
dD_{ini} = \begin{cases} \dfrac{d\varepsilon_{eq}^{p}}{\chi \cdot e^{-\frac{3}{2}T}} & \left(T \geqslant -\dfrac{1}{3}\right) \\ 0 & \left(T < -\dfrac{1}{3}\right) \end{cases} \tag{12}
$$

where χ is a material constant which can be obtained from a coupon test. Ductile fracture initiation is assumed to occur when D_{ini} reaches 1.0.

3.2 Ductile crack propagation model

Eq. (12) assumes that the crack propagates immediately after crack initiation when D_{ini} reaches 1.0. However, A ductile fracture model with only a crack initiation rule generally overestimates the crack propagation rate, especially for the cases where non-uniform strain distribution prevails[21].

A crack propagation rule based on an energy balance approach[13] is proposed in a previous study. A crack propagation index, D_{prop}, is defined as:

$$
D_{prop} = \frac{G}{G_c} \tag{13}
$$

Where G is current absorbed energy of a unit area since fracture initiation, and G_c is the threshold value for absorbed energy of a unit area of crack surface. G_c is a material constant which can be obtained from a coupon test. The deterioration of stress-carrying capacity is considered based on the concept of effective stress σ_e:

$$
\sigma_e = (1 - D_{prop}) \cdot \sigma \tag{14}
$$

The loading /unloading tangent modulus of a damaged material, E_d, is reduced to:

$$
E_d = (1 - D_{prop}) \cdot E \tag{15}
$$

where E is the Young's modulus of undamaged materials. Finally, an analytical model combining plasticity, crack initiation and propagation is illustrated in Fig. 1[22,23].

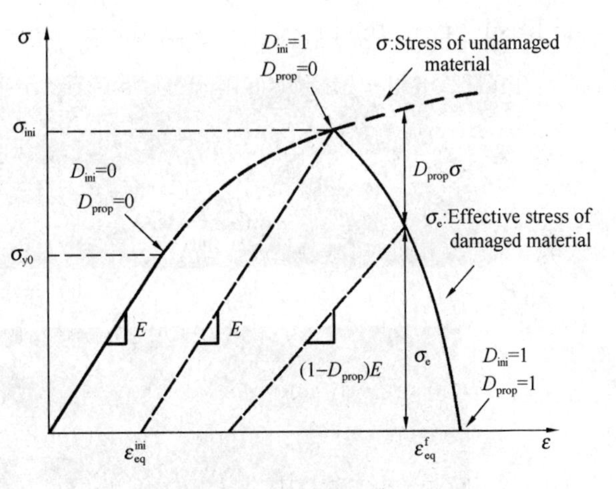

Fig. 1 Illustration of ductile fracture model

4. Metallic dampers proposed

Based on plasticity models and ductile fracture models listed above, a series of novel damping devices based on topology optimization are proposed.

Fish-Bone shaped Buckling-Restraint Braces (FB-BRBs)[24,25] aim to achieve large maximum ductility, cumulative ductility and energy dissipation capacities. The core plate of FB-BRB is divided into several segments by multiple pairs of stoppers as shown in Fig. 2. Post-necking straining of a FB-BRB is more significant through generating several necks in several segments, compared with a single neck for conventional BRBs. The stoppers control the development of the maximum necking deformation, and lead to necking and restraining of the segments subsequently.

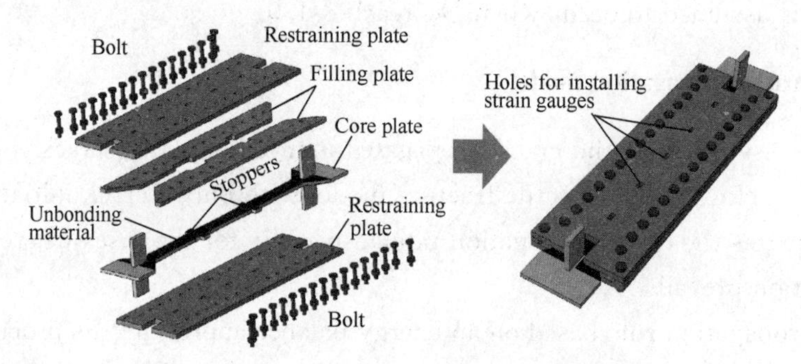

Fig. 2 Configuration of a FB-BRB

To analyze the FB-BRBs, 8 specimens were tested under cyclic loading. By the combined Chaboche model and the ductile fracture model, numerical simulation showed good agreement with experiment results, as shown in Fig. 3 and Fig. 4. Thus, both accuracy and feasibility of the proposed methods are testified.

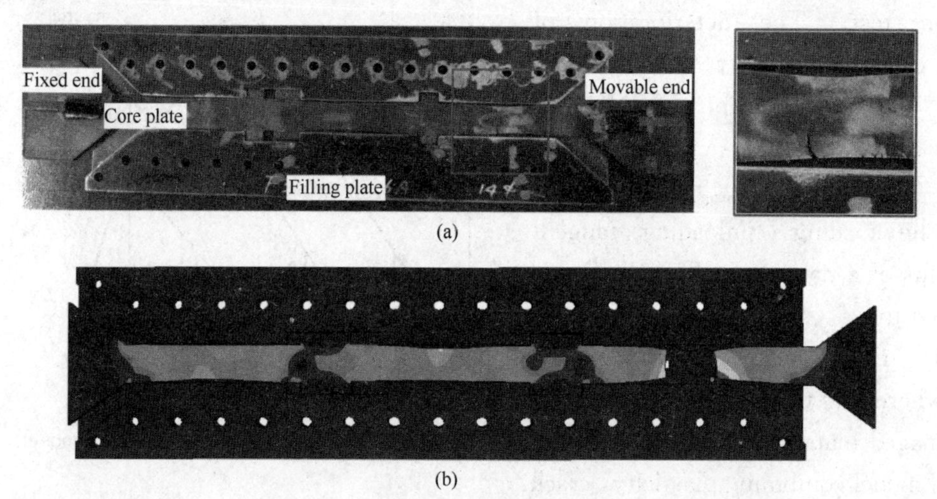

Fig. 3 Experimental results and numerical simulation of a FB-BRB

(a) Necking and fracture for a FB-BRB; (b) Numerical simulation of necking and fracture for a FB-BRB

There have been other related metallic damping devices proposed:

Perforated Buckling-Restraint Braces (PBRBs)[26] refer to BRBs with perforated core plates, as shown in Fig. 5. The deformation mechanism of the PBRBs under compression is shortening of the core plate and in-plane shear deformation of the perforated core plate. The openings of the core plates enable PBRBs to yield rapidly under earthquake strikes and meanwhile enhance energy dissipation capacity.

Brace-type Shear Fuses (BSFs)[27] are a type of all-steel dampers consisting of a core plate, two filling plates, two perforated cover plates, as shown in Fig. 6. It is designed to dissipate seismic energy via yielding mainly under shear. Perforated cover plates also make damage of the core plate visible.

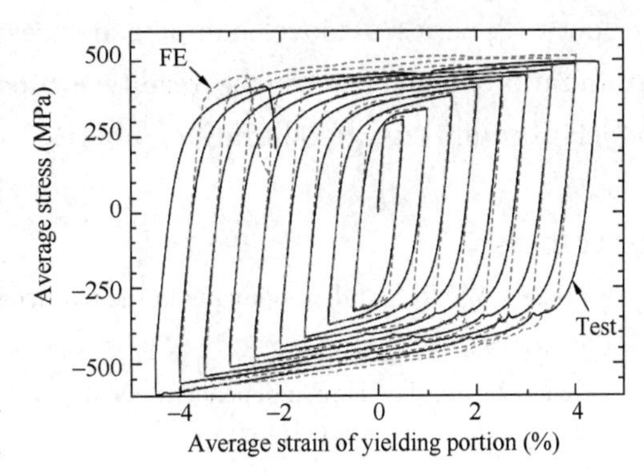

Fig. 4 Comparison between experimental result and numerical modeling

Mini-BRBs[28,29] are all-steel BRBs with small bearing capacities employed at beam-to-column connections, as shown in Fig. 7. The core bars of mini-BRBs are three-side-weakened, and the remaining arcs of core bar were designed to contact the outer tube with minimum displacements, thus mitigating buckling deformation.

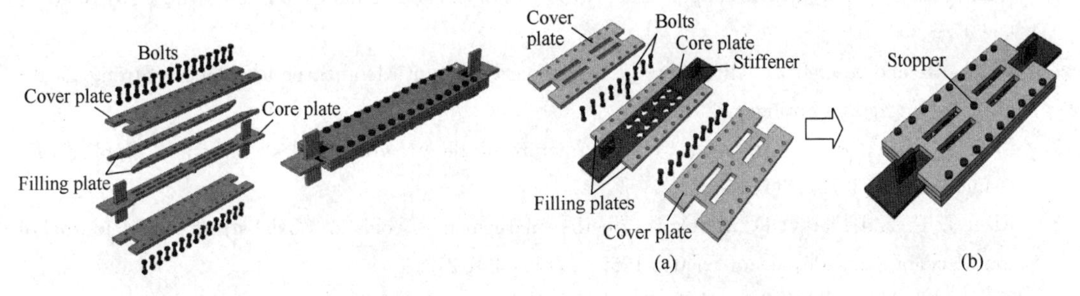

Fig. 5 Configuration of a PBRB

Fig. 6 Configuration of a BSF

(a) Components; (b) Assembled specimen

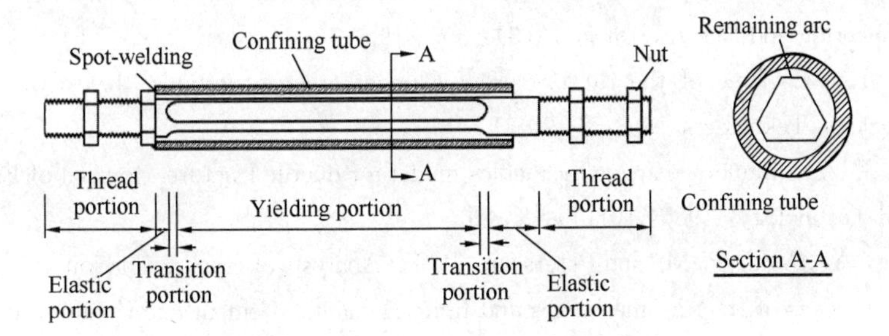

Fig. 7 Configuration of a mini-BRB

5. Conclusions

This paper presents several plasticity models and mesoscopic based ductile fracture models. The proposed ductile fracture models are capable to simulate the whole process of crack initiation, crack propagation and final fracture. The combined models are both simple in their forms and accurate to predict and simulate hysteretic behaviors of metallic dampers. A series of novel damping devices are proposed under the concept of topology optimization, and they can be readily evaluated by the combined plasticity and fracture models through simple calibration.

References

[1] Prager, W. Recent developments in the mathematical theory of plasticity. Journal of Applied Physics, 1949, 20(3): 235-241.

[2] Chaboche, J. L. Time-independent constitutive theories for cyclic plasticity. International Journal of Plasticity, 1986, 2(2): 149-188.

[3] Jia, L. J. and Kuwamura, H. Prediction of cyclic behaviors of mild steel at large plastic strain using coupon test results. Journal of Structural Engineering, 2014, 140(2): 04013056.

[4] Yoshida, F. and Uemori, T. A model of large-strain cyclic plasticity describing the Bauschinger effect and workhardening stagnation. International Journal of Plasticity, 2002, 18(5-6): 661-686.

[5] Jia, L. J., Koyama, T. and Kuwamura, H. Prediction of cyclic large plasticity for prestrained structural steel using only tensile coupon tests. Frontiers of Structural and Civil Engineering, 2013, 7(4): 466-476.

[6] Jia, L. J. and Ge, H. B. Ultra-low-Cycle Fatigue Failure of Metal Structures Under Strong Earthquakes. Singapore: Springer, 2019.

[7] McClintock, F. A. A criterion for ductile fracture by the growth of holes. Journal of Applied Mechanics, 1968, 35(2): 363-371.

[8] Rice, J. R. and Tracey, D. M. On the ductile enlargement of voids in triaxial stress fields. Journal of the Mechanics and Physics of Solids, 1969, 17(3): 201-217.

[9] Tvergaard, V. and Needleman, A. Analysis of the cup-cone fracture in a round tensile bar. Acta Metallurgica, 1984, 32(1): 157-169.

[10] Chaboche, J. L. Anisotropic creep damage in the framework of continuum damage mechanics. Nuclear Engineering and Design, 1984, 79(3): 309-319.

[11] Kachnov, L. M. Time of the rupture process under creep conditions. Izvestiya Akademii Nauk SSSR Otdelenie Tekniches, 1958, 8: 26-31.

[12] Lemaitre, J. A continuous damage mechanics model for ductile fracture. Journal of Engineering Materials and Technology, 1985, 107(1): 83-89.

[13] Hillerborg, A., Modéer, M. and Petersson, P. E. Analysis of crack formation and crack growth in concrete by means of fracture mechanics and finite elements. Cement and Concrete Research, 1976, 6(6): 773-781.

[14] Ling, Y. Uniaxial true stress-strain after necking. AMP Journal of Technology, 2004, 5(1): 37-48.

[15] Jia, L. J. and Kuwamura, H. Ductile fracture simulation of structural steels under monotonic tension. Journal of Structural Engineering, 2014, 140(5): 1-12.

[16] Zaverl Jr, F. and Lee, D. Constitutive relations for nuclear reactor core materials. Journal of Nuclear Materials, 1978, 75(1): 14-19.

[17] Kuwamura, H. and Yamamoto, K. Ductile crack as trigger of brittle fracture in steel. Journal of Structural Engineering, 1997, 123(6): 729-735.

[18] Miner, M. A. Cumulative fatigue damage. Journal of Applied Mechanics, 1945, 12(3):

A159-A164.
[19] Bao, Y. and Wierzbicki, T. On the cut-off value of negative triaxiality for fracture. Engineering Fracture Mechanics, 2005, 72(4): 1049-1069.
[20] Jia, L. J. and Kuwamura, H. Ductile fracture model for structural steel under cyclic large strain loading. Journal of Constructional Steel Research, 2015, 106: 110-121.
[21] Jia, L. J., Ikai, T., Shinohara, K. and Ge, H. B. Ductile crack initiation and propagation of structural steels under cyclic combined shear and normal stress loading. Construction and Building Materials, 2016, 112: 69-83.
[22] Kang, L., Ge, H. B. and Kato, T. Experimental and ductile fracture model study of single-groove welded joints under monotonic loading, Engineering Structures, 2015, 85(2): 36-51.
[23] Kang, L., Ge, H. B., and Fang, X. An improved ductile fracture model for structural steels considering effect of high stress triaxiality, Construction and Building Materials, 2016, 115: 634-650.
[24] Jia, L. J., Ge, H. B., Maruyama, R. and Shinohara, K. Development of a novel high-performance all-steel fish-bone shaped. Engineering Structures, 2017, 138: 105-119.
[25] Jia, L. J., Ge, H. B., Xiang, P. and Liu, Y. Seismic performance of fish-bone shaped buckling-restrained braces with controlled damage process. Engineering Structures, 2018, 169: 141-153.
[26] Jia, L. J., Dong, Y., Ge, H. B., Kondo, K. and Xiang, P. Experimental study on high-performance buckling-restrained braces with perforated core plates. International Journal of Structural Stability and Dynamics, 2019, 19(1): 1-23.
[27] Jia, L. J., Xie, J. Y., Wang, Z., Kondo, K. and Ge, H. B. Initial studies on brace-type shear fuses. Engineering Structures, 2020, 208: 110318.
[28] Yang, S., Guan, D. Z., Jia, L. J., Guo, Z. X. and Ge, H. B. Local bulging analysis of a restraint tube in a new buckling-restrained brace. Journal of Constructional Steel Research, 2019, 161: 98-113.
[29] Guan, D. Z., Yang, S., Jia, L. J. and Guo, Z. X. Development of miniature bar-type structural fuses with cold formed bolted connections. Steel and Composite Structures, 2020, 34(1): 53-73.

5 高性能钢结构若干重要概念及实现方法*

李国强[1,2]
（1. 同济大学土木工程防灾国家重点实验室，上海；2. 同济大学土木工程学院，上海）

摘　要：本文介绍了高性能钢结构若干重要概念，指出高性能钢结构应满足承载性好、承灾性好和施工性好要求，而承载性好的钢结构应在确定的条件下具有较好的使用功能和较大的承载能力，承灾性好的钢结构应在罕遇的大地震下不发生整体结构倒塌，以及在可能发生的火灾、爆炸作用下产生的结构局部破坏不引起结构的连续性倒塌，而施工性好的钢结构应便于制作与安装，减少施工工期与成本。本文还提出了高性能钢结构相关重要概念的实现方法，指出高承载性钢结构可以从采用高强结构钢材料、高承载结构构件和高效结构体系三个方面加以实现；另将传统钢结构地震下易屈服的构件（部件）替换为消能-承载双功能构件是提高钢结构抗震性能的有效方法，而加强结构的整体性（包括保证楼板的整体性和刚度、保证梁柱连接的强度和转动能力、加强角柱和边柱等），可以大大提高钢结构抵抗火灾、爆炸、撞击等偶然作用下连续性倒塌的能力；此外，通过采用轧制型钢构件、现场螺栓连接可提高钢结构的施工性，并可通过抗侧结构与承重结构分离的方式提高钢结构现场螺栓连接的可实现性。

Some Important Concepts on High-Performance Steel Structures and Realization

Li Guoqiang[1,2]
(1. National Key Lab for Disaster Reduction in Civil Engineering, Tongji University, Shanghai;
2. College of Civil Engineering, Tongji University, Shanghai)

Abstract: A number of important concepts on the high-performance steel structures are introduced in this paper. It is indicated that the high-performance steel structures need possessing the characteristics of high capacity to bear loads and to resist disasters, and high feasibility for construction. High load-bearing steel structures should bear high capacity for service and supporting loads. High performance steel structures for disaster-resistance should bear high capacity for collapse prevention on overall structures subjected to extreme earthquakes, fire and explosion. While high construction-feasibility steel structures should be easily fabricated and erected. The realization of high performance steel structures is also proposed in this paper. It is pointed that the realization of high load-bearing steel structures can be realized with employing high-strength steel, high load-bearing structural elements and high-efficiency structural system. The effective way to realize high performance steel structures for earthquake-resistance is to replace the structural elements

* 论文发表于《建筑钢结构进展》2020 年第 5 期；国家重点研发计划项目"高性能钢结构体系研究与示范应用（2016YFC07012）"资助。

with bi-functional elements of energy dissipation and load-bearing. While the effective way to realize high performance steel structures for fire- and blast-resistance is to guarantee the structural integrality with ensuring the integrality of floor slabs and strengthening beam-to-column connections, perimeter columns and corner columns. Finally, the hot-rolled steel for fabrication and bolted connection for erection is recommended to increase the construction-feasibility of steel structures and separation of structural systems to respectively resist lateral and vertical actions is also recommended to increase the applicability of bolted connection for erection of steel structures.

1. 引言

在 1999 年第 2 届钢结构进展国际会议上（The 2nd International Conference on Advances in Steel Structures，Hong Kong，China），美国里海大学 L. W. Lu 教授做了一个有关高性能钢结构的大会主题报告“High-Performance Steel Structures：Recent Research”[1]，然而这个报告介绍的重点是采用高性能钢材的钢结构，提出高性能钢应具有强度高、延性和韧性大、可焊性和可加工性好的特性。然而，钢材强度越高，通常其延性和可焊性变得越差，因此制作强度、延性、可焊性和可加工性均好的钢材是一个极大的挑战，会增加钢材生产成本。由于用于工程的钢结构用钢量大，仅通过采用成本较高的高性能钢做成高性能钢结构不应是工程结构追求的目标。

工程结构一般是整体结构发挥功能作用，而整体结构的性能不仅与结构材料有关，还与结构构件形式和结构体系有关，因此提高结构性能不能仅限于采用高性能结构材料，而应从整体结构性能的角度，考虑高性能钢结构问题。

2. 高性能钢结构的一些重要概念

2.1 结构的功能需求

工程结构是工程设施（建筑、桥梁、隧道、地下工程等）的骨架，其主要功能是维持工程设施在预定设计使用期内的完整性，使工程设施能正常运行。

一般工程结构应具有以下功能[2,3]：

（1）正常使用功能：结构在设计使用期内经常出现的荷载（永久荷载、楼面荷载、风等）作用下产生的变形、振动等，不应影响结构的正常使用；

（2）承载安全功能：结构在设计使用期内会出现但出现概率较小的荷载（概率较小的永久荷载、活荷载、地震等）的作用下，在结构连接、节点、构件中产生的内力不应超过其极限承载力，对整体结构的作用不应超过其极限承载力；

（3）承灾安全功能：结构在设计使用期不一定出现，而一旦出现对结构性能影响很大的灾害（罕遇地震、火灾、爆炸等）作用下，结构应保持完整、不倒塌。

2.2 高性能结构的要求

对于高性能结构，应具有高安全性能、高使用性能、高施工性能、高环保性能、高维护性能、高耐久性能等特征[4]，对于高性能钢结构应至少满足以下重要要求[5-8]：

（1）承载性好：在同样的条件下，结构具有更好的使用功能和更大的承载能力；或在满足同样的正常使用功能和承载安全功能的条件下，使用的结构材料较少或结构的成本更低；

（2）承灾性好：在罕遇的大地震下整体结构不倒塌，在可能发生的火灾、爆炸或撞击作用下产生的结构局部破坏不会引起结构的连续性倒塌，且灾后结构中产生的破坏能方便修复，使结构快速恢复使用；

（3）施工性好：结构应便于制作与安装，施工工期短，施工成本低。

3. 高承载性钢结构的实现方法

高承载性钢结构可以从结构材料、结构构件和结构体系三个方面加以实现。

3.1 采用高强度钢材料

目前我国有从 Q235 到 Q960 强度等级的结构钢材标准[9-11]。然而我国钢结构仍以采用 Q235 和 Q345 钢材为主[12]，如采用更高强度的钢材，在同样的荷载作用下对于相同的结构构件，可以减小构件截面，从而减少用钢量。例如对于受相同压力的钢柱，采用 Q460 钢与采用 Q235 钢相比，钢材强度提高了约 50%，可以减小钢柱的截面，但截面减小后，柱的长细比增大，柱的稳定承载力会减小，因此对于受压钢构件采用 Q460 钢替换 Q235 钢，构件截面或用钢量的节省要小于强度增大的比例，但仍可节省可观的用钢量。而对于受拉钢构件，由于没有稳定问题，采用高强钢可取得更好的节材效益。

国家体育场（鸟巢）为跨度超过 300m 的大跨度钢结构，设计时采用 Q345 钢，构件的最大钢板厚度需要 220mm，而采用 Q460 钢，构件的最大钢板厚度仅为 110mm[13]（图 1）。尽管 Q460 钢比 Q345 钢的强度只大约 33%，但构件截面减小后结构的自重减轻，可进一步减小结构构件的内力，因此使构件的用钢量减小 50%左右。可见，对于自重内力效应敏感的结构，采用高强钢可以取得非常好的节材效益。

图 1　国家体育场（鸟巢）结构构件的最大钢板厚度

Fig. 1　The maximum thickness of steel plates for National Stadium (Bird Nest)

3.2 采用高承载构件形式

采用不同的构件的形式，对于同样的荷载作用，承载力将不同。例如，对于简支梁，在具有同样的承载力条件下，波纹腹板远比平腹板的抗剪稳定性好，在保证相同抗剪稳定性的条件下，波纹腹板可比平腹板薄，因此波纹腹板工字形钢梁的用钢量要小于平腹板工字形钢梁[14]；如采用钢桁架梁可以利用上下弦杆抵抗弯矩，其腹杆的用钢量比波纹腹板钢梁的实腹截面用钢量又可以更小。因此在同样的用钢量条件下，波纹腹板工字形钢梁承载力要大于平腹板工字形钢梁[14]，而钢桁架梁的承载力又将大于波纹腹板工字形钢梁(图 2)。

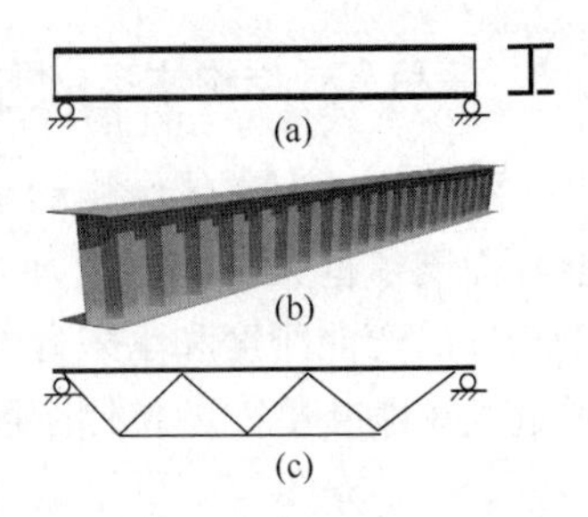

图 2 同样承载条件下不同形式简支梁的用钢量对比

Fig. 2 Steel consumption comparison of various simply-supported beams under the same loading

(a) 实腹梁：用钢量大；
(b) 波纹腹板梁：用钢量较小；
(c) 桁架梁：用钢量小

纽约世贸大楼的楼面梁跨度达 18.3m[15]，如采用实腹梁用钢量大，因而实际工程采用了桁架梁（图 3)。烟台机场屋盖梁的跨度达 55m，采用了波纹腹板工字形钢梁，与平腹板工字形钢梁相比，用钢量节省 15%[14]（图 4)。

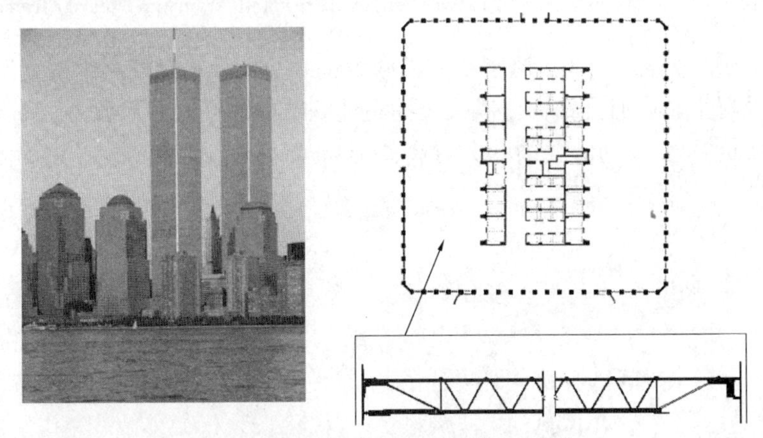

图 3 纽约世贸大楼的楼面桁架梁[15]

Fig. 3 The truss-type girder of World Trade Centre[15]

图 4 烟台机场屋盖波纹腹板梁[14]

Fig. 4 Corrugated web steel beams for terminal building of Yantai Airport[14]

3.3 采用高效结构体系

结构体系对整体结构的承载性能影响很大。刚接框架结构抵抗侧力靠梁柱的抗弯性能，而支撑框架结构抗侧力主要靠支撑抗轴力性能，而钢构件的轴向刚度和承载力远大于钢构件的弯曲刚度和承载力，因此在用钢量相同的条件下，支撑框架的抗侧刚度和承载力要远大于刚接框架的抗侧刚度和承载力（图 5）。

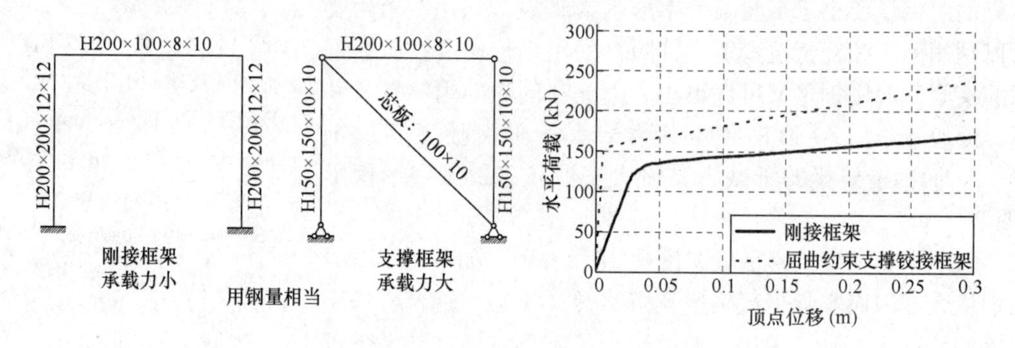

图 5 刚接框架与支撑框架抗侧能力对比

Fig. 5 Capacity comparison between moment-resistant frame and braced frame

混凝土剪力墙或核心筒的抗侧刚度和承载力也非常大，在钢框架中设置混凝土核心筒构成钢框架-混凝土芯筒混合结构可以大大提高纯钢结构的刚度和承载力（图 6），这种钢-混凝土混合结构用于 100m 至 200m 的高层建筑，可以减少结构用钢量和成本 30%左右[16,17]。

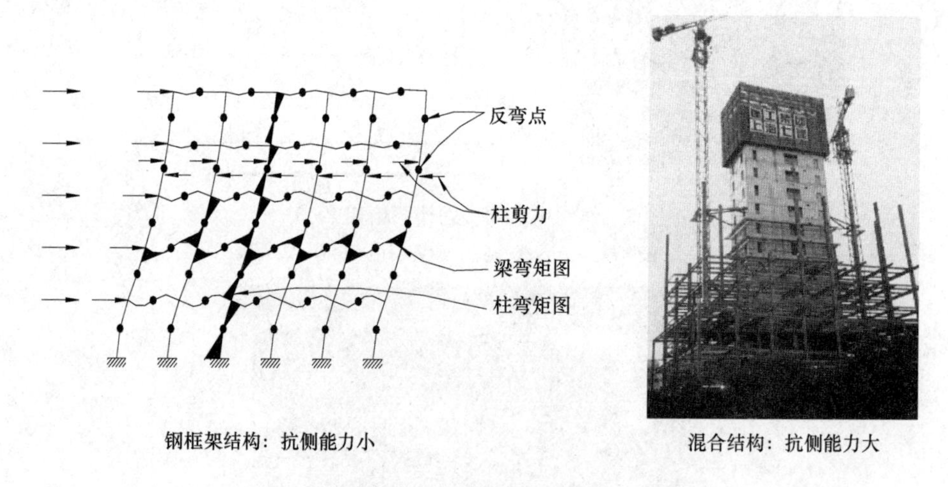

图 6 钢框架结构与混合结构抗侧能力对比[18]

Fig. 6 Capacity comparison between steel frame and steel-concrete hybrid structure[18]

美国纽约帝国大厦 381m 高，采用钢框架-支撑结构体系，用钢量为 206kg/m²，而纽约世贸大楼 412m 高，采用框筒结构体系，用钢量为 186.6kg/m²（图 7）。可见，尽管世贸大楼比帝国大厦高度大，但采用了抗侧力效率更高的框筒结构体系，用钢量更少[18]。

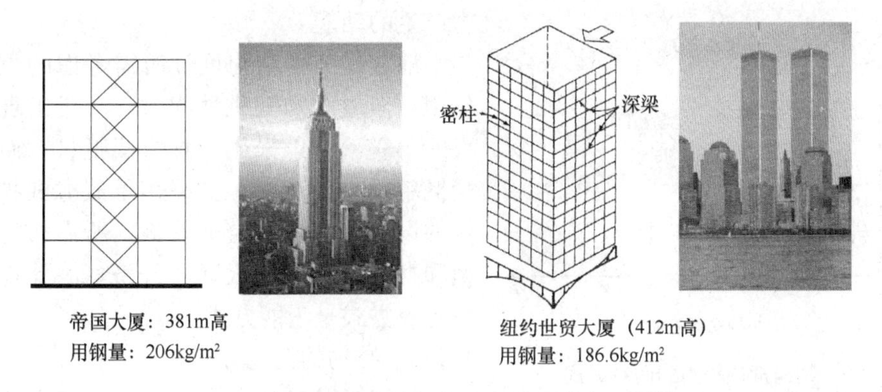

图 7 纽约世贸大楼与帝国大厦用钢量对比[18]

Fig. 7 Comparison of steel consumption between WTC and Empire Tower[18]

4. 高承灾性钢结构的实现方法

4.1 结构承受静力和动力的方法

4.1.1 静力和动力的区别

结构在设计使用期内会遭受静力和动力的作用（图 8）。静力作用随时间（大于结构基本周期 10 倍的范围内）基本保持恒定，例如结构的自重、承受的人员、设备的重量、土压力等；而动力作用随时间（小于结构基本周期的范围内）发生变化，例如冲击、地震等，动力将引起结构振动，产生惯性力和动能。由于风的脉动周期较长，因此对于周期较小的低矮建筑、小跨度桥等工程结构产生的动力振动效应较小，而对于周期较大的高层建筑、大跨度桥等工程结构产生的动力振动效应较大。

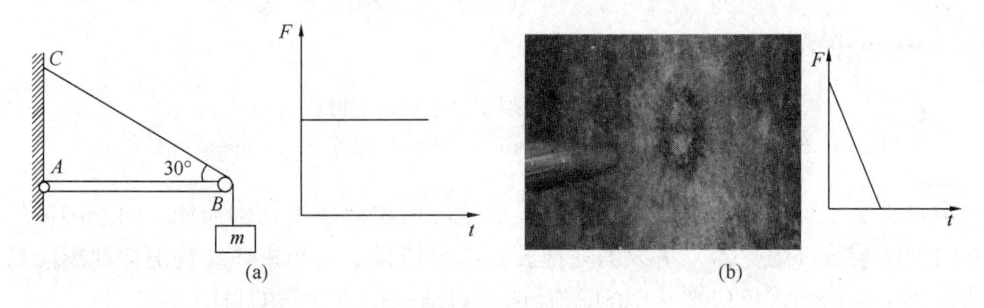

图 8 结构承受的典型静力与动力

Fig. 8 Typical static and dynamic action on structures

（a）静力（重力）；（b）动力（冲击力）

4.1.2 结构抵抗静力和动力的方式

对于静力，结构只能以承载方式抵抗，即整体结构及所有结构构件的承载力需大于静力作用和在构件中产生的内力。但对于动力，结构可以采用承载和消能两种方式抵抗[19]（图 9）。采用承载方式抵抗动力，要求结构或抵抗动力构件的承载力大于动力产生的作用力或作用效应；而采用消能方式抵抗动力，则要求结构或抵抗动力构件的消能能力大于动

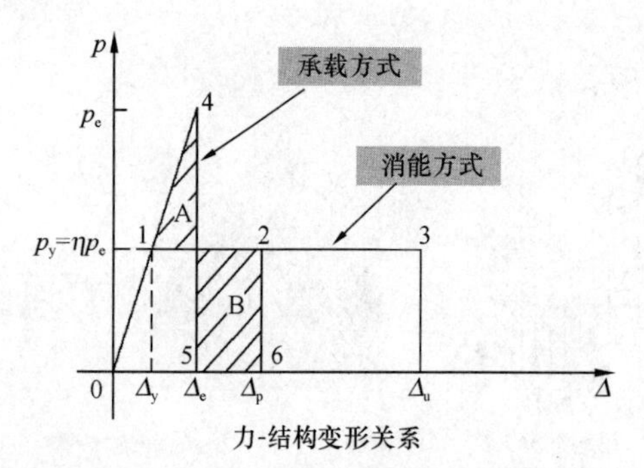

图 9　结构抵抗动力的两种方式

Fig. 9　Two ways to resist dynamic action

力产生的动能。

地震是工程结构设计时需考虑的重要动力作用，但地震的随机性很大，对于出现概率较大的小震，可以采用承载方式抵抗；而对于出现概率很小的大震，如采用承载方式抵抗，结构的设计地震作用会非常大，所需结构构件截面和材料用量会很大而不经济，因此抵抗大震更有效方式是消能方式。

4.1.3　结构的消能方法

结构可以通过自身构件屈服产生塑性变形消能，也可以额外附加专门装置（阻尼器）消能。

图 10 是钢结构和混凝土结构在地震反复作用下屈服后进入塑性变形状态以后的典型力-位移滞回曲线，滞回曲线包围的面积就是结构的消能[20]。显然，由于钢结构滞回曲线饱满、消能能力比混凝土结构强，故在同样的地震条件也即同样的地震能量消耗需求下，钢结构的地震位移反应要比混凝土结构小，因此钢结构的抗震性能比混凝土结构好。

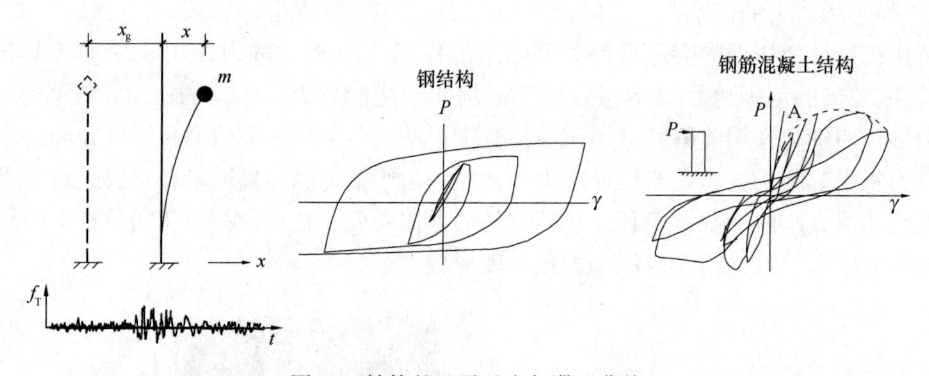

图 10　结构的地震反应与滞回曲线

Fig. 10　Seismic response and hysteretic curves of steel and concrete structures

结构屈服发生塑性变形虽然消耗了地震能量，但也造成了结构损伤。而采用额外设置附加专门消能装置（阻尼器）可以消耗地震的部分能量，减小主体结构消耗地震能量的需求，从而减小结构地震反应[20]，达到减轻地震对结构造成损伤的目的（图 11）。

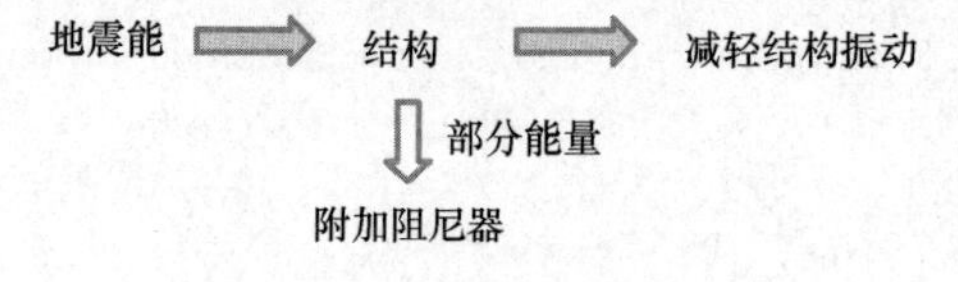

图 11　附加阻尼器消能减震原理

Fig. 11　The principle of seismic reduction with energy-dissipating dampers

消能装置或阻尼器主要有两种类型，一种是位移型阻尼器（图 12），另一种是速度型阻尼器[21]。一般位移型阻尼器采用金属材料制成，通过塑性变形消能；而速度型阻尼器采用黏稠材料，通过在这种材料中的相对运动产生阻尼，阻尼力与运动速度成比例（一般为非线性比例），从而消耗能量。

但传统金属位移型或黏稠材料速度型阻尼器有以下不足：

（1）额外设置消能阻尼器，平时闲置；

（2）消能能力不强，阻尼力最大为百吨级。

为克服传统阻尼器的不足，可采用消能-承载双功能构件[22]（图 13），这种构件是利用金属阻尼器的原理，通过特殊设计，使传统承载钢构件同时具备金属阻尼器的性能。由于消能-承载双功能构件是利用钢构件的承载力，使其屈服转化成塑性阻尼力，可很容易使阻尼力达千吨甚至万吨级，从而大大提高结构的消能能力和抗震性能。

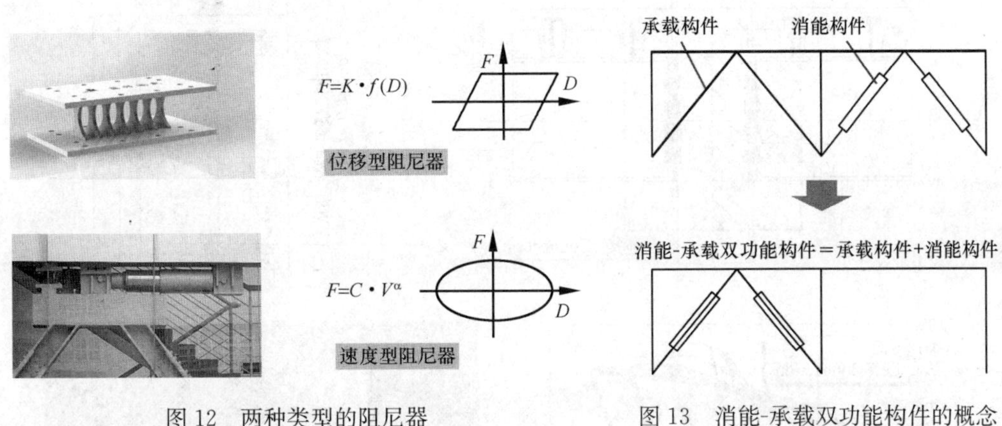

图 12　两种类型的阻尼器

Fig. 12　Two types of dampers

图 13　消能-承载双功能构件的概念

Fig. 13　The concept of bi-functional elements of energy-dissipation and load-bearing

4.2　提高钢结构抗震性能的方法

4.2.1　提高钢框架结构抗震性能的方法

刚接框架是钢结构常用的一种结构形式，但刚接框架梁柱节点在地震中容易破坏[23]（图 14），其原因是梁柱刚性节点一般需焊接，节点的塑性转动变形能力不强，难以满足

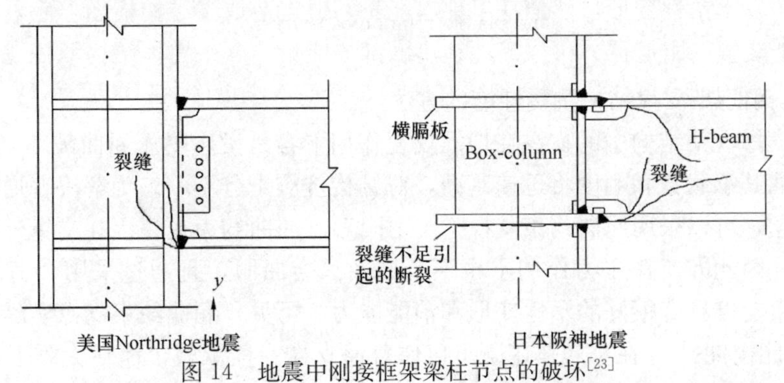

图 14　地震中刚接框架梁柱节点的破坏[23]

Fig. 14　Damages of beam-to-column connections in Northridge EQ and Kobe EQ[23]

消耗大震时在结构中产生的能力消耗需求。

解决刚性梁柱节点在地震下易破坏的问题，可以采用端板式半刚性梁柱节点形式[24]（图 15），这种节点可以通过设计控制节点仅端板屈服，而梁端端板由于发生的是弯曲变形，塑性变形能力很强，因此可使节点的转动能力很大（超过 0.06rad），满足结构 1/20 层高以上的层间侧移变形需求[25]。可见，端板式半刚性梁柱节点中的端板，实际上可看作是一种消能-承载双功能构件（部件）。

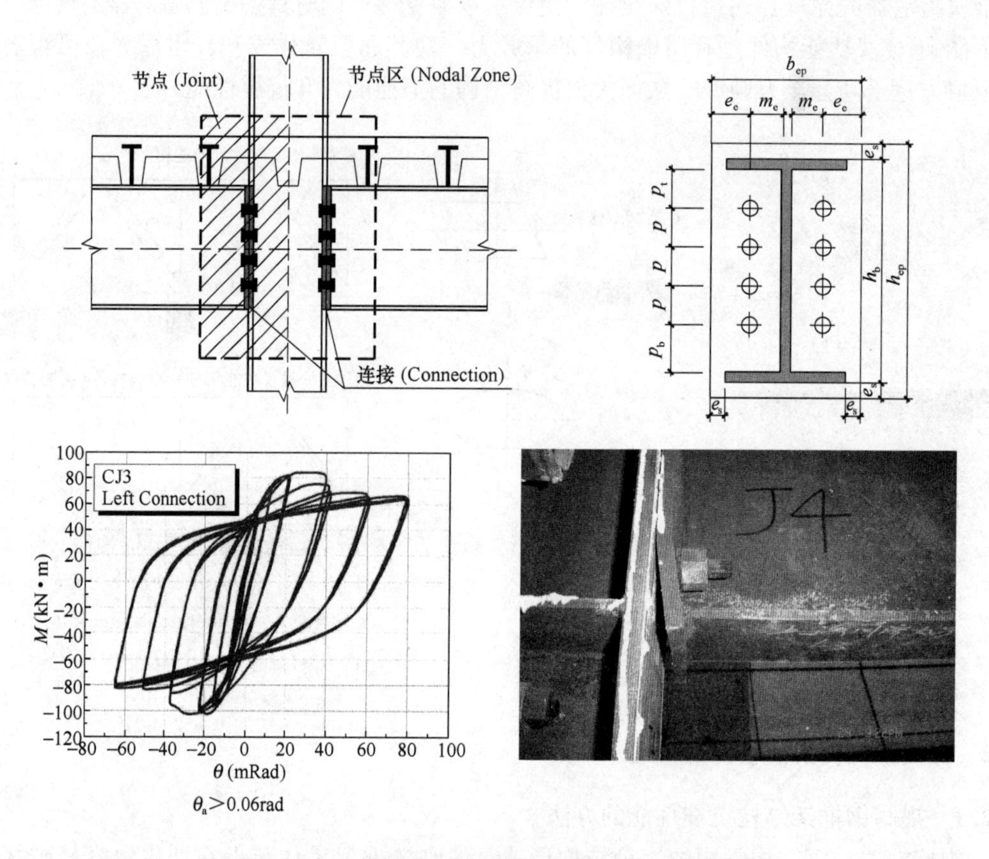

图 15 端板式半刚性梁柱节点及其塑性转动变形[25]

Fig. 15 Semi-rigid end-plate beam-to-column connection and plastic rotation capacity[25]

4.2.2 提高钢框架-支撑结构抗震性能的方法

地震下支撑框架结构中的支撑在地震反复作用下容易发生受压屈曲破坏（图 16），而支撑一旦屈曲其承载力和刚度将迅速衰减，对结构抗震十分不利。为解决普通钢支撑易屈曲破坏的问题，可以采用屈曲约束支撑[26]（图 17）。屈曲约束支撑因在支撑外围设有套管约束其屈曲，因此即使在压力作用下也不会屈曲只会屈服，且通过采用合适的钢材和构造，屈曲约束支撑具有很好的塑性变形和消能能力。可见，屈曲约束支撑实际上是一种消能-承载双功能构件[27]，在小震作用下可以像普通支撑一样承载，而在大震下可以像金属阻尼器一样消能减震。

图 16　地震下支撑的破坏

Fig. 16　Damage of steel brace Under earthquake

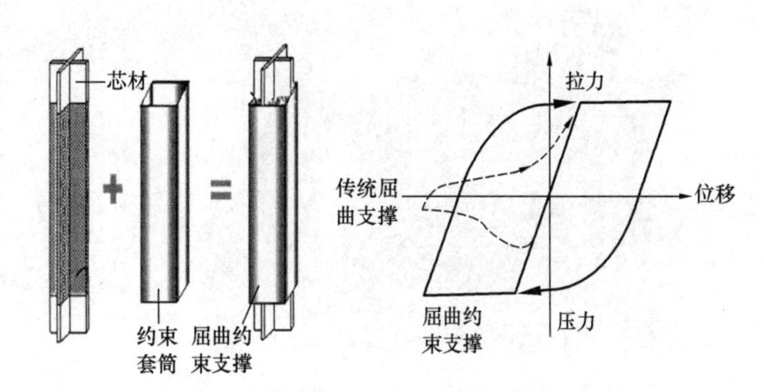

图 17　屈曲约束支撑及其滞回性能[26]

Fig. 17　Buckling-restrained brace and hysteretic behavior[26]

在结构中设置支撑有时会妨碍建筑功能，例如通道、门窗等，此时可以设置消能-承载双功能钢板墙[27-29]（图 18）。设置消能-承载双功能钢板墙可使钢框架承受小震的抗侧承载力和刚度以及抵御大震的消能能力大大提高[30]，而结构延性仍很好（图 19）。

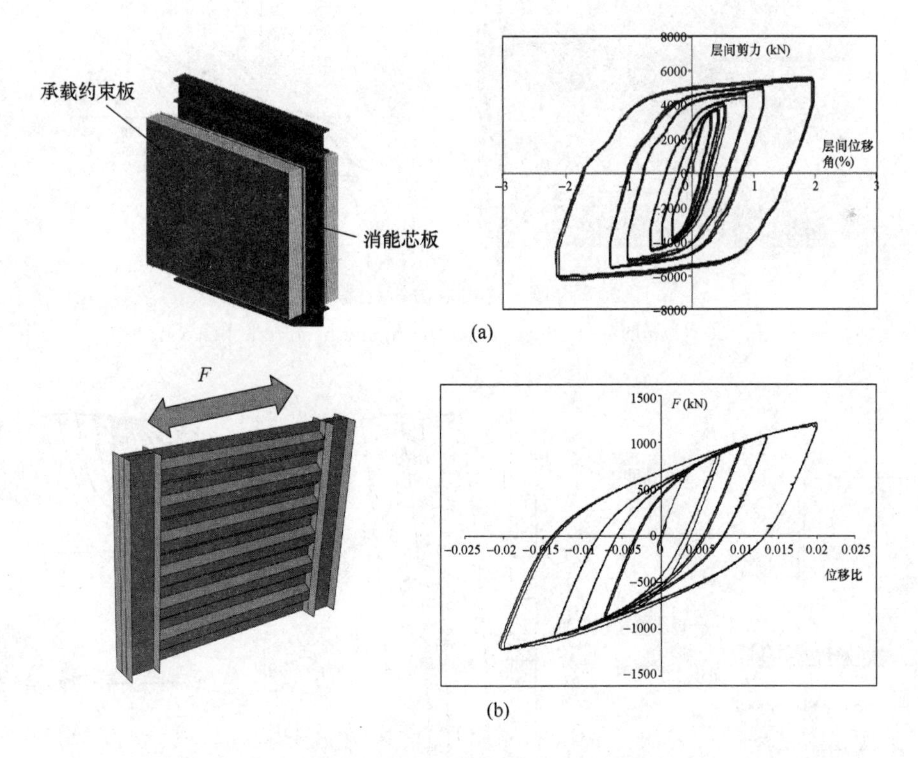

图 18　两种形式的消能-承载双功能钢板墙[29]

Fig. 18　Two types of bi-functional walls with energy-dissipating and load-bearing[29]

(a) 屈曲约束钢板墙[28]；(b) 无屈曲波纹钢板墙

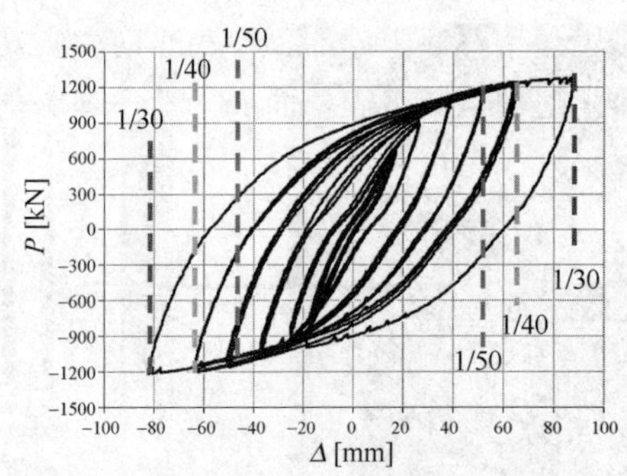

图 19　设置消能-承载双功能钢板墙的钢框架的滞回性能[30]

Fig. 19　Steel frame braced with bi-functional wall and hysteretic behavior[30]

4.2.3　提高钢-混凝土混合结构抗震性能的方法

钢框架-混凝土芯筒混合结构（图 6）抵抗地震侧力作用以芯筒为主，芯筒因需开门洞通常实际为联肢剪力墙结构。而传统混凝土连梁延性差，地震下易破坏，消能能力不强[31]（图 20）。为克服混凝土连梁的缺点，在联肢墙中可以采用钢连梁（图 21），而钢连

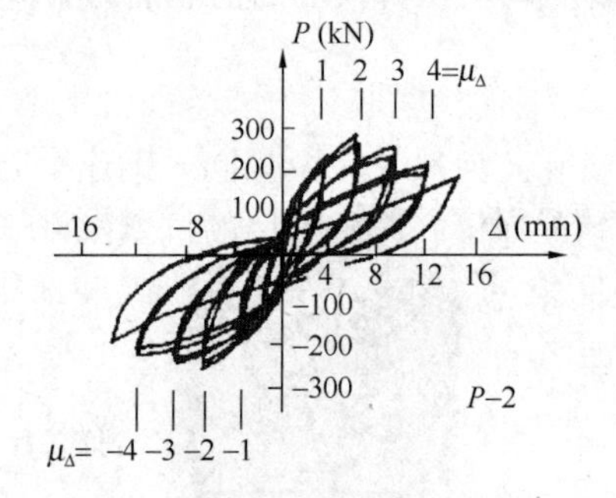

图 20　混凝土连梁的破坏及滞回性能

Fig. 20　Damage of concrete link beam and hysteretic behavior

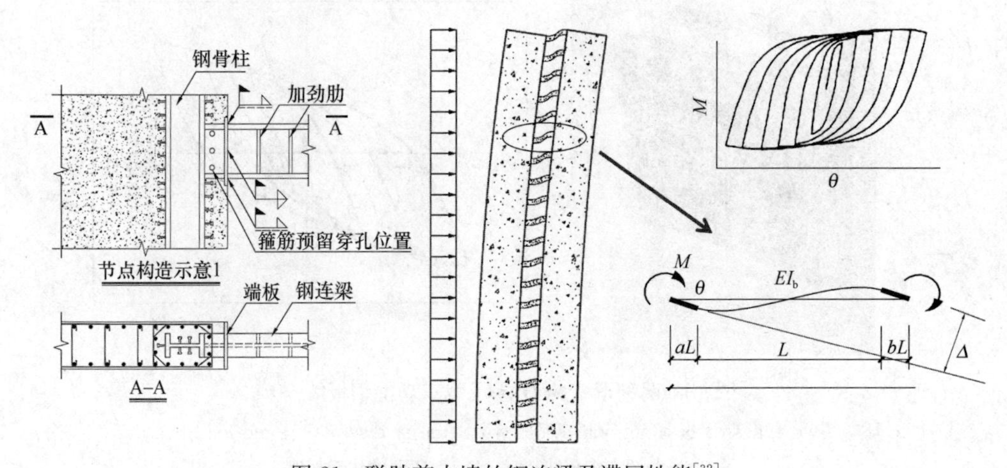

图 21　联肢剪力墙的钢连梁及滞回性能[32]

Fig. 21　Steel link beams in coupled shear walls and hysteretic behavior[32]

梁实际上也是一种消能-承载双功能构件[32]。由于普通钢连梁小震下承载不消能，还可以额外设置一个金属阻尼器，与普通钢连梁复合构成双阶屈服钢连梁[33]（图 22），这样小震下金属阻尼器消能减震，而大（中）震下普通钢连梁消能减震。

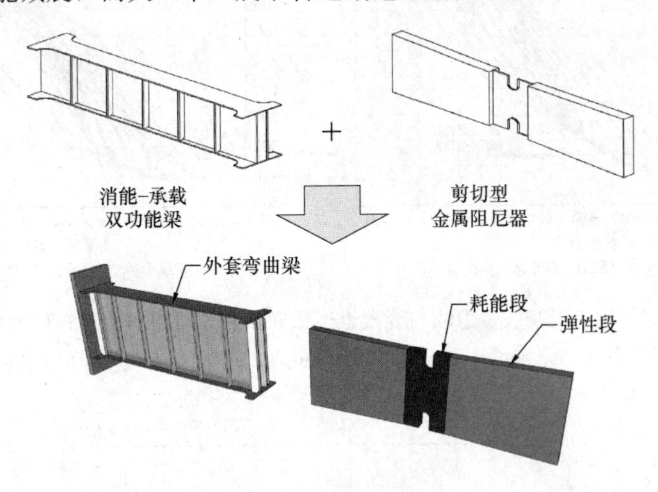

图 22　双阶屈服钢连梁[33]

Fig. 22　Configuration of double yielding steel beam[33]

图 23 为钢连梁联肢剪力墙与混凝土连梁联肢剪力墙对比试验模型[34]，两个试验模型的墙肢材料和几何尺寸完全相同，钢连梁与混凝土连梁的刚度也接近，因此钢连梁联肢剪力墙与混凝土连梁联肢剪力墙的抗侧刚度一致，但在相同的反复荷载作用下，两者的滞回性能差异非常大（图 24）。钢连梁联肢剪力墙的承载力比混凝土连梁联肢剪力墙约大 1 倍，消能能力约大 5.5 倍（图 25）；且在同样的侧移条件下，混凝土连梁的破坏远远严重与钢连梁（图 26），并难于修复。

钢连梁联肢剪力墙

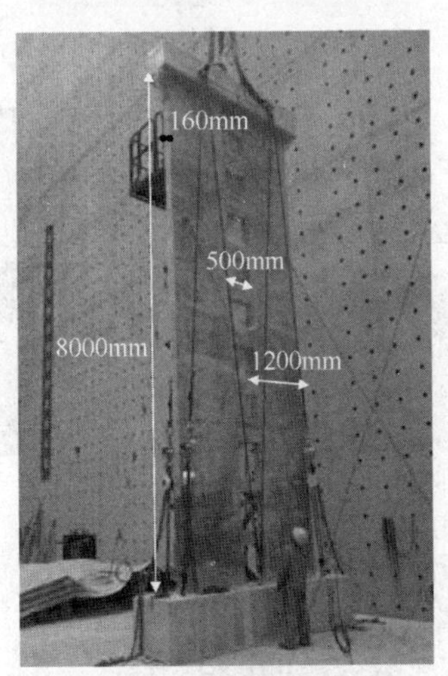

混凝土连梁联肢剪力墙

图 23　钢连梁联肢剪力墙与混凝土连梁联肢剪力墙对比试验模型[34]

Fig. 23　Coupled shear wall specimens with steel link beams and concrete link beams[34]

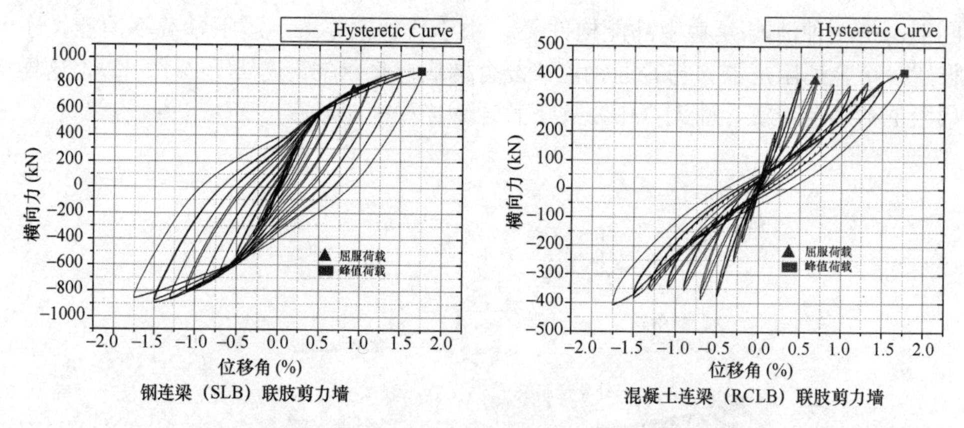

图 24　钢连梁联肢剪力墙与混凝土连梁联肢剪力墙滞回性能的对比[34]

Fig. 24　Hysteretic behavior comparison between coupled shear walls with different link beams[34]

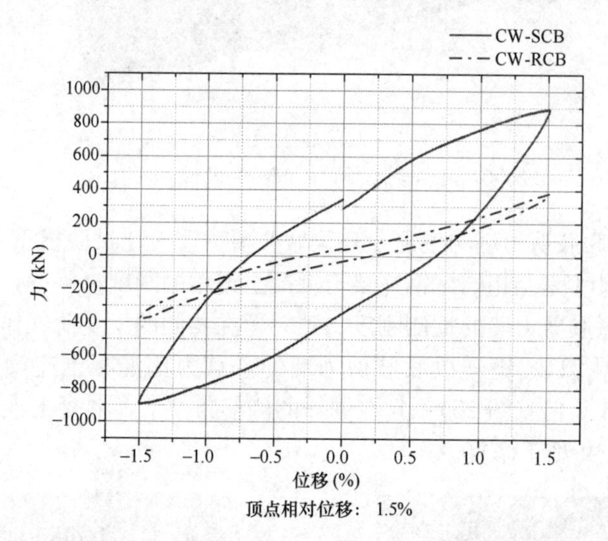

图 25　钢连梁联肢剪力墙与混凝土连梁联肢剪力墙消能能力的对比[34]

Fig. 25　Energy dissipation comparison between coupled shear walls with different link beams[34]

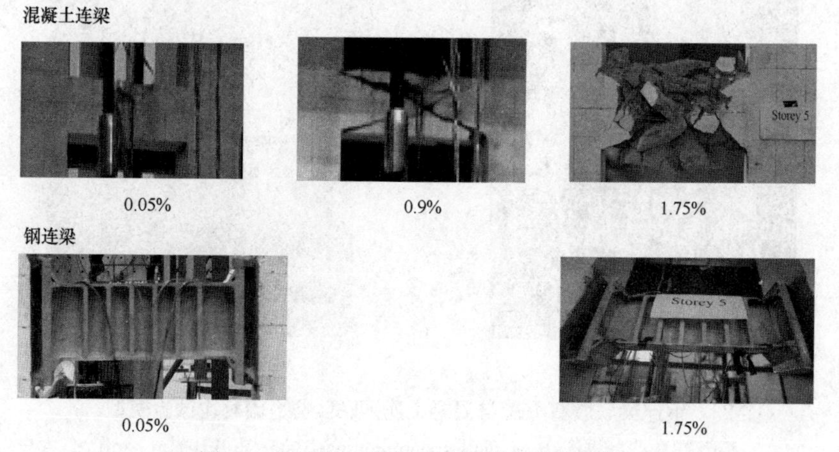

图 26　钢连梁与混凝土连梁破坏的对比[34]

Fig. 26　Comparison of damages of steel link beam and concrete link beam[34]

4.2.4 高强钢在抗震钢结构中的应用方法

随着钢材强度的提高，钢材的延性（塑性变形能力）将降低（图 27），钢材很难做到强度很高，延性也很好。因此，高强钢用于抗震结构时，可通过设计用于在地震下不发生屈服（或屈服后发生塑性变形很小）的结构部位或构件[12]（图 28）。

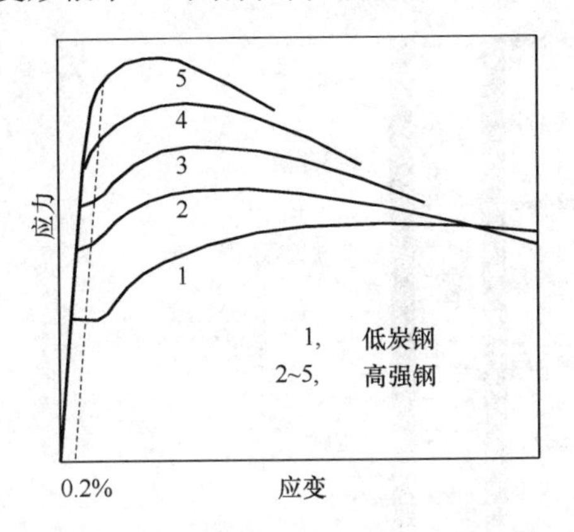

图 27　钢材强度与塑性变形能力的关系

Fig. 27　Relationship between strength and ductility of steels

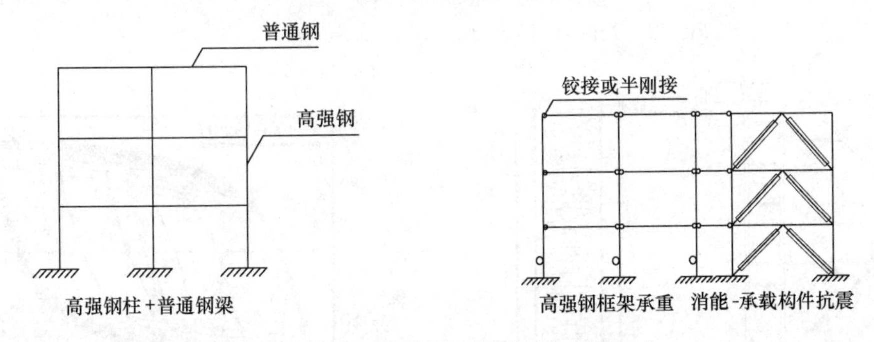

图 28　高强钢在钢结构抗震中的应用方式

Fig. 28　The method to apply high-strength steel in structures against earthquakes

4.2.5 提高钢结构抗震可恢复性的方法

结构抗震可恢复性是指地震后结构不需修复或可方便快速修复使结构功能恢复的特性[35]，如地震下结构即使不倒塌，但发生有很大的震后残余变形（图 29），或主要承重结构构件有严重破坏，则结构很难修复而不能继续使用，则这种结构不具有可恢复性。

图 29　地震后建筑的残余变形

Fig. 29　The residual deformation of building after earthquake

为保证结构的可恢复性，可采用下列方法：

（1）应避免主要承重结构竖向构件（承重柱、承重墙等）发生严重破坏，而主要采用消能-承载钢支撑、消能-承载钢板墙，或框架梁、联肢墙连梁等抵抗地震作用抗震；

（2）应采取措施，减小结构震后残余变形。

图 30 是一种消能摇摆柱（墙），这种构件有

两个柱（墙）肢，在之间设置消能部件，通过其发生塑性变形消能[36]。试验证明这种消能摇摆构件具有很好的塑性变形能力和滞回消能能力（图 31），且这种构件两个承重的柱（墙）肢始终保持弹性，仅消能部件屈服消能，本身具有较好的可恢复性。

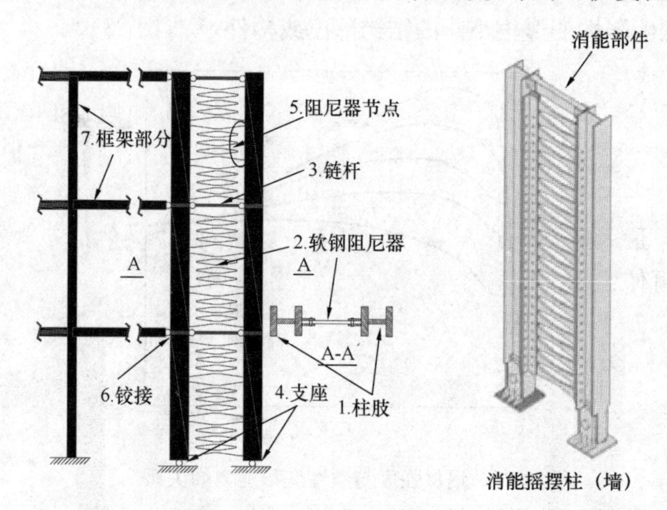

图 30　消能摇摆柱（墙）[36]

Fig. 30　Energy-dissipating rocking column (wall)[36]

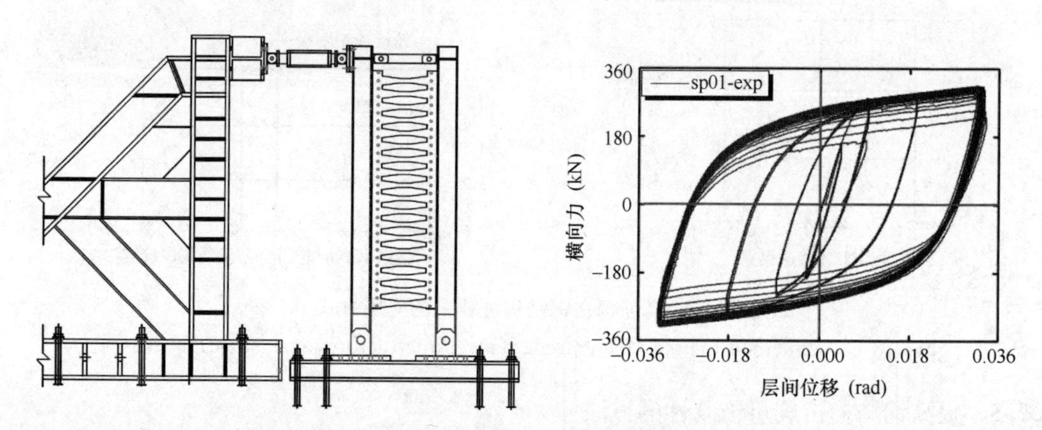

图 31　消能摇摆柱（墙）的滞回性能[36]

Fig. 31　Hysteretic behavior of energy-dissipating rocking column (wall)[36]

消能摇摆柱（墙）用于结构抗震的方式如图 32 所示，设计目标为[37]：（1）小震下，

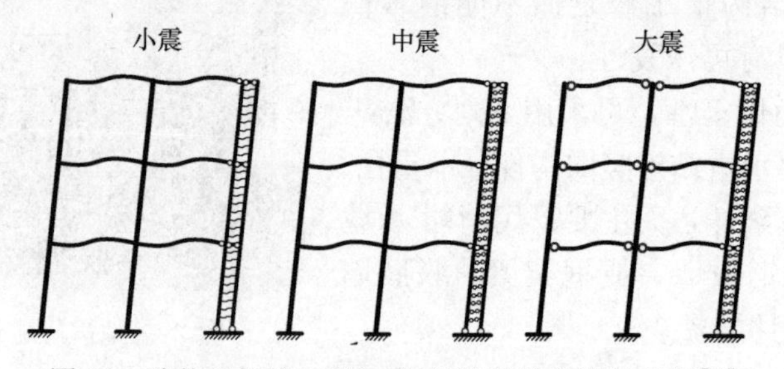

图 32　消能摇摆柱（墙）-框架结构抗震设计目标[37]

Fig. 32　Aseismic design target of frame structure with energy-dissipating rocking column[37]

消能摇摆柱（墙）和主体结构保持弹性；（2）中震下，消能摇摆柱（墙）中的消能部件屈服消能减震，而主体结构保持弹性；（3）大震下，消能摇摆柱（墙）和主体结构均屈服，但不倒塌。由于消能摇摆柱（墙）具有很大的平面内刚度，通过摇摆可使主体结构沿竖向的层间变形均匀，从而防止主体结构薄弱层的发生。

为证明消能摇摆柱（墙）对提高框架结构可恢复性的有效性，进行了消能摇摆柱框架与普通框架结构模型的拟动力对比试验[37]（图 33），按照抗侧刚度和承载力相同的原则设计两个对比试验模型，除其中一个模型采用消能摇摆柱外，两个模型其他部分的结构材料相同，构件几何尺寸基本一致。试验表明，在相同的大震条件下，采用消能摇摆柱的框架与普通框架相比，最大层间位移减小 40%，残余层间变形减小 90%（图 34）。可见，消能摇摆柱（墙）可使主体结构的可恢复性大大提高。

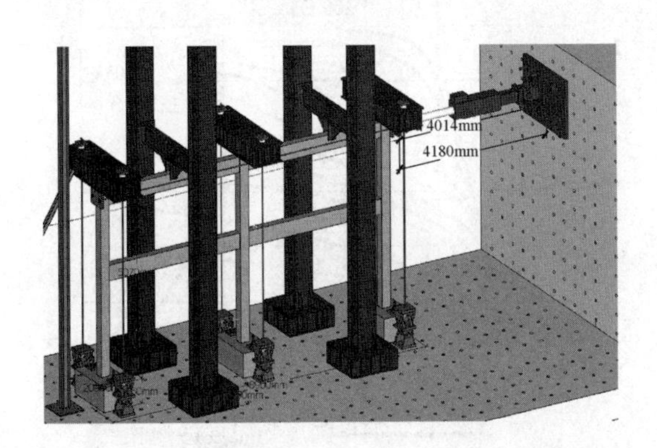

图 33　普通框架与消能摇摆柱框架的对比试验[38]

Fig. 33　Comparative tests on conventional frame and frame with energy-dissipating rocking column[38]

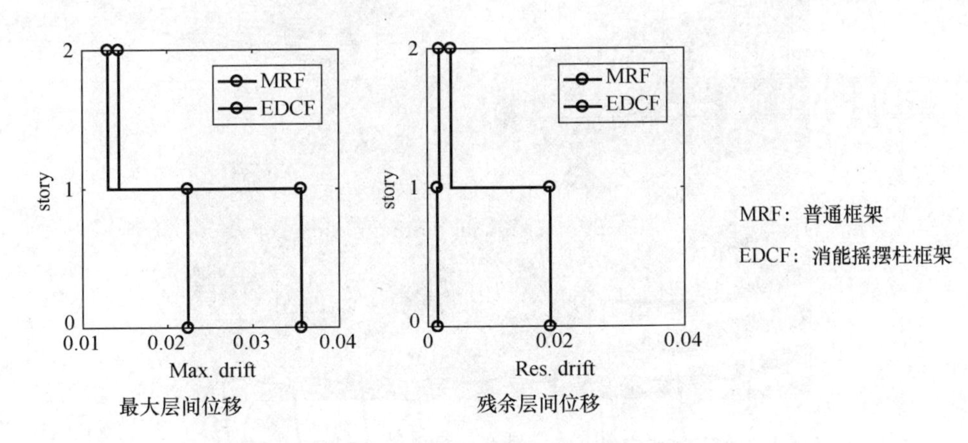

图 34　大震下普通框架与消能摇摆柱框架最大层间变形和残余层间变形的对比[38]

Fig. 34　Comparison of maximum story drift under rare earthquake and residual story drift after rare earthquake between conventional frame and frame with energy-dissipating rocking column[38]

4.3 提高钢结构抗火、抗爆性能的方法

4.3.1 火灾和爆炸作用

火灾是建筑中最常发生的灾害。建筑中一旦发生火灾，温度将迅速升高（图 35），一般 10 分钟内温度将达到 600℃以上，30 分钟内温度会超过 800℃。而钢材强度随温度升高会降低（图 36），钢材温度为 600℃时的强度，仅为常温时强度的 30%左右，因此火灾下钢结构的承载力会大幅降低，而承受不了其上的荷载而破坏[39]（图 37）。

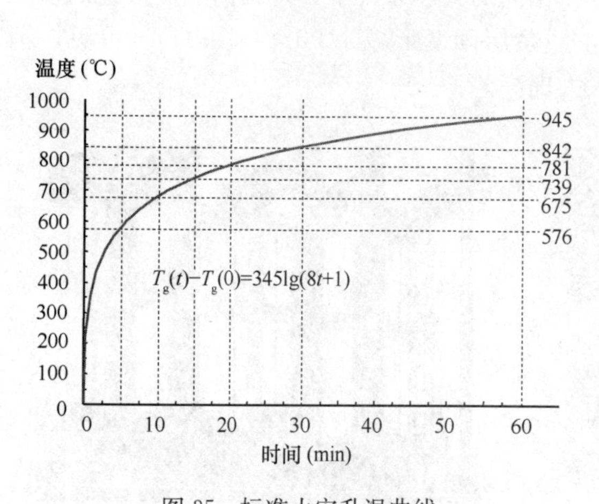

图 35　标准火灾升温曲线

Fig. 35　Temperature elevation of standard fire

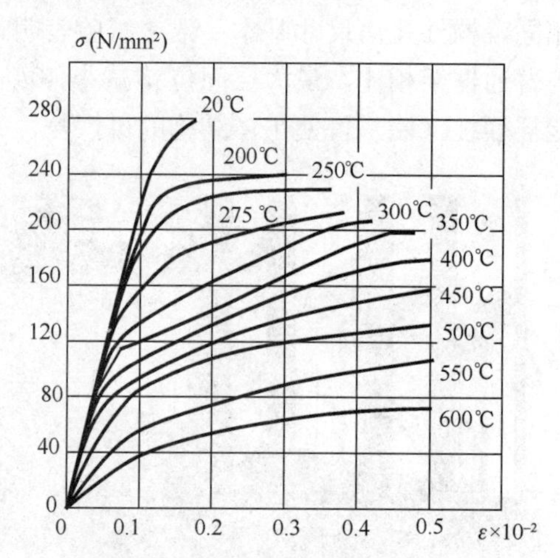

图 36　钢材强度随温度的变化

Fig. 36　Strength reduction of steel with temperature elevation

图 37　火灾中钢结构的破坏

Fig. 37　Damage of steel structure in fire

爆炸在建筑中也会发生，有意外事故爆炸（如煤气泄漏）和人为爆炸（恐怖行为）。爆炸会产生很大的冲击作用[40]（图 38），这种冲击作用虽然持续时间很短（毫秒级），但压力值很大（kPa～MPa 级），会造成结构构件的严重破坏（图 39）。

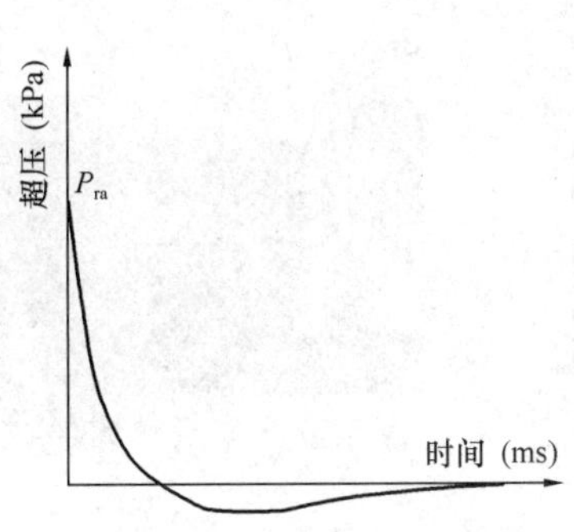

图 38 爆炸冲击作用

Fig. 38 Impact action of blast

图 39 爆炸造成的结构破坏

Fig. 39 Structural damage due to blast

4.3.2 提高结构构件抗火和抗爆炸能力的方法

火灾和爆炸对结构的直接影响往往是局部的，针对可能直接受火和爆炸作用的结构构件采取相关措施，可提高其抗火和抗爆能力。

火灾温度虽然很高，但可以采用防火涂料和防火板包裹钢构件[39]（图 40），隔离火的热量向钢构件的传播，延缓火灾下钢构件的升温速度，延长钢构件的耐火时间，达到具有足够时间进行人员疏散和灭火救援的目的。

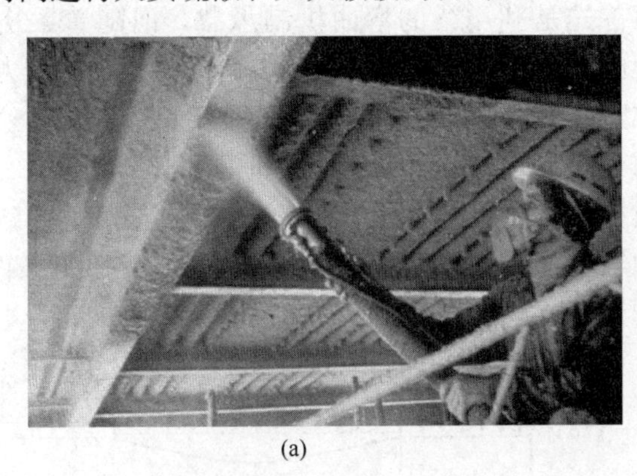

(a)

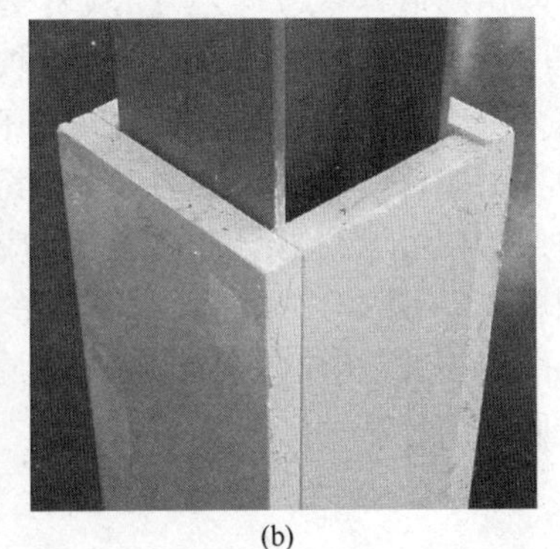

(b)

图 40 钢结构防火措施

（a）防火涂料；（b）防火板

Fig. 40 Steel fire protection measurements

(a) Fire insulation coating; (b) Fire insulation board

爆炸作用是一种持续时间很短的冲击作用，采用承载方式抵抗爆炸的最大冲击力是不经济和不必要的，应采用消能方式抵冲击作用的冲量，因此对于受爆炸作用的构件，最有效的抵抗方式是采用延性好的材料和构件形式[7]。

4.3.3 提高整体结构抗连续性倒塌的方法

火灾和爆炸作用对结构造成的直接破坏尽管是局部的，但结构部分构件破坏后，其原来承受的荷载要分散到其他构件来承担，如果其他构件不能承受这种增加的荷载，就会引起其他构件相继破坏，这种破坏如果连续发生，最终会造成整体结构的连续性倒塌[41]。

1995 年美国俄克拉荷马城 Murrah Federal 建筑，因爆炸造成结构连续性倒塌（图 41），导致 167 人死亡，592 人受伤。

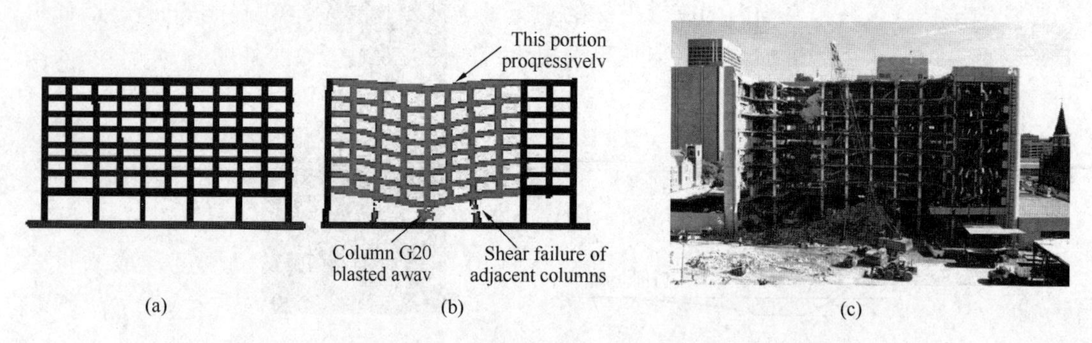

图 41 美国俄克拉荷马城 Murrah Federal 建筑的连续性倒塌

（a）破坏前；（b）破坏原因；（c）破坏后

Fig. 41 Progressive collapse of Murrah Federal building in Oklahoma，USA

（a）before damage；（b）reason of damage；（c）after damage

提高整体结构抗连续性倒塌能力的最有效方法是加强结构的整体性，从而可以利用以下三个有用的效应。

（1）梁的悬链线效应

两端受轴向约束的梁，随着挠度的增大会产生一种类似于悬索的悬链线效应[42]，承受梁上的横向荷载（图 42）。即约束梁首先是利用梁的弯曲机制承载，但随着梁挠度增大，梁的两端有向内缩的趋势，但这种内缩由于受到梁两端的约束使梁产生轴向拉力，梁挠度越大梁中拉力也越大，而梁拉力在竖向（梁横向）上的分量可承受梁上荷载，这种效应即为悬链线效应[43,44]。梁的这种悬链线机制承载有可能比弯曲机制承载大几倍。

框架结构可以利用梁的悬链线效应抵抗由于火灾或爆炸造成的少量柱破坏而导致的连续性倒塌[45]（图 43）。但是，框架梁的悬链线效应要能充分发挥作用需满足以下条件[46]：一是梁两端需有足够的轴向约束；二是梁柱节点需有足够的转动能力，以使梁能产生悬链线效应所需的足够大挠度。

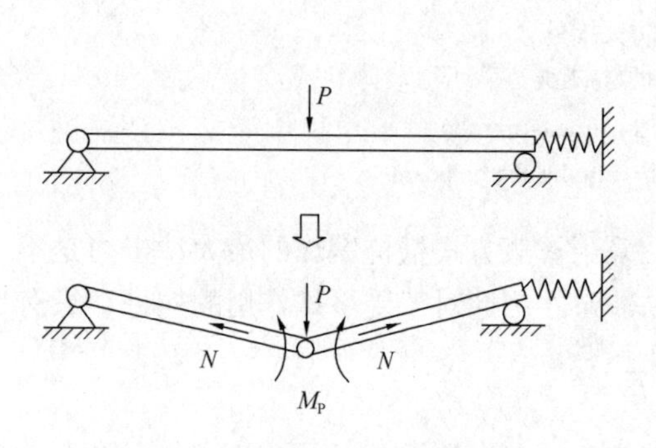

图 42 约束梁的悬链线效应

Fig. 42 Catenary action of restrained beam

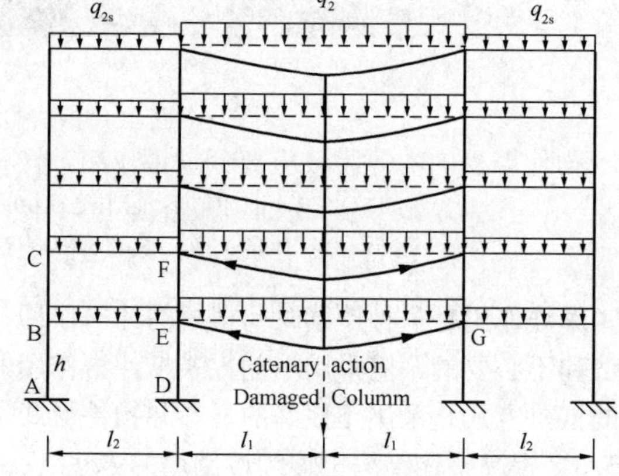

图 43 框架结构抗连续性倒塌的悬链线效应

Fig. 43 The catenary action of frame for prevention of progressive collapse

对于框架中间跨内的柱失效情况，虽然梁的跨度增大一倍（假定梁的跨度均相等），但失效跨梁两端会受到同一平面未失效跨的框架约束（图 43），以及不在失效跨同一平面的未失效结构通过楼板平面内的刚度提高的约束（一般结构是三维和有楼板的），因此失效跨梁两端会有很大的轴向约束。但如果框架边跨内柱失效，失效跨外侧的梁端，主要由边柱的抗弯刚度提供梁的轴向刚度，因此边柱抗弯刚度宜设计得较大。

（2）框架的空腹效应

两层以上框架结构的边柱（特别是角柱）如果发生破坏，可以利用空腹效应抵抗连续性倒塌（图 44）。空腹效应可以理解为旋转 90°的框架承受竖向荷载的一种效应，显然，要充分发挥框架空腹效应的作用，梁柱节点需有足够的转动刚度，成为抗弯框架。

（3）楼板的薄膜效应

板的薄膜效应类似于梁的悬链线效应，是利用板在大挠度下板内张力的竖向分量承受板上荷载的一种效应（图 45）。楼板内一般布置有钢筋网，这种钢筋网随着楼板挠度的增大也会产生薄膜效应[47,48]（图 46），显然，可以利用这种楼板的薄膜效应抵抗框架结构的连续性倒塌[49]（图 47）。

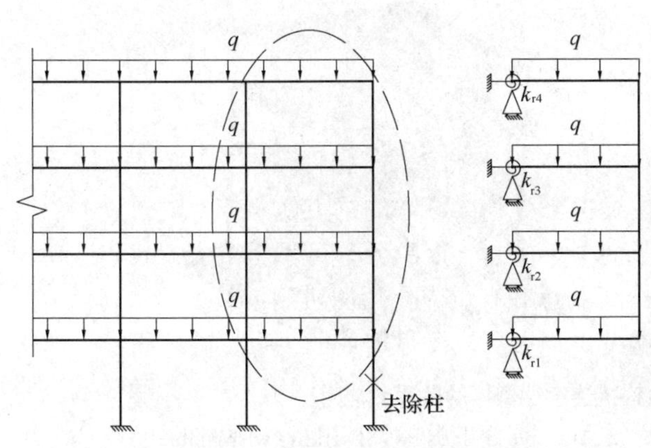

图 44　框架结构抗连续性倒塌的空腹效应

Fig. 44　The effect of moment frame for prevention of progressive collapse

图 45　板的薄膜效应

Fig. 45　Membrane action of slab

为使楼板内的薄膜效应充分发挥作用，楼板内的钢筋网应连续布置，对于四边支承的楼板（内柱失效情况），板四周的混凝土会自然形成一个压力环承受板中间部位钢筋网中的拉力（图 46f）；对于三边支承的楼板（边柱失效情况），如果边柱不靠近角柱，可以依靠失效柱相邻跨的楼板平面内刚度对失效柱跨楼板内钢筋网的薄膜效应拉力提供支承，但如果失效边柱靠近角柱，楼板的薄膜效应拉力在角柱这一边，则主要靠角柱的抗弯刚度来支承；而对于两边支承的楼板（角柱失效情况），则不能形成楼板薄膜效应，只能利用框架的空腹效应抵抗连续性倒塌。

从以上讨论可知，为使框架梁的悬链线效应、框架的空腹效应和楼板的薄膜效应充分发挥，以提高整体结构抗连续性倒塌的能力，应保证结构的整体性，包括保证楼板的整体性和刚度、保证梁柱连接的强度和转动能力、加强角柱和边柱。

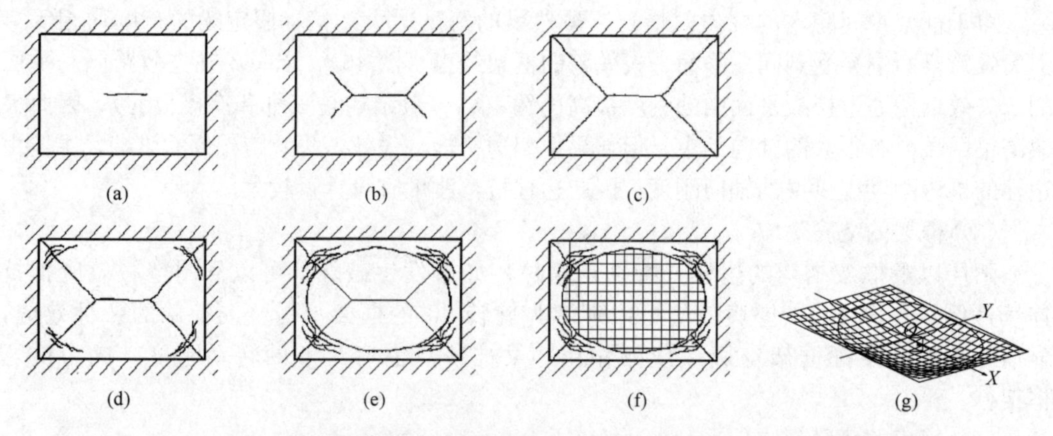

图 46　楼板内薄膜效应随挠度增大的发展过程

Fig. 46　Formation of membrane action in reinforced concrete slab

(a) 开始屈服；(b) 屈服线进一步发展；(c) 形成破坏机构；(d) 薄膜效应的产生；
(e) 薄膜效应充分发展；(f) 薄膜效应的极限状态；(g) 楼板内的钢筋网

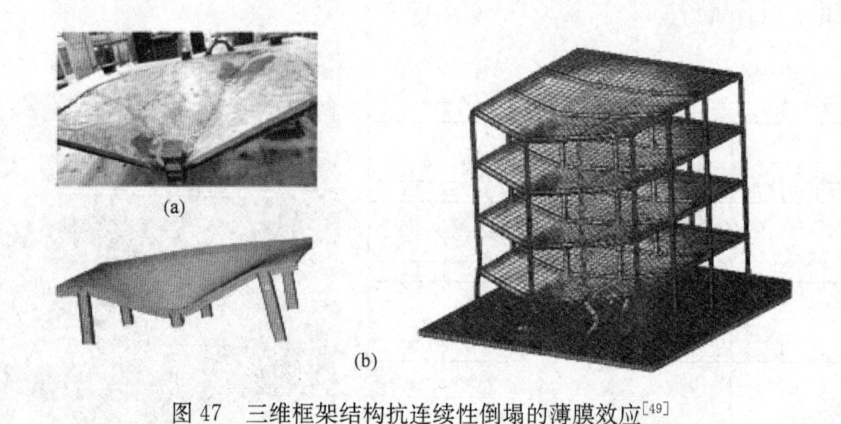

图 47　三维框架结构抗连续性倒塌的薄膜效应[49]

Fig. 47　Membrane action to prevent progressive collapse of 3D frame with slabs[49]

(a) 试验结果；(b) 有限元结果

5. 高施工性钢结构的实现

5.1　提高钢结构制作效率的方式

钢结构施工分工厂制造和现场安装两个阶段。

钢结构构件一般均在工厂制作，为提高钢结构构件的制作效率，降低制造成本，可采用以下方式：

(1) 钢结构设计尽量采用轧制型钢，少采用钢板焊接构件，减小钢构件制作工作量；

(2) 一个钢结构工程采用的型钢规格、钢板厚度、钢材等级尽可能归并，在合理用钢量的条件下减少钢材种类数，以便批量采购，降低成本，提高生产效率；

（3）采用自动化制造设备、信息化管理手段，提升制造智能化水平（图 48），提高钢构件制作精度和效应，减低成本。

5.2 提高钢结构现场安装效率的方式

钢结构的现场安装连接方式主要有焊接和螺栓连接两种形式（图 49）。

钢结构现场焊接的优点是材料省，对钢结构制造和安装的精度要求低；而缺点是对工人的技术要求高，工作量大，人工成本高，施工质量受环境影响大，质量保障难度大，且对钢结构的防腐涂装影响大，钢结构易生锈（图 50）。

图 48 钢结构智能化制造

Fig. 48 Smart fabrication of steel elements

而钢结构现场采用螺栓连接可克服焊接的缺点，可采用工具施工，对工人的技术要求低，工作量小，施工速度快，效率高，人工成本较低，且对钢结构的防腐涂装影响小，钢结构不易生锈（图 50）；但缺点是材料用量大，材料成本较高，另对钢结构制造和安装的精度要求高。

然而，随着我国人工成本的进一步提高，钢结构现场螺栓连接的总成本优势将逐渐凸显，且随着钢结构制造和安装技术水平的提高，将较易满足螺栓连接对钢结构制造和安装的高精度要求。因此，为提高钢结构现场安装的效率，应优先采用螺栓连接。

图 49 钢结构的现场焊接与螺栓连接

Fig. 49 Welded and bolted connections in site

图 50 焊接与螺栓连接对钢结构防腐的影响

Fig. 50 Effect of different connections on corrosion prevention of steel structures

5.3 钢结构现场螺栓连接的实现方法

钢结构现场梁柱连接节点数量多，而梁柱刚性连接一般需焊接，而螺栓连接方便用于梁柱铰接或半刚性连接[18]（图 51）。对于闭口箱型截面柱，梁柱也可以采用单向螺栓连

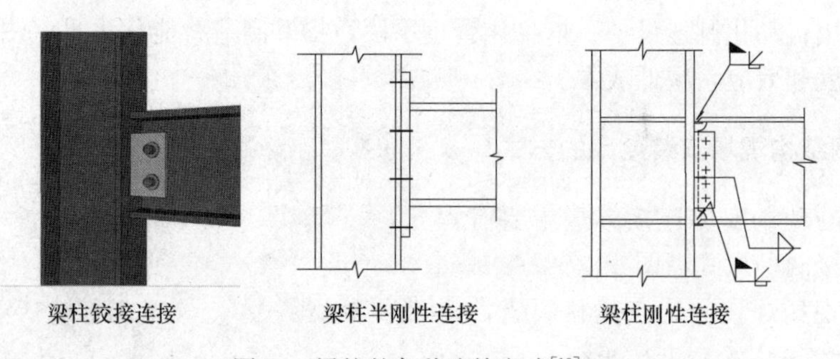

图 51　梁柱的各种连接方式[18]

Fig. 51　Various beam-to-column connections[18]

接[50]（图 52）。

然而，框架结构如采用梁柱铰接则不能提供抗侧刚度，框架结构如采用半刚性连接提供的抗侧刚度也较小，且结构设计计算较麻烦。解决这一问题可采用整体结构体系抗侧结构与承重结构分离的方法（图 53），在整体结构中专门设计抗侧结构体系承受地震和风等侧向力，而整体结构的其他部分为承重结构，由于不承受侧力，因此梁柱节点无需采用刚性连接。

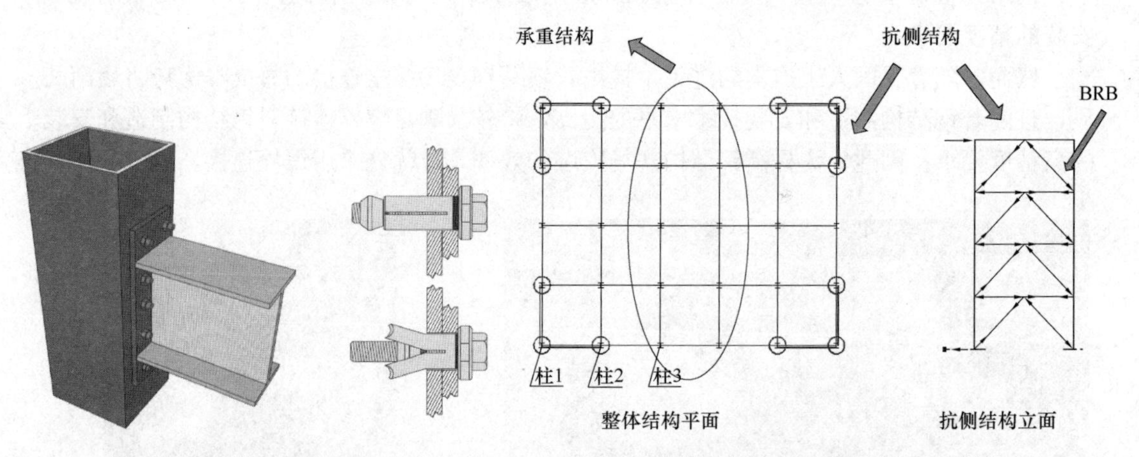

图 52　梁柱单向螺栓连接[50]

Fig. 52　Blind bolts for beam to box column connection[50]

图 53　整体结构体系抗侧结构与承重结构分离

Fig. 53　Separation of structural systems to respectively resist lateral forces and vertical forces

采用整体结构体系抗侧结构与承重结构分离的设计方法，应注意保证结构楼板的整体性和楼板平面内较大刚度，通过楼板将整体结构承受的侧向力传递到抗侧结构部分。如果结构需抗震，则整体结构中的抗侧结构体系宜采用消能-承载双功能构件抵抗地震作用。

如承重结构采用半刚性梁柱连接，可以偏安全地忽略其承受的侧向力，仅考虑半刚接框架承受重力，除可以简化半刚接框架的设计外，与铰接框架和刚接框架相比，在相同重力作用下，半刚接框架梁端的弯矩比刚接框架梁小，而跨中弯矩比铰接框架梁小（图 54），还可以考虑跨中正弯矩由钢-混凝土组合梁承受，而梁端负弯矩仅由钢梁承受的特点，调整半刚接框架梁跨中弯矩与梁端弯矩绝对值的比值，优化半刚接框架的设计，与传统刚接框架相比，可节省用钢量 10%以上[25]。

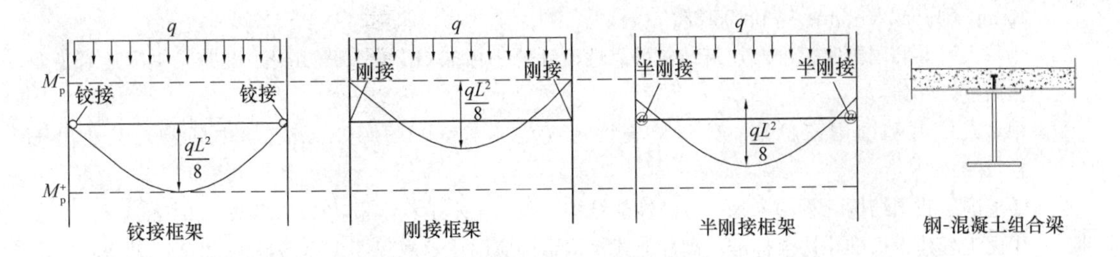

图 54 重力荷载作用下铰接、刚接、半刚接框架梁弯矩的对比[25]

Fig. 54 Comparison of moments in the beams of pined, rigid and semi-rigid connected frames under gravity loads[25]

6. 总结

本文介绍了高性能钢结构的若干重要概念及实现方法，可归纳总结如下：

（1）高性能钢结构应至少满足承载性好、承灾性好和施工性好的要求。

（2）高承载性钢结构可以采用高强度结构钢材、高承载构件形式和高承载结构体系的方式实现。

（3）高承灾性钢结构应在罕遇的大地震下和可能发生的火灾或爆炸作用下保持完整、不倒塌，且产生的破坏能方便修复。

（4）结构抵抗地震作用的方式有承载方式和消能方式两种，对于出现概率较大的小震可以采用承载方式抵抗，而对于出现概率较小的大震应优先采用消能方式抵抗。

（5）高强钢用于抗震结构的有效方式是，在地震下不发生屈服或屈服后发生塑性变形很小的结构部位或构件采用高强钢。

（6）对于地震下钢结构抗侧力的主要构件采用消能-承载双功能构件，可起到大震下消能减震作用，提高结构抗地震整体倒塌能力。

（7）在结构中采用消能摇摆柱（墙）可大大减小大震后的残余变形，提高主体结构的可恢复性。

（8）保证楼板的整体性和刚度、保证梁柱连接的强度和转动能力、加强角柱和边柱等，可以充分发挥框架梁的悬链线效应、框架的空腹效应和楼板的薄膜效应，大大提高结构抵抗火灾、爆炸等偶然作用下连续性倒塌的能力。

（9）可通过采用轧制型钢构件和现场螺栓连接提高钢结构的施工性。

（10）通过抗侧结构与承重结构分离的方式提高可提高钢结构现场螺栓连接的可实现性，即在整体结构中专门设计抗侧结构体系承受地震和风等侧向力，而整体结构的其他部分仅为不承受侧力的承重结构，使梁柱节点可较方便地采用螺栓铰接连接或螺栓半刚性连接形式。

参考文献

[1] L. W. Lu, R. Sause, J. M. Ricles. High-Performance Steel Structures: Recent Research. Proceedings of The Second International Conference on Advances in Steel Structures, 15-17 December 1999, Hong

Kong, China, Volume 1, pp 75-86.
[2] 中华人民共和国国家标准. 建筑结构可靠性设计统一标准 GB 50068—2018. 北京：中国建筑工业出版社，2019.
[3] 中华人民共和国国家标准. 建筑抗震设计规范 GB 50011—2010. 北京：中国建筑工业出版社，2010.
[4] 聂建国. 我国结构工程的未来——高性能结构工程. 土木工程学报，2016，46(9)：1-8.
[5] 中国工程建设标准化协会标准. 高性能建筑钢结构应用技术规程 T/CECS 599—2019. 北京：中国计划出版社，2019.
[6] 中华人民共和国国家标准. 建筑钢结构防火技术规范 GB 51249—2017. 北京：中国计划出版社，2018.
[7] 中国工程建设标准化协会标准. 民用建筑防爆技术规程 T/CECS 736—2020. 中国计划出版社，2021.
[8] 中国工程建设标准化协会标准. 建筑结构抗倒塌设计规范 CECS 392：2014. 北京：中国计划出版社，2014.
[9] 中华人民共和国国家标准. 优质碳素结构钢 GB/T 699—2015，2015.
[10] 中华人民共和国国家标准. 低合金高强度结构钢 GB/T 1591—2008，2008.
[11] 中华人民共和国国家标准. 高强度结构用调质钢板 GB/T 16270—2009，2009.
[12] 李国强，王彦博，陈素文，孙飞飞. 高强度结构钢研究现状及其在抗震设防区应用问题. 建筑结构学报，2013，34(1)：1-13.
[13] 范重. 国家体育场(鸟巢)结构设计. 北京：中国建筑工业出版社，2011.
[14] 李国强，张哲，范昕. 波纹腹板钢结构性能、设计与应用. 北京：中国建筑工业出版社，2018.
[15] FEMA 403 (2002). World Trade Centre Building Performance Study: Data Collection. Preliminary Observation and Recommendations. Federal Emergency Management Agency and Federal Insurance and Mitigation Administration.
[16] 李国强，张洁. 上海地区高层建筑采用钢结构与混凝土结构的综合经济比较分析. 建筑结构学报，2000，(02)：75-79.
[17] 李国强，陈素文，丁翔，陆烨. 高层建筑钢-混凝土混合结构设计实例. 建筑钢结构进展，2005，(06)：40-48.
[18] 李国强. 多高层建筑钢结构设计. 北京：中国建筑工业出版社，2004.
[19] 朱伯龙，屠成松，许哲明. 工程结构抗震设计原理. 上海：上海科学技术出版社，1982.
[20] 李杰，李国强. 地震工程学导论. 北京：地震出版社，1992.
[21] 李爱群. 工程结构减振控制. 北京：机械工业出版社，2007.
[22] Guo-Qiang LI, Enhancing Performance of Buildings in Seismic Zone with Structural Metal Dampers, Proceedings of Second International Conference on Performance-based and Lifecycle Structural Engineering (PLSE 2015), Brisbane, Australia, 9-11 December 2015.
[23] 李国强，孙飞飞，沈祖炎. 强震下钢框架梁柱焊接连接的断裂行为. 建筑结构学报，1998，(04)：19-28.
[24] 中国工程建设标准化协会标准. 端板式半刚性连接钢结构技术规程 CECS 260：2009. 北京：中国计划出版社，2009.
[25] 李国强，石文龙，王静峰. 半刚性连接钢框架结构设计. 北京：中国建筑工业出版社，2009.
[26] Akira Wada and Masayoshi Nakashima. From infancy to maturity of buckling restrained brace research, Proceedings of 13 WCEE, Vancouver, BC, Canada, 2004.
[27] 孙飞飞，李国强，宫海，王新娣. 耗能钢支撑及钢板剪力墙结构设计指南. 上海：同济大学出版

社，2015.
[28] 李国强，金华建，孙飞飞，陆烨. 屈曲约束钢板剪力墙约束板研究(Ⅱ)——承载力需求及试验验证. 土木工程学报，2016，(07)：53-60+71.
[29] 金华建，孙飞飞，李国强. 无屈曲波纹钢板墙抗震性能与设计理论. 建筑结构学报，2020，(05)：53-64.
[30] 刘青，陆烨，李国强. 钢板墙束柱受力性能试验研究及理论分析. 建筑结构学报，2014，(02)：59-66.
[31] 王亚勇. 汶川地震建筑震害的启示——抗震概念设计. 建筑结构学报，2008，(04)：20-25.
[32] Li Guo-Qiang，Pang Mengde，Sun Feifei，Jiang Jian，Hu Dazhu. Seismic behavior of coupled shear wall structures with various concrete and steel coupling beams. Structural design of tall and special buildings，2018，27(1)：e1405.
[33] Li，Guo-Qiang，Pang，Meng-de，Sun，Fei-fei，Jiang，Jian. Study on two-level-yielding steel coupling beams for seismic-resistance of shear wall systems. Journal of constructional steel research，2018，144：327-343.
[34] Li Guo-Qiang，Pang Meng-De，Li Yan-Wen，Li Liu-Lian，Sun Fei-Fei，Sun Jian-Yun. Experimental comparative study of coupled shear wall systems with steel and reinforced concrete link beams. Structural Design of Tall and Special Buildings，2019，28(18)：e1678.
[35] 吕西林，陈云，毛婉君. 结构抗震设计的新概念——可恢复功能结构. 同济大学学报(自然科学版). 2011(7)：941-948.
[36] Li Yan-Wen，Li Guo-Qiang，Sun Fei-Fei，Jiang Jian. Experimental study on continuous energy-dissipative steel columns under cyclic loading. Journal of constructional steel research，2018，141：104-117.
[37] Li Yan-Wen，Li Guo-Qiang，Jiang Jian，Sun Fei-Fei. Mitigating seismic response of rc moment resisting frames using steel energy-dissipative columns. Engineering structures，2018，174：586-600.
[38] Li Yan-Wen，Li Guo-Qiang，Jiang Jian，Wang Yan-Bo. Experimental study on seismic performance of RC frames with Energy-Dissipative Rocking Column system. Engineering Structures，2019，194：406-419.
[39] 李国强，韩林韩，楼国彪，蒋首超. 钢结构及钢-混凝土组合结构抗火设计. 北京：中国建筑工业出版社，2006.
[40] 李忠献，师燕超. 建筑结构抗爆分析理论. 北京：科学出版社，2015.
[41] 姜健，李国强. 建筑钢结构防连续性倒塌各国规范设计及分析[J]. 解放军理工大学学报(自然科学版). 2014，15(6)，540-551.
[42] Y. Z. Yin，Y. C. Wang. Analysis of catenary action in steel beams using a simplified hand calculation method，Part 1：theory and validation for uniform temperature distribution. Journal of Constructional Steel Research，2005，61(2)：pp183-211 .
[43] 王开强，李国强，杨涛春. 考虑悬链线效应的约束钢梁在分布荷载作用下的性能(Ⅰ)——理论模型. 土木工程学报，2010，(01)：9-15.
[44] 李国强，王开强，杨涛春. 考虑悬链线效应的约束钢梁在分布荷载作用下的性能(Ⅱ)——数值算例验证. 土木工程学报，2010，(01)：16-20.
[45] B. A. Izzuddin，A. G. Vlassis，A. Y. Elghazouli，D. A. Nethercot. Progressive collapse of multi-storey buildings due to sudden column loss — Part I：Simplified assessment framework [J]. Engineering Structures，2008，30：pp1308-1318.
[46] Li Liu-Lian，Li Guo-Qiang，Jiang Binhui，Lu Yong. Analysis of robustness of steel frames against

progressive collapse. Journal of constructional steel research，2018，143：264-278.

[47] C. G. Bailey. Membrane action of unrestrained lightly reinforced concrete slabs at large displacements. Engineering Structures，2001，23，pp470-483.

[48] C. G. Bailey，D. B. Moore. The Structural Behaviour of Steel Frames with Composite Floor Slabs Subjected to Fire，Part 1：Theory. The Structural Engineer，2000，78(11)：pp19-27.

[49] J Z Zhang，G Q Li. Collapse resistance of steel beam-concrete slab composite substructures subjected middle column loss. Journal of Constructional Steel Research，2018，145：471-488.

[50] 中国工程建设标准化协会标准. 矩形钢管构件自锁式单向螺栓连接设计规程 T/CECS 605—2019. 北京：中国计划出版社，2019.

6 摇摆桥墩的研究综述*

杜修力[1]，周雨龙[1]，韩　强[1]，王智慧[2]
（1. 北京工业大学 城市与工程安全减灾教育部重点实验室，北京 100124；
2. 北京市市政工程设计研究总院有限公司，北京 100082）

摘　要：桥梁作为交通生命线的枢纽工程，是交通基础设施震后功能恢复的关键所在。摇摆结构可将地震损伤控制在摇摆界面以避免主体结构破坏，且具有较好的抗震性能和自复位能力，因此在提高结构震后恢复能力方面具有显著优势。本文首先简要介绍了摇摆桥墩的基本原理，回顾了摇摆理念在桥梁工程应用的发展历史，综述了摇摆桥墩的研发和试验研究、摇摆桥墩与上部结构的连接方式、摇摆结构分析方法等发展现状，总结了目前摇摆桥墩研究的不足和发展趋势。

关键词：桥梁结构；震后恢复能力；摇摆桥墩；综述
中图分类号：U442.55　　　　**文献标志码**：A

State-of-the-art on Rocking Piers

Du Xiuli[1], Zhou Yulong[1], Han Qiang[1], Wang Zhihui[2]
(1. Key Laboratory of Urban Security and Disaster Engineering of Ministry of Education,
Beijing University of Technology, Beijing 100124, China
2. Beijing General Municipal Engineering Design & Research Institute Co., Ltd., Beijing 100082, China)

Abstract: As important joints of transportation lifeline systems, bridges have a significant effect on the resilience of post-earthquake serviceability. Rocking structures can limit damage in rocking interface to void damage in main structure, with fine seismic performance and self-centering capacity. Therefore, rocking structures have significant advantages in improving post-earthquake resilience. This paper introduced the basic mechanism of rocking piers and its application history briefly, then reviewed the state of the art of rocking pier development and experiment, joint construction of rocking piers and superstructures, and analytical method of rocking structures. The shortages and development tendencies of existing rocking piers were summarized.

Keywords: bridge structure; post-earthquake resilience; rocking pier; state-of-the-art

1. 引言

经过多年的研究和实践，现有桥梁抗震设计规范[1-3]基本能达到“小震不坏、大震不

* 杜修力，周雨龙，韩强等．摇摆桥墩的研究综述［J］．地震工程与工程振动，2018，038（005）：1-11.

倒”的抗震设防目标，但对于震后桥梁结构残余交通功能和恢复能力没有明确量化标准。然而桥梁震后丧失交通功能造成的交通中断严重影响灾后救援和灾区重建工作，造成的间接经济损失更是难以估计。因此，在地震作用下，有效控制桥梁结构的地震损伤，缩短桥梁结构受损后的恢复时间，已成为现阶段桥梁结构抗震设计中非常重要的理念及追求目标。

目前，我国的桥梁抗震设计采用延性抗震和减隔震设计方法。延性抗震设计方法利用桥墩塑性铰区的塑性变形来耗散地震能量，因此塑性铰区在地震中损伤破坏严重，且震后墩顶残余位移可能较大。减隔震设计方法可避免桥墩明显塑性损伤，但在罕遇地震作用下桥梁有发生落梁或倒塌的潜在风险，且不易控制主梁残余位移，致使震后桥梁的交通功能受限。因此，目前我国桥梁抗震设计规范中给出的抗震设计方法不能有效控制桥梁结构在受损后的恢复时间，震后桥梁交通通行能力的快速恢复难以实现。

近年来，秉承减隔震理念的摇摆桥墩在具有较好耐久性和稳定性的同时，可提供优良的抗震能力和震后恢复能力，且极适用于预制拼装技术的应用，使桥梁结构建造速度更快、质量更可控、环境影响更小，是现代桥梁结构抗震设计方法和先进建造技术的前沿领域[4]。迄今，有关摇摆桥墩抗震性能的研究在国外开展较多，我国在这一领域的研究则相对较少。国际上摇摆桥墩的工程应用还并不多见，主要原因是摇摆桥墩合理的构造形式、抗震性能等应用基础问题还缺少系统成熟的成果。

本文简要介绍了摇摆桥墩的基本原理，回顾了摇摆桥梁结构的工程应用，综述了摇摆桥墩的抗震性能和分析方法的发展现状，总结了目前摇摆桥墩研究的不足和发展趋势。

2. 摇摆桥墩的基本原理及工程应用

2.1 摇摆桥墩的基本原理

图1所示为典型单柱摇摆桥墩，摇摆桥墩与承台之间设置摇摆界面，通过无粘结预应力束和耗能装置将摇摆桥墩和承台连接成整体［如图1(b) 所示］。摇摆桥墩利用其摇摆

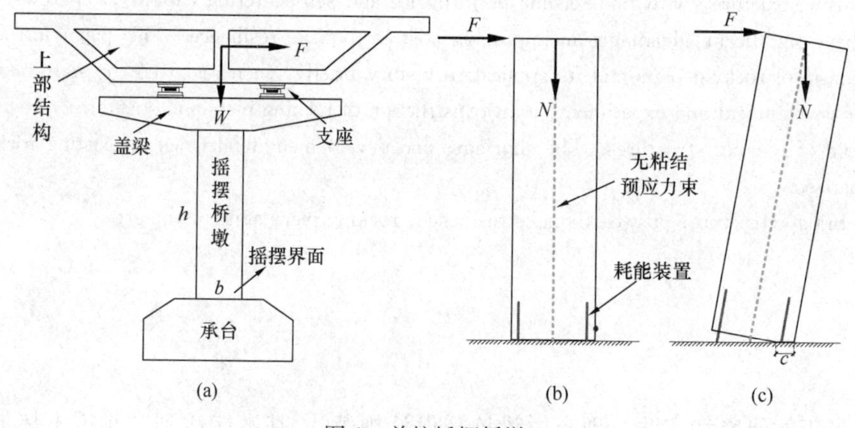

图1 单柱摇摆桥墩

Fig. 1 Single-column rocking pier

(a) 单柱摇摆桥墩横断面；(b) 初始位置；(c) 摇摆变形

界面的变形来实现桥梁结构的位移需求，桥墩与承台的连接部分在地震作用下不断“提离”和“闭合”，来实现桥墩的摇摆行为［如图 1(c) 所示］，以降低桥墩抗侧刚度，延长桥梁结构自振周期从而起到隔震作用。摇摆桥墩主要为受压构件，因此可将损伤控制在摇摆界面上，以避免桥墩主体发生损伤破坏，且可在变形位置采用不易受损的控制元件（如无粘结预应力束）和易于修复的消能装置，来提高摇摆桥墩的自复位能力和耗能能力。

当上部结构自重与无粘结预应力束产生的复位弯矩大于耗能钢筋的抵抗弯矩时，桥墩发生可控制的摇摆，塑性变形仅发生在摇摆界面的受压区（受压区高度为 c），结构的整体性得以保持。在预应力束和上部结构重力的作用下，摇摆桥墩具有极佳的自复位能力。摇摆桥墩利用上部结构自重和无粘结预应力束的自复位特性［如图 2(a) 所示］，以及耗能装置的耗能能力［如图 2 (b) 所示］，提供摇摆桥墩以旗帜形滞回方式运动［如图 2 (c) 所示］。

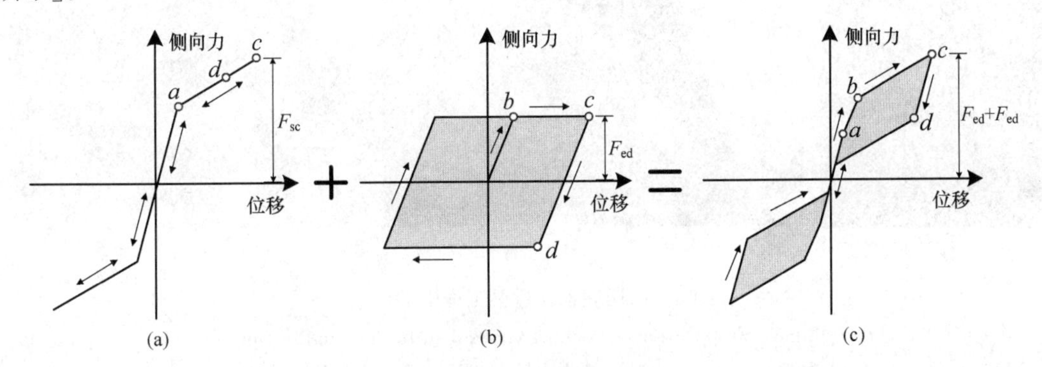

图 2　摇摆桥墩的滞回行为

Fig. 2　Hysteretic behavior of rocking pier

(a) 自复位组件；(b) 耗能组件；(c) 旗帜形滞回行为

2.2　摇摆桥墩的工程应用

1974 年 Beck 和 Skinner[5] 首次开展了摇摆理念在桥梁抗震设计中的研究，成果在新西兰 South Rangitikei 铁路桥中［如图 3(a) 所示］得到了应用[6]。在 20 世纪 90 年代，Priestley 等[7] 提出摇摆理念可作为桥梁抗震加固的一种有效方法，随后 Astaneh-Asl 和 Shen[8] 研究了桥墩与基础间的有限摇摆，并应用于美国旧金山-奥克兰海湾大桥改造的抗震加固设计中；Dowdell 和 Hamersley[9] 也将桥墩摇摆构造应用到温哥华狮门大桥的北引高架桥的抗震加固中［如图 3(b) 所示］。将桥墩与基础之间的弱化处理作为一种抗震加固措施还应用在加州的卡齐尼兹大桥和金门大桥的抗震加固中[10,11]。尽管摇摆桥墩在桥梁工程中已有应用，但其发展较为滞后且处于初步阶段，主要工程应用集中在钢桁架桥梁的抗震加固方面。

3. 摇摆桥墩发展现状

3.1　摇摆桥墩的研发和试验研究

目前摇摆桥墩的主要研究内容集中在将无粘结预应力技术和耗能装置与摇摆桥墩的联

(a)

(b)

图 3　摇摆理念在桥梁工程中的应用

Fig. 3　Application of rocking concept in bridge engineering

(a) 新西兰 South Rangitikei 铁路桥；(b) 温哥华狮门大桥的北引高架桥

合应用上，并在预制拼装技术中有一定涉及。无粘结预应力束对提高摇摆桥墩的抗倾覆能力和减小残余位移起到更为重要的作用。Mander 和 Cheng[12] 最早提出无损伤破坏的预应力摇摆桥墩设计理念［如图 4(a) 所示］，以采用无粘结预应力技术的摇摆单柱桥墩为研究对象，进行了低周循环荷载试验研究，并给出该类摇摆桥墩抗震设计方法和简化计算公

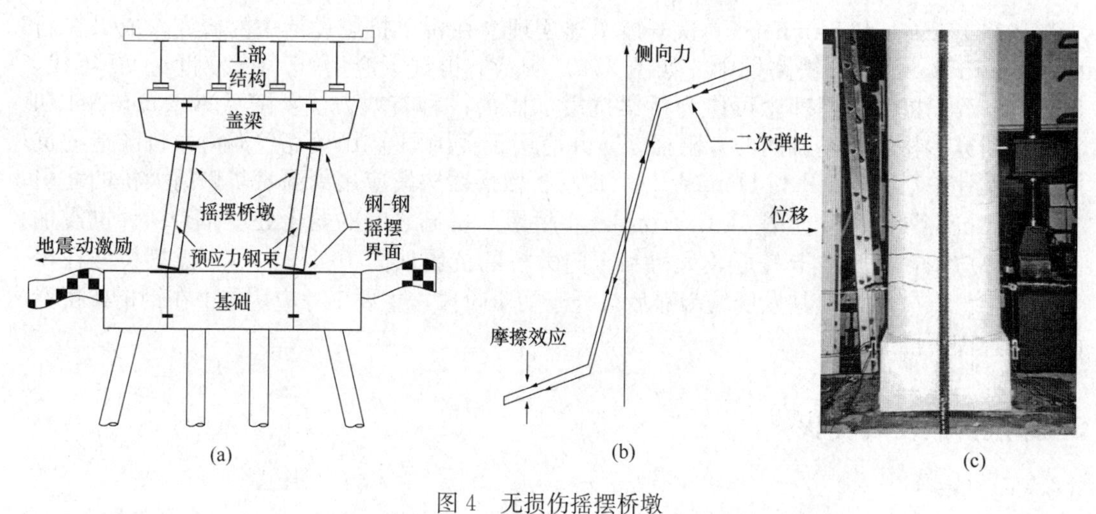

(a)　(b)　(c)

图 4　无损伤摇摆桥墩

Fig. 4　Rocking pier with damage avoidance concept

(a) 设计理念；(b) 分析模型；(c) 最后损伤状态

式；为避免摇摆界面出现显著损伤破坏，桥墩底部与承台顶部之间的摇摆界面均设置76mm厚钢板，试验结果表明：该类摇摆桥墩的抗侧承载能力主要由桥墩轴力（由上部结构和预应力束提供）决定，且具有显著自复位能力［如图 4(b) 所示］，试验后该摇摆桥墩无明显损伤而仅有细微裂缝［如图 4(c) 所示］。Cheng[13]在此基础上进行了无损双柱墩的振动台试验，验证了该类桥墩在地震作用下的无损理念和分析模型的有效性。同样采用预应力技术，夏修身和陈兴冲[14]针对铁路高墩提出利用高墩底部提离进行隔震的抗震设计方案，并验证了该摇摆高墩设计方案的有效性。

与梭形滞回关系的 RC 桥墩相比［如图 5(a) 所示］，无粘结预应力摇摆桥墩可避免桥墩主体发生较大损伤破坏且具有较好的自复位能力［如图 5(b) 所示］。但由于预应力束在耗能方面的贡献基本可以忽略，因此该类摇摆桥墩的耗能能力不足，致使位移需求较大。依托美国预制结构抗震计划项目[15]，Palermo 等[16]提出无粘结预应力技术与内置耗能钢筋联合应用的摇摆桥墩，耗能钢筋埋置在承台并延伸至桥墩。该类摇摆桥墩的力-位移滞回关系为旗帜形［如图 5(c) 所示］，与仅设置预应力束的摇摆桥墩（S 形滞回关系）相比，其侧向承载能力和耗能能力显著增加，位移需求明显减小，使桥墩具有更好的抗震性能。随后，内置不同类型耗能装置的联合摇摆桥墩的抗震能力又通过拟静力和拟动力试验进一步验证和研究[17-20]，并与常规现浇 RC 桥墩对比，该类摇摆桥墩具有较好自复位能力，并显著减小桥墩损伤。

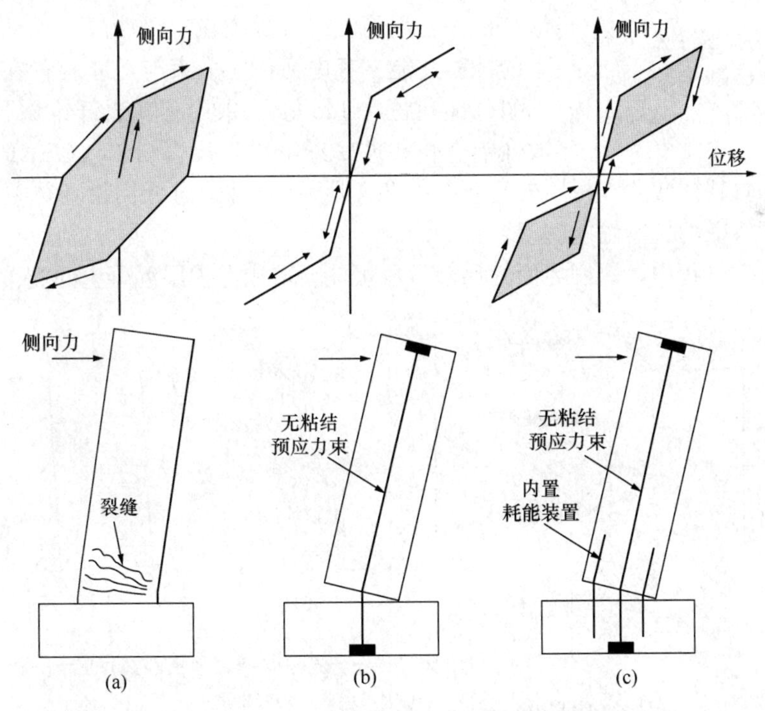

图 5　不同类型桥墩及其滞回关系

Fig. 5　Different type and hysteresis relationship of pier

(a) 梭形 RC 桥墩；(b) S 形摇摆桥墩；(c) 旗帜形摇摆桥墩

虽然在地震作用下内置耗能装置的预应力摇摆桥墩损伤较小，但耗能钢筋损伤严重，且在震后难以更换，致使该类摇摆桥墩在震后需要较为复杂的修复工作。因此，代替内置耗能装置的外置耗能装置（如防屈曲软钢阻尼器，角钢阻尼器和斜撑阻尼器等）被应用在无粘结预应力摇摆桥墩中（如图 6 所示），这些外置耗能装置作为保险丝单元可对桥墩提供有效的保护，且在震后更换极为方便[21-25]。国内学者针对联合应用的预应力摇摆桥墩的研究相对滞后，葛继平等[26]、布占宇等[27]研究了无粘结预应力联合内置耗能钢筋的预制节段拼装桥墩抗震性能，郭佳等[28]、谭真等[29]研究了外置耗能装置的预应力摇摆桥墩的抗震性能。

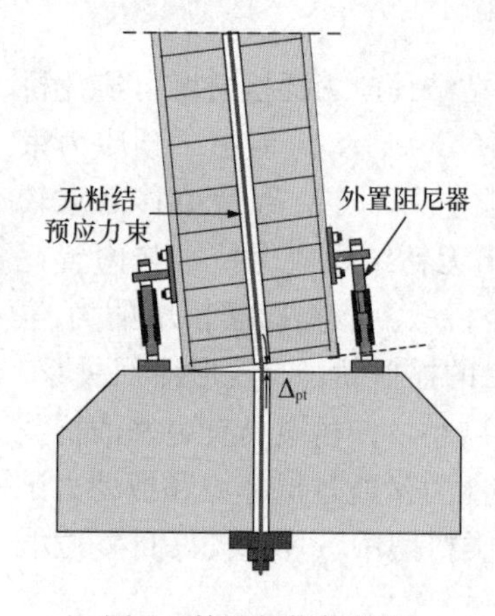

图 6　外置耗能装置的预应力摇摆桥墩

Fig. 6　Post-tensioned rocking pier with external dissipaters

近年来，可提供更优性能的新型材料在摇摆桥墩中的应用得到初步研究。Elgawady 等[24, 25, 30]对纤维增强复合材料（FRP）包裹混凝土的预应力摇摆桥墩进行低周循环往复荷载试验研究，得出该类摇摆桥墩可提供较好延性，且残余位移可忽略，桥墩和承台仅出现较小损伤。Dawood 等[31, 32]通过有限元的方法对 Elgawady 等的试验进行了验证，并进一步研究 FRP 套筒高度，初始预应力和桥墩几何特性等参数对该类摇摆桥墩抗震性能的影响，并给出该类桥墩基于性能的抗震设计方法。Trono 等[33]通过对联合应用摇摆桥墩的振动台试验验证了复合纤维增强混凝土材料可减小摇摆桥墩的地震损伤。Nikbakht 等[34]研究了采用形状记忆合金筋的摇摆桥墩在非线性静力和动力荷载作用下的抗震性能。考虑到预应力束与耗能装置在氯盐环境中的耐久性问题，Guo 等[35]采用玄武岩纤维增强复合材料、改性铝合金和玻璃纤维增强聚合材料（GFRP）来制作预应力钢筋、外置耗能装置和桥墩外套筒（如图 7 所示），并通过低周循环往复荷载试验来研究该类摇摆桥墩的抗震性能。

在桥墩摇摆过程中，摇摆界面的局部承压可能会造成保护层混凝土的剥落和脚点混凝

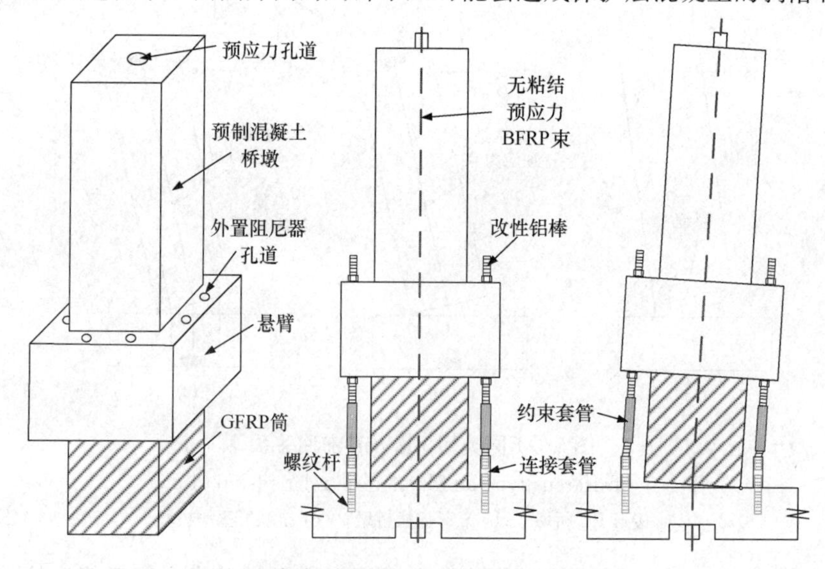

图 7　考虑耐久性问题的预应力摇摆桥墩

Fig. 7　Post-tensioned rocking pier considering durability

土的压碎[16-20]。采用钢材料对摇摆界面进行加强可限制局部损伤，若干学者[12, 13, 19, 21]通过拟静力、伪动力和振动台试验验证了此类方法的有效性。纤维增强复合材料和复合纤维增强混凝土的应用同样可限制摇摆界面的损伤[24, 25, 30, 33, 35]。

平齐端面对接方式使桥墩-承台界面的剪力传递仅依赖于界面摩擦。采用该种对接方式可能会造成摇摆桥墩在地震作用下发生滑动[36]，致使体系稳定性较差，且存在造成高应力状态的预应力束剪切破坏的危险。何铭华等[37, 38]提出嵌合式接头可为摇摆桥墩提供更为可靠的剪力传递机制；郭佳等[28]通过试验验证了球入式嵌合接头的抗剪有效性；Cheng[13]和 Marriott[22]分别采用角钢和剪力键作为抗剪措施同样避免了摇摆桥墩发生显著滑动。

3.2 摇摆桥墩与上部结构的连接方式

桥墩与上部结构的连接方式主要分为支座支撑和墩梁固结。对于采用支座支撑的连接方式，摇摆桥墩与常规桥墩无显著差别[16-20]。但是在连续刚构桥梁中的应用往往受限，虽然摇摆桥墩可以避免墩底出现塑性破坏，却会显著增加桥墩与上部结构连接处的弯矩需求，致使墩顶附近出现更严重的塑性破坏[39]［如图 8(a) 所示］。因此，采用仅传递剪力而不能传递弯矩的销栓作为连接桥墩与上部结构的构造方式，成为潜在解决摇摆桥墩与桥梁结构变形不协调而导致墩顶严重塑性破坏的方案［如图 8(b) 所示］。

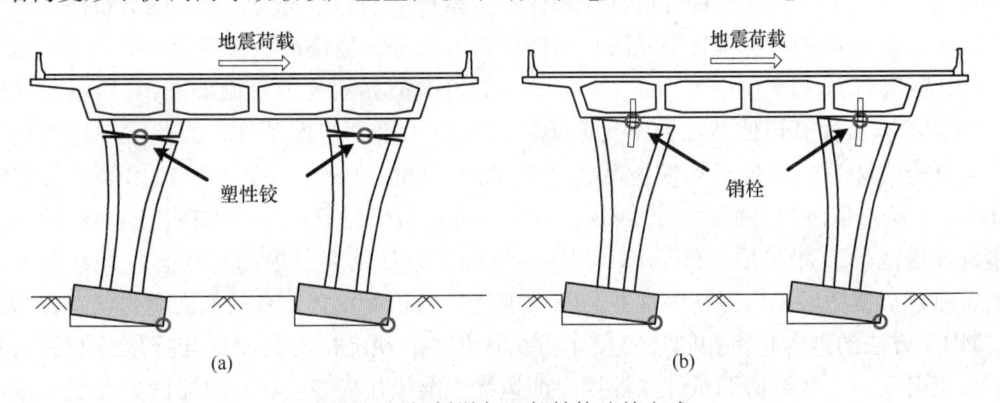

图 8　摇摆桥墩与上部结构连接方式

Fig. 8　Joint construction of rocking pier and superstructure

(a) 墩梁固结；(b) 销栓连接

20 世纪初，类似销栓的柔性连接方式已开始应用于钢筋混凝土桥墩与上部结构和承台的连接，以避免桥墩在连接处的损伤破坏。2006 年，Keever 等[40]提出一种连接桥墩与上部结构的销栓构造形式［如图 9(a) 所示］，并将该销栓连接构造应用于位置关键、抗震性能要求极高的旧金山-奥克兰海湾大桥引桥的更换工程中［如图 9(b) 和 (c) 所示］，且该销栓连接方式逐渐在美国加州新建桥梁工程中得到推广。该销栓连接方式的具体构造为：钢管作为销栓部分埋入桥墩，并伸入上部结构中由钢罩预留出的孔洞，钢管与钢罩之间预留的空隙可使钢管自由转动；螺旋箍筋布置于埋置桥墩的钢管外围以增强钢管周围混凝土的约束强度。Zaghi 和 Saiidi 通过静载试验研究作为销栓的钢管剪切屈服强度和极限强度[41]以及拟静力试验研究采用销栓连接上部结构的桥墩力学性能和销栓作用[42]，并提

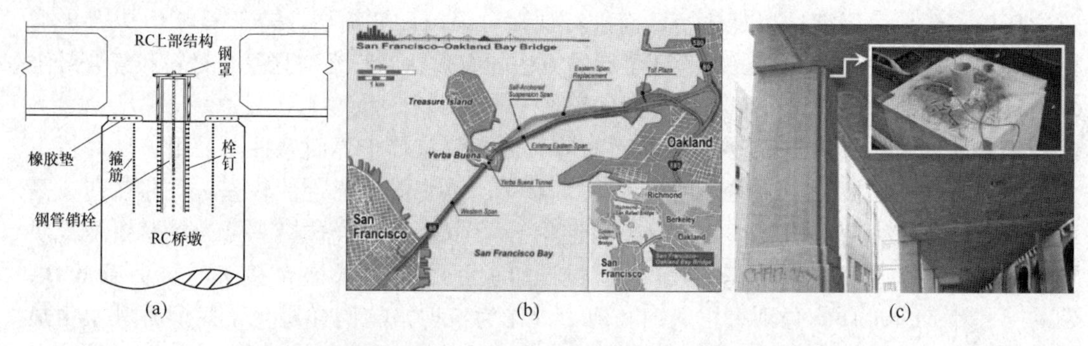

(a) (b) (c)

图 9 销栓在旧金山-奥克兰海湾大桥引桥的应用

Fig. 9 Utilization of pipe-pin in the approach ramps of San Francisco-Oakland Bay

(a) 销栓构造；(b) 旧金山-奥克兰海湾大桥；(c) 引桥工程应用

出销栓连接的可能破坏模式和分析模型[41]，给出具有销栓构造的双柱式桥墩的简化数值计算方法[43]。研究结果表明：在地震作用下这种销栓连接构造可有效消减弯矩的传递，而仅将剪力和轴力传递至下部结构，显著减小桥墩顶部损伤而避免塑性铰的形成。

3.3 摇摆结构的分析方法

目前研究人员已提出多种分析方法来计算摇摆桥墩的力学行为，如截面分析法、集中塑性铰法，多弹簧法，实体有限元法和刚体分析方法等。摇摆桥墩在侧向水平力 F、轴力 N、阻尼器反力 F_d 和预应束反力 F_{pt} 共同作用下的摇摆界面受力机制如图 10 所示。从图 10 中可知摇摆界面开口使截面曲率无限大，而违背了平面假设条件，因此常规现浇桥墩的截面分析方法不能实现对摇摆桥墩的弯矩-曲率分析。Pampanin 等[44]提出等效悬臂梁（MBA）方法，以解决平截面假设问题。该方法随后被广泛验证并分别在美国和新西兰的抗震设计规范[45, 46]中采用。Palermo 等[17]在 MBA 的基础上考虑消压状态，即在摇摆界面消压前，摇摆桥墩反应保持弹性并与等效现浇桥墩相等，且二者的截面应力状态保持一致。MBA 方法的理念是将相同几何尺寸和配筋形式的现浇桥墩的变形与摇摆桥墩进行等效，从而引入一个变形协调条件（摇摆界面边缘混凝土压应变 $\varepsilon_c(\theta)$ 和摇摆界面受压区高度 c 的关系）：

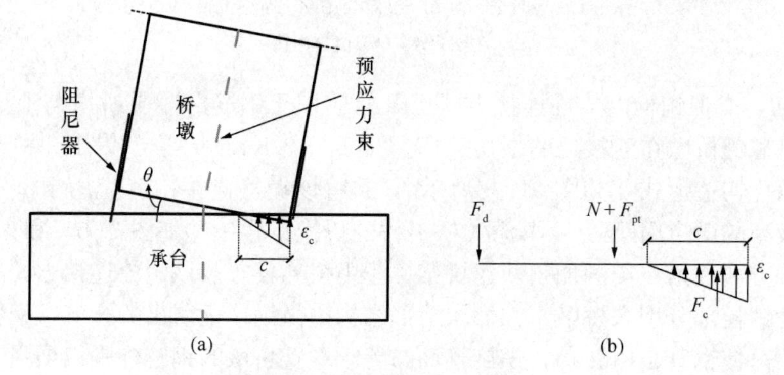

(a) (b)

图 10 摇摆界面的力学机制

Fig. 10 Mechanical mechanism of rocking interface

(a) 摇摆界面变形模式；(b) 摇摆界面受力分析

$$\varepsilon_c(\theta)=\left[\frac{\theta\cdot h}{(h-0.5h_p)h_p}+\phi_y\right]c \tag{1}$$

式中，θ 为摇摆界面的截面转角；h 为桥墩计算高度；h_p 为桥墩塑性铰高度；ϕ_y 为现浇桥墩屈服曲率。

图 11 所示为 MBA 方法的计算流程：首先给出截面转角 θ 并假设对应的中性轴高度 c，再根据静力平衡条件迭代出近似精确的中性轴高度 c，从而计算摇摆桥墩顶部承受的侧向水平力 F。

集中塑性铰法是分析摇摆结构反应的方法中最简单的一种，Pampanin 等[47]、Palermo 等[16]采用集中塑性铰法来模拟摇摆结构体系反应。与传统集中塑性铰法有所区别的是，这里的塑性铰转动特性由两个转动弹簧并联表征，以分别模拟自复位体系和耗能体系的弯矩-转角滞回行为。

多弹簧法（如图 12 所示）是将摇摆界面接触特性由一组受压弹簧来表征，桥墩采用弹性梁柱单元，预应力束和耗能装置均采用桁架单元并分别考虑预应力束和耗能装置的力学特性。Marriott 等[22, 23]、Guo 等[35]均通过拟静力试验验证了该方法可较为精确地模拟摇摆桥墩的滞回行为。与多弹簧法相比，实体有限元法可更精细化考虑摇摆界面开合和混凝土局部承压，因此能更准确地模拟摇摆结构的力学行为，但使用该方法来分析摇摆结构反应的耗时较大[31, 32]。

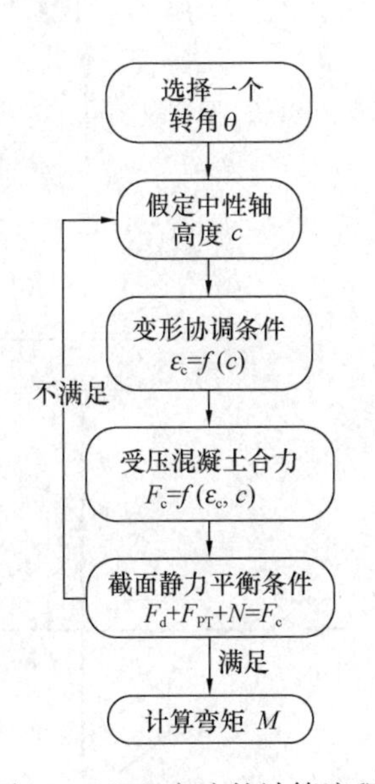

图 11　MBA 方法的计算流程

Fig. 11　Calculated flow of MBA method

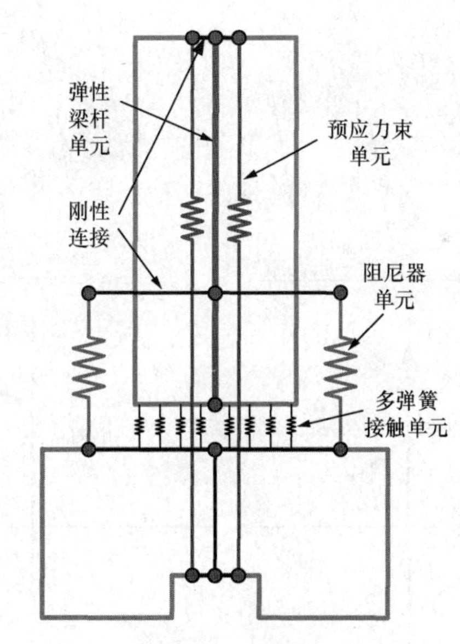

图 12　多弹簧法

Fig. 12　Multi-spring method

在地震作用下，摇摆桥墩主要作为受压构件，如摇摆界面刚度较大则可将摇摆桥墩简化成摇摆刚体。1963 年 Housner[48]首次给出分析水塔摇摆行为的刚体动力分析模型［如图 13(a) 所示］。后续学者对刚体摇摆的动力行为进行了更深入的研究，给出了刚体摇摆模型的适用性，并对 Hounser 提出的模型进行了改进。Tso 和 Wong[49]指出刚体高度非线性摇摆行为会使其在地震作用下的动力特性极为复杂。Aslam 等[50]提出对刚体摇摆反应与传统弹性或延性结构反应一致性的质疑，并进一步通过试验验证其质疑的正确性。

Makris 和 Konstantinidis[51, 52]提出摇摆刚体动力特性与双线性或弹性-阻尼单自由度体系反应谱不相匹配的质疑，并采用数值方法验证了摇摆刚体动力反应与传统单自由度体系的反应规律严重不相符，如当激励振幅增大 1.6%时，摇摆刚体位移反应增大了 125%，因此建议将摇摆反应谱用于摇摆刚体的设计。Aslam 等[50]针对单纯摇摆刚体的不稳定性，提出预应力束在摇摆刚体中应用的方案和分析模型［如图 13(b) 所示］，指出在摇摆刚体中采用预应力束可显著提高摇摆刚体的抗倾覆能力和屈服后的二次刚度，尤其在预应力束刚度足够大的情况下，摇摆刚体的二次刚度可变为正值。虽然具有预应力束的摇摆刚体的力学特征仍保持高度非线性，但其动力特性变得与传统延性体系更为一致，因此采用预应力技术的摇摆刚体设计更适用采用常规反应谱，而摇摆反应谱则不再适用[50]。为提高摇摆刚体抗侧强度、耗能能力和稳定性，Makris 和 Zhang[53]提出一种在摇摆界面边缘布置延性耗能元件的改进形式［如图 13(c) 所示］，并研究发现地震作用下延性耗能元件可显著提高摇摆刚体的耗能能力，以及弯矩-转角骨架曲线的初始刚度和抗侧强度。

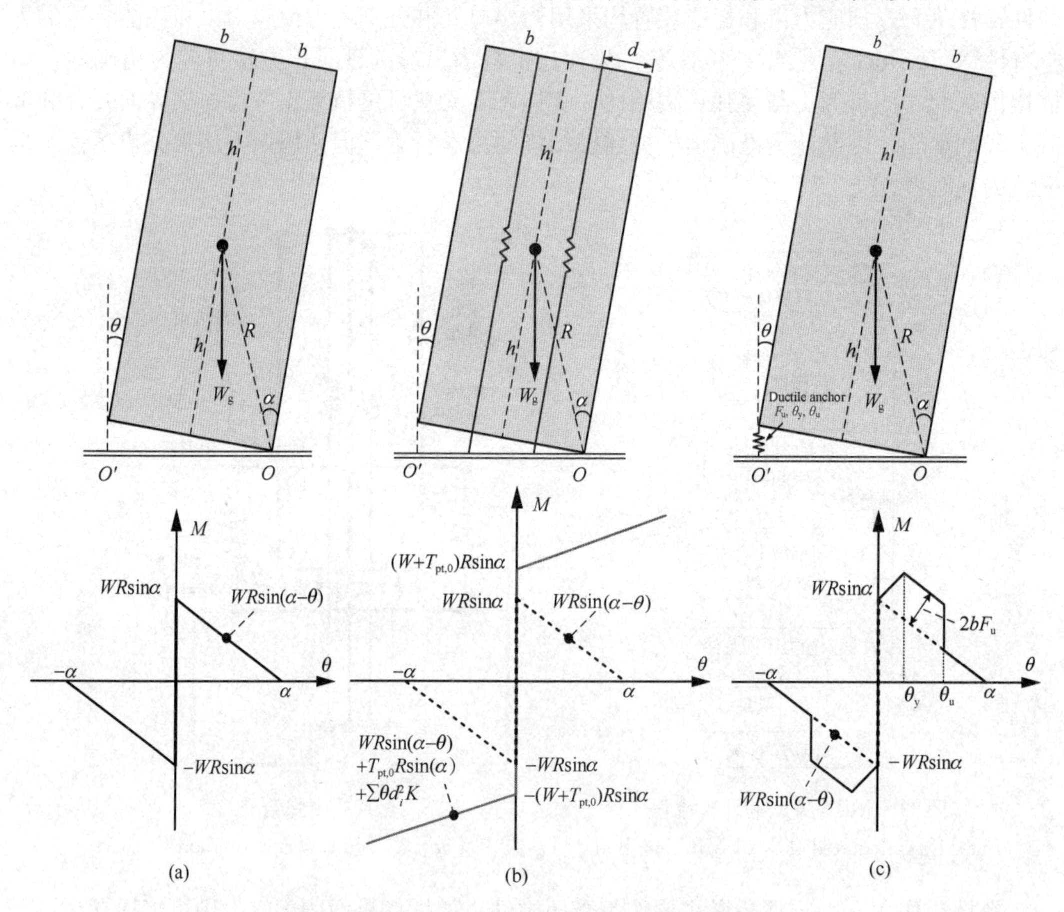

图 13　摇摆刚体的分析模型及弯矩-转角骨架曲线

Fig. 13　Analytical models and moment-rotation envelope curves of rocking rigid blocks

(a) Housner 模型；(b) Aslam 模型；(c) Makris 模型

Makris 和 Vassiliou[54, 55]以古希腊神庙为研究对象，在 Housner 模型的基础上提出神庙结构的刚体分析模型，并将该方法引入到摇摆框架结构的应用中，得出该类体系比单柱摇摆桥墩的稳定性更好，且上部结构越重该体系越稳定的结论。Dimitrakopoulos 和 Paraskeva[56]采用该分析模型对在近场地震作用下的摇摆框架桥进行无量纲化的易损性分析，并得出该类摇摆结构发生倒塌受速度脉冲影响显著的结论。随后 Makris 和 Vassiliou[57]提出预应力摇摆框架结构的刚体分析模型并分析其反应规律。

4. 现有研究的不足和发展趋势

摇摆结构最早可追溯到公元前 510 年的希腊神庙，且该建筑在 2500 年内经历多次大地震而未发生倒塌，目前作为古建筑仍保存较好[52]。但摇摆结构的良好抗震能力在当时并未受到重视，直至 1963 年，Housner[48]首次提出水塔的摇摆可有效防止其在大地震作用下的发生倒塌。该摇摆理念在 20 世纪 70 年代首次应用于摇摆高墩铁路桥[5, 6]，虽然随后已有大量学者针对摇摆桥墩进行研究，但摇摆桥墩的应用发展滞后，且主要局限在钢桁架桥的抗震加固领域。目前，摇摆桥墩的研究不足以及其未来发展趋势可总结为：

（1）预应力束和消能减震装置（如：黏滞阻尼器、软钢阻尼器等）在摇摆桥墩中的应用可提供其自复位能力和耗能能力。嵌合式接头或剪力键的应用可避免摇摆界面滑移，采用钢材料、纤维增强聚合材料和复合纤维增强混凝土对摇摆界面加强可限制摇摆界面的局部损伤。

（2）目前国内外研究手段多采用拟静力或拟动力试验测试摇摆桥墩，而采用振动台试验的摇摆桥梁结构动力性能研究较少，因此，在真实地震动激励下摇摆桥梁结构的抗震能力和震后恢复能力，以及连接节点的地震反应（如：桥墩与承台接缝的碰撞效应）的真实评估是需要解决的关键研究问题。

（3）目前国内外研究主要集中于单柱摇摆桥墩的研究，而对于我国城市桥梁和公路桥梁中常见、抗倒塌能力更好的双柱式摇摆桥墩的研究则较少。针对双柱式摇摆桥墩的桥墩-上部结构连接节点合理构造形式的研究暂存不足，为更好地控制桥墩在地震作用下的摇摆幅值，可在桥墩-上部结构节点中引入消能减震元件。

（4）目前国内外针对摇摆桥墩的研究对象以二次刚度为正的摇摆桥墩为主，而负二次刚度摇摆桥墩未受到足够重视。在地震作用下负二次刚度摇摆桥墩可避免共振发生且能起到显著隔震效果，且不依赖于预应力束的应用而易于设计、施工和维护。因而，负二次刚度摇摆桥墩的抗震性能和抗倒塌能力应开展充分研究。

（5）适用的摇摆桥墩抗震设计方法暂无翔实依据，为使摇摆桥墩在工程中推广应用，需建立成体系的摇摆桥墩及其桥梁结构设计方法。基于摇摆桥墩力学特性，基于直接位移的抗震设计方法是摇摆桥墩及其桥梁结构抗震设计的未来发展趋势。在该方法的研究中，需对摇摆桥梁整体及局部变形限值、等效阻尼比等关键问题展开充分研究。

5. 结语

摇摆桥墩可将损伤控制在摇摆界面以避免主体结构的严重损伤，且具有较好的自复位能力和耗能能力，可在满足常规抗震需求的同时，提供较好的震后可快速恢复能力，保证交通生命线的震后畅通，给快速救灾抢险以及灾后重建工作提供便利，最大限度地减少地震造成的经济损失。但目前摇摆桥墩的工程应用主要局限在钢桁架桥的抗震加固领域，摇摆桥梁的新建工程应用极少。其主要原因是摇摆桥梁结构的研究还处于发展阶段，其合理构造形式尚需进一步研发，适用的摇摆桥梁抗震设计方法无翔实依据，值得我国桥梁工程抗震研究人员和技术人员开展研究。

参考文献

[1] 公路桥梁抗震设计细则 JTG/T B02-01—2008[S]. 北京：人民交通出版社，2008.
Guidelines for seismic design of highway bridges JTG/T B02-01—2008[S]. Beijing: China Communications Press, 2008.

[2] 城市桥梁抗震设计规范 CJJ 166—2011[S]. 北京：中国建筑工业出版社，2011.
Code for seismic design of urban bridges CJJ 166—2011[S]. Beijing: China Building Industry Press, 2011.

[3] 公路工程抗震规范 JTG B02—2013[S]. 北京：人民交通出版社，2013.
Specification of seismic design for highway engineering JTG B02—2013[S]. Beijing: China Communications Press, 2013.

[4] Shahawy M. Prefabricated bridge elements and systems to limit traffic disruption during construction [M]. Washington, DC: Transportation Research Board, 2003.

[5] Beck J L, Skinner R I. The seismic response of a reinforced concrete bridge pier designed to step[J]. Earthquake Engineering & Structural Dynamics, 1973, 2(4): 343-358.

[6] Cormack L G. The design and construction of the major bridges on the Mangaweka rail deviation[J]. Transactions of the Institution of Professional Engineers New Zealand: Civil Engineering Section, 1988, 15(1): 17-23.

[7] Priestley M J N, Seible F, Calvi G M. Seismic design and retrofit of bridges[M]. State of New Jersey: John Wiley & Sons, 1996.

[8] Astaneh-Asl A, Shen J H. Rocking behavior and retrofit of tall bridge piers[C]//Structural Engineering in Natural Hazards Mitigation. ASCE, 1993: 121-126.

[9] Dowdell D J, Hamersley B A. Lions' Gate Bridge North Approach: Seismic retrofit[C]//Behaviour of Steel Structures in Seismic Areas: Proc., 3rd Int. Conf.: STESSA 2000. Balkema, 2000: 319-326.

[10] Jones M H, Holloway L J, Toan V, et al. Seismic retrofit of the 1927 Carquinez Bridge by a displacement capacity approach [C]//Second National Seismic Conference on Bridges and Highways. 1997.

[11] Ingham T J, Rodriguez S, Nader M N, et al. Seismic retrofit of the golden gate bridge[C]//Proc., National Seismic Conf. on Bridges and Highways: Progress in Research and Practice. Federal Highway Administration, 1995.

[12] Mander J B, Cheng C T. Seismic resistance of bridge piers based on damage avoidance design[R]. New York: US National Center for Earthquake Engineering Research (NCEER), 1997.

[13] Cheng C T. Shaking table tests of a self-centering designed bridge substructure[J]. Engineering Structures, 2008, 30(12): 3426-3433.

[14] 夏修身，陈兴冲. 铁路高墩桥梁基底摇摆隔震与墩顶减震对比研究[J]. 铁道学报，2011，33(9)：102-107.
XIA Xiushen, CHEN Xingchong. Controlled rocking and pier top seismic isolation of railway bridge

with tall piers[J]. Journal of the China Railway society, 2011, 33(9): 102-107. (in Chinese)

[15] Stanton J F, Nakaki S D. Design guidelines for precast concrete seismic structural systems[M]. Washington, DC: University of Washington, 2002.

[16] Palermo A, Pampanin S, Calvi G M. Concept and development of hybrid solutions for seismic resistant bridge systems[J]. Journal of Earthquake Engineering, 2005, 9(06): 899-921.

[17] Palermo A, Pampanin S, Marriott D. Design, modeling, and experimental response of seismic resistant bridge piers with posttensioned dissipating connections[J]. Journal of Structural Engineering, 2007, 133(11): 1648-1661.

[18] Palermo A, Pampanin S. Enhanced seismic performance of hybrid bridge systems: Comparison with traditional monolithic solutions[J]. Journal of Earthquake Engineering, 2008, 12(8): 1267-1295.

[19] Solberg K, Mashiko N, Mander J B, et al. Performance of a damage-protected highway bridge pier subjected to bidirectional earthquake attack[J]. Journal of structural engineering, 2009, 135(5): 469-478.

[20] Ou Y C, Wang P H, Tsai M S, et al. Large-scale experimental study of precast segmental unbonded posttensioned concrete bridge columns for seismic regions[J]. Journal of structural engineering, 2010, 136(3): 255-264.

[21] Chou C C, Chen Y C. Cyclic tests of post-tensioned precast CFT segmental bridge columns with unbonded strands[J]. Earthquake engineering & structural dynamics, 2006, 35(2): 159-175.

[22] Marriott D, Pampanin S, Palermo A. Quasi-static and pseudo-dynamic testing of unbonded post-tensioned rocking bridge piers with external replaceable dissipaters[J]. Earthquake Engineering & Structural Dynamics, 2009, 38(3): 331-354.

[23] Marriott D, Pampanin S, Palermo A. Biaxial testing of unbonded post-tensioned rocking bridge piers with external replacable dissipaters[J]. Earthquake Engineering & Structural Dynamics, 2011, 40(15): 1723-1741.

[24] ElGawady M, Booker A J, Dawood H M. Seismic behavior of posttensioned concrete-filled fiber tubes[J]. Journal of Composites for Construction, 2010, 14(5): 616-628.

[25] ElGawady M A, Sha'lan A. Seismic behavior of self-centering precast segmental bridge bents[J]. Journal of Bridge Engineering, 2010, 16(3): 328-339.

[26] 葛继平，魏红一，王志强. 循环荷载作用下预制拼装桥墩抗震性能分析[J]. 同济大学学报(自然科学版)，2008，36(7)：894-899.
GE Jiping, WEI Hongyi, WANG Zhiqiang. Seismic performance of precast segmental bridge column under cyclic loading[J]. Journal of Tongji University (Natural Science), 2008, 36(7): 894-899.

[27] 布占宇，唐光武. 无黏结预应力带耗能钢筋预制节段拼装桥墩抗震性能研究[J]. 中国铁道科学，2011，32(3)：33-40.
BU Zhanyu, TANG Huangwu. Seismic performance investigation of unbonded prestressing precast segmental bridge piers with energy dissipation bars[J]. China Railway Science, 2011, 32(3): 33-40.

[28] 郭佳，辛克贵，何铭华，等. 自复位桥梁墩柱结构抗震性能试验研究与分析[J]. 工程力学，2012，29(A01)：29-34.
GUO Jia, XIN Kegui, HE Minghua, et al. Experimental study and analysis on the seismic performance of a self-centering bridge pier[J]. Engineering Mechanics, 2012, 29(A01): 29-34.

[29] 谭真. 设置粘弹性阻尼器的预应力节段拼装桥墩抗震性能研究[D]. 哈尔滨：哈尔滨工业大学，2013.

TAN Zhen. Seismic performance of post-tensioned precast concrete segmental bridge columns with viscoelastic dampers[D]. Haerbin: Harbin Institute of Technology, 2013.
[30] ElGawady M A, Dawood H M. Analysis of segmental piers consisted of concrete filled FRP tubes [J]. Engineering Structures, 2012, 38: 142-152.
[31] Dawood H, ElGawady M, Hewes J. Behavior of segmental precast posttensioned bridge piers under lateral loads[J]. Journal of Bridge Engineering, 2011, 17(5): 735-746.
[32] Dawood H M, ElGawady M. Performance-based seismic design of unbonded precast post-tensioned concrete filled GFRP tube piers[J]. Composites Part B: Engineering, 2013, 44(1): 357-367.
[33] Trono W, Jen G, Panagiotou M, et al. Seismic response of a damage-resistant recentering posttensioned-HyFRC bridge column[J]. Journal of Bridge Engineering, 2015: 04014096.
[34] Nikbakht E, Rashid K, Hejazi F, et al. Application of shape memory alloy bars in self-centring precast segmental columns as seismic resistance[J]. Structure and Infrastructure Engineering, 2015, 11 (3): 297-309.
[35] Guo T, Cao Z, Xu Z, et al. Cyclic load tests on self-centering concrete pier with external dissipators and enhanced durability[J]. Journal of Structural Engineering, 2015: 04015088.
[36] Barthes C B. Design of earthquake resistant bridges using rocking columns[D]. California: University of California, Berkeley, 2012.
[37] 何铭华，辛克贵，郭佳. 新型自复位桥梁墩柱节点的局部稳定性研究[J]. 工程力学，2012，29 (4)：122-127.
HE Minghua, XIN Kegui, GUO Jia. Local stability study of new bridge piers with self-centering joints[J]. Engineering Mechanics, 2012, 29(4): 122-127.
[38] 何铭华，辛克贵，郭佳，等. 自复位桥墩的内禀侧移刚度和滞回机理研究[J]. 中国铁道科学，2012，33(5)：22-28.
HE Minghua, XIN Kegui, GUO Jia, et al. Research on the intrinsic lateral stiffness and hysteretic mechanics of self-centering pier[J]. Zhongguo Tiedao Kexue, 2012, 33(5): 22-28.
[39] Deng L, Kutter B L, Kunnath S K. Centrifuge modeling of bridge systems designed for rocking foundations[J]. Journal of Geotechnical and Geoenvironmental Engineering, 2012, 138 (3): 335-344.
[40] Keever M D, Chung P, Holden T, et al. Innovative solution to seismic design challenges on the Mococo overhead project[J]. Technical Memorandum of Public Works Research Institute, 2006, 4009: 257-269.
[41] Zaghi A E, Saiidi M S. Bearing and shear failure of pipe-pin hinges subjected to earthquakes[J]. Journal of Bridge Engineering, 2010, 16(3): 340-350.
[42] Esmaili Zaghi A. Seismic design of pipe-pin connections in concrete bridges[M]. Nevada: University of Nevada, Reno, 2009.
[43] Zaghi A E, Saiid S M, El-Azazy S. Shake table studies of a concrete bridge pier utilizing pipe-pin two-way hinges[J]. Journal of Bridge Engineering, 2010, 16(5): 587-596.
[44] Pampanin S. Emerging solutions for high seismic performance of precast/prestressed concrete buildings[J]. Journal of Advanced Concrete Technology, 2005, 3(2): 207-223.
[45] ACI T1. 2-03. Special hybrid moment frames composed of discretely jointed precast and post-tensioned concrete members[S]. Michigan: American Concrete Institute, 2003.
[46] NZS3101. The designing of concrete structures and commentary[S]. Wellington: Standards New Zealand, 2006.

[47] Pampanin S, Priestley M J N, Sritharan S. Analytical modelling of the seismic behaviour of precast concrete frames designed with ductile connections[J]. Journal of Earthquake Engineering, 2001, 5 (03): 329-367.

[48] Housner G W. The behavior of inverted pendulum structure during earthquakes[J]. Bulletin of the Seismic of America, 1963, 2(53): 403-417.

[49] Tso W K, Wong C M. Steady state rocking response of rigid blocks part 1: Analysis[J]. Earthquake engineering & structural dynamics, 1989, 18(1): 89-106.

[50] Aslam M, Scalise D T, Godden W G. Earthquake rocking response of rigid bodies[J]. Journal of the Structural Division, 1980, 106(2): 377-392.

[51] Makris N, Konstantinidis D. The rocking spectrum and the shortcomings of design guidelines[R]. California: Pacific Earthquake Engineering Research Center, 2001.

[52] Makris N, Konstantinidis D. The rocking spectrum and the limitations of practical design methodologies[J]. Earthquake engineering & structural dynamics, 2003, 32(2): 265-289.

[53] Makris N, Zhang J. Rocking response and overturning of anchored equipment under seismic excitation[R]. California: Pacific Earthquake Engineering Research Center, 1999.

[54] Makris N, Vassiliou M F. Planar rocking response and stability analysis of an array of free-standing columns capped with a freely supported rigid beam[J]. Earthquake Engineering & Structural Dynamics, 2013, 42(3): 431-449.

[55] Makris N, Vassiliou M F. Are some top-heavy structures more stable? [J]. Journal of Structural Engineering, 2014, 140(5): 06014001.

[56] Dimitrakopoulos E G, Paraskeva T S. Dimensionless fragility curves for rocking response to near-fault excitations[J]. Earthquake Engineering & Structural Dynamics, 2015, 44(12): 2015 - 2033.

[57] Makris N, Vassiliou M F. Dynamics of the rocking frame with vertical restrainers[J]. Journal of Structural Engineering, 2014, 141(10): 04014245.

7 数据驱动的减振高层建筑全寿命性能推演

——减振系统破坏、再生及精准量化评估

薛松涛[1,2]　谢丽宇[1]　唐和生[1]　杨朋超[1]

（1. 同济大学土木工程学院，上海；2. 日本东北工业大学建筑系，日本仙台）

摘　要：环境侵蚀及极端灾害会诱发减振高层建筑及其减振系统的性能退化和突变，数据驱动的结构健康监测是实现减振高层结构全寿命性能评估和风险管控的可靠途经。以一栋减振高层建筑结构为例，利用结构的地震响应监测数据，重演了日本“3.11”大地震对减振系统的破坏，介绍了减振系统的再生设计。总结了减振高层建筑及其减振系统在极端灾害下的灾变机理的研究现状，提出了基于三个极限状态的研究思路，提炼了极限破坏机理、抗灾性能演变规律、设计和再生性能控制理论的三个关键问题。未来需要开展系统、深入的研究，探索考虑极端灾害效应的性能控制设计理论，建立减振系统可修复、可替换的高层建筑一体化初始和再生设计方法。

关键词：减振高层建筑；减振系统；三极限状态；性能再生；日本“3.11”地震

Data driven life-cycle performance estimation of vibration-controlled, high-rise building: vibration-controlled system's failure, rehabilitation, and precise performance estimation

Xue Songtao[1,2] Xie Liyu[1] Tang Hesheng[1] Yang Pengchao[1]

(1. College of Civil Engineering, Tongji University, Shanghai, China;
2. Department of Architecture, Tohoku Institute of Technology, Sendai, Japan)

Abstract: Environmental erosion and extreme loading can significantly damage the performance of both vibration-controlled, high-rise building and its vibration-controlled system. Data driven structural health monitoring provides such a reliable approach for the life-cycle performance estimation and risk management of vibration-controlled, high-rise building. A steel-framed building equipped with oil dampers is utilized to exemplify the fundamental ideas of data driven life-cycle performance estimation. The failure event of oil dampers of the building during the 2011 Great East Japan earthquake is simulated and replayed based on structural earthquake measurements. To rehabilitate the energy dissipation system and to enhance the structural performance, a novel hybrid retrofit strategy was proposed and is introduced here. The state-of-the-art of research in the failure process of vibration-controlled system and building under extreme loading is summarized. And a three limit states based research framework is originally proposed, which highlights the failure mechanism, the performance evolution and the performance-based design and retrofit philosophy for vi-

bration-controlled building. In the near future, more efforts need to be spent in developing the performance-based design theory for vibration-controlled, high-rise building under extreme loading, and in establishing an integrated design and retrofit methodology, which can incorporate the idea of recoverable and replaceable vibration controlled system.
Keywords: vibration-controlled building, vibration-controlled system, three limit states, performance rehabilitation, 3. 11 earthquake

1. 引言

高层建筑仍是我国未来城市建设的主要发展方向，是人多地少的城市朝集约化发展所不可替代的建筑形式。然而，高层建筑在强震或强台风作用下可能出现安全性或舒适性问题，由于其容纳人员众多、功能多样、结构复杂，一旦出现极限状态下的破坏，将导致超出一般建筑的灾难性后果。目前我国的设计规范和方法仍然是以满足预期可能发生的理想状态为目标，但面对频发、超预期发生的重（特）大自然灾害缺乏充分的应对策略和技术储备，当前的规范可能会对国家的防灾策略带来重大挑战。因此，有必要深入研究高层建筑在强震、强台风等极端灾害下的极限破坏机理及控制理论和方法，从战略角度重新规划和制定高层建筑的设计规范和方法，以确保极端灾害发生时国民经济不致遭到灾难性打击。

目前从安全性和舒适度考虑，高层建筑的发展趋势是在结构体系中加装减振（震）系统，有效控制由地震或风振引起的结构振动，提高结构的抗风、抗震性能。早在 1972 年，为了控制美国世贸大厦的风致响应舒适度，结构中就已经安装了一万余个黏弹性阻尼器；而强台风和地震频发的日本，现在已经拥有 4000 栋以上的减振高层建筑，如日本东京新建地标 Sky Tree，为了最大限度地减少地震响应，安装了大量的油阻尼器。国内现有减振建筑还不是很多，但发展势头强劲，新建减振建筑数量迅猛增长，住房和城乡建设部于 2014 年 2 月发文大力推广减振建筑。目前我国已建成的减振建筑如上海环球金融中心，顶部安装了两台风阻尼器，上海中心则安装摆式电涡流调谐质量阻尼器；汶川大地震后，四川新建了大量减振公共建筑。

事实证明，在现有规范设计范围内，减振系统在高层建筑的安全性及舒适度控制方面有着优异的表现。但非常遗憾的是，从世界范围来看，很少有研究考虑过极端荷载作用下减振系统的失效问题和失效后的应对方法，亦没有考虑过失效以后高层建筑本体结构的性能，以及主灾害之后结构性能迅速再生的设计方法。实际上，即使没有受到极端荷载的作用，减振系统作为一个机械构件，其使用寿命也远远小于建筑本身的寿命。基于此，在建筑物的有效寿命期内，有必要考虑减振系统的失效状态极其应对方法。

2011 年 3 月 11 日的东日本大地震，导致地震中心仙台市内一栋建筑的减振系统失效，安装于 1 层的油阻尼器全部遭到破坏，同时另一栋隔震建筑的隔震器也出现了裂缝，这是高层建筑减振系统在世界范围内首次出现的破坏现象。当时造成了房屋所有者、使用者对结构安全的不信任，也让我们必须面对以下不容回避的问题：减振系统作为一种机械系统，除了其老化失效以外，在极端灾害下有可能突然破坏。

因此，有必要形成应对此类灾害的新方法和新思路，包括寻找减振系统的破坏原因，评价减振系统损伤后建筑本体结构的性能，同时研究崭新的减振结构再生设计方法。这是确保减振结构安全，有效控制自然灾难造成的经济损失，保障城市、社会和环境可持续发展的重要和紧迫研究内容。

2. 减振结构的震害案例

2.1 油阻尼器减振系统破坏案例

薛松涛等[1]首次报道了日本“3.11”地震对一栋8层钢结构建筑所安装的消能减震装置造成的破坏，如图1所示。这是位于日本宫城县仙台市东北工业大学校园内的一栋行政楼，建于2003年，高34.2m、长48m、宽9.6m，地下1层为钢筋混凝土结构，地上8层为混凝土预制楼板的钢结构，第1层和第2层构成底部大空间（该合并层在下文中均用第1层指代），层高为8m。该结构在无消能构件布置的情况下，已满足日本对学校建筑的抗震规范要求，但为了验证该大学自主研发的一类油阻尼器的抗震性能，同时也为了提高该栋行政楼的结构抗震性能，在每层的两个方向上各安装了4个油阻尼器，一共安装了56个油阻尼器，油阻尼器在结构中的布置如图1（b）所示。

“3.11”地震造成第1层8组油阻尼器被完全破坏，油液完全泄漏，如图1（e）所示。由于强烈的地震作用，位于第1层的油阻尼器经历了远远超出其工作行程及缓冲行程范围的冲击，致使固定于V形斜撑上的中部缸体与固定阻尼器的U形支座发生了直接的碰撞，造成了位于地面U形支座的张开以及两侧活塞的脱落。位于第3、4层油阻尼器的黏弹性密封材料发生严重的磨损，导致油液的完全泄漏，不能提供阻尼恢复力，但油阻尼器的机械构件并未发生任何损伤和塑性变形。除了油阻尼器的受损，该栋结构的其他部分并未发生任何的结构损伤，在经过震后安全评估后重新投入使用。

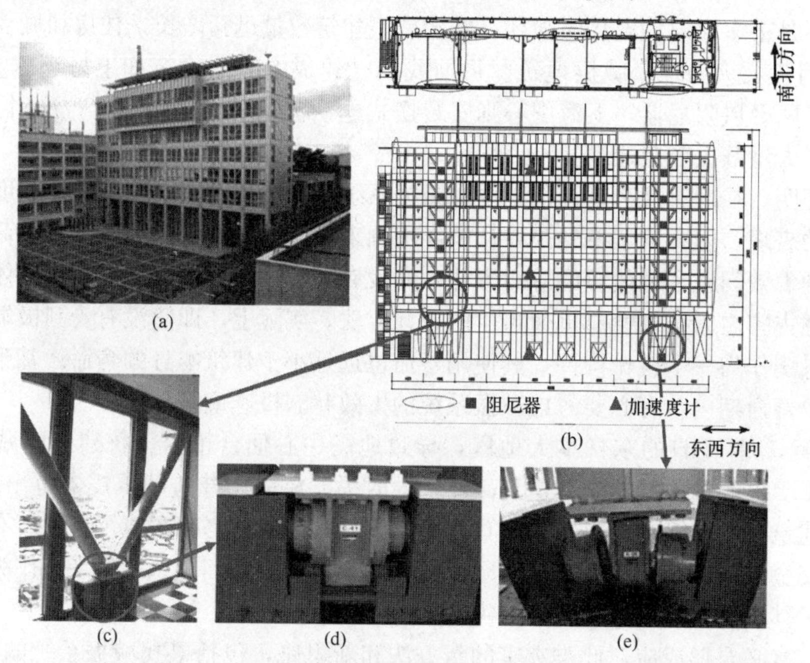

图1 油阻尼器钢框架建筑结构

Fig. 1 A steel-framed building equiped with oil dampers

（a）8层钢结构建筑外观；（b）油阻尼器的布置图；（c）位于3楼的油阻尼器支撑；（d）3楼的油阻尼器；（e）在“3.11”地震中遭到破坏的油阻尼器

2.2 其他减隔震系统损伤案例

日本“3.11”地震后，日本隔震协会立即成立了振动控制结构的调查委员会对隔震结构和消能减震结构进行了震害调查[2,3]，调查对象包括了日本境内327栋隔震结构（60%分布在东京附近，25%位于宫城县）和130栋消能减震结构。有30栋装配有铅芯隔振垫的结构上发现了问题。一些隔震结构的铅芯隔振垫在“3.11”地震后受到了损伤（出现了裂缝），另有一栋隔震结构的铅芯隔振垫在“3.11”地震前已有深度为8mm的裂缝，地震后裂缝深度扩展至32mm。

日本隔震协会在“3.11”地震的震后调查中也发现了金属阻尼器受到了损伤[3]，如金属阻尼器的残余变形，以及用于固定金属阻尼器的高强螺栓发生了松动。一些结构的软钢阻尼器进入了屈服状态，软钢阻尼器正是利用其材料在屈服之后的耗能特性来增加结构阻尼，在地震之后可能已经进入屈服阶段，造成耗能能力的降低，需要在震后进行检查、评估，确定是否需要更换阻尼器。

速度型减振器、位移型金属阻尼器以及隔震垫的损伤案例改变了“结构布置了减振系统就安全无患”的传统看法，警示我们，无论是突发极端灾害作用还是长期性能退化，减振系统都有可能失效。在正常使用过程中阻尼器也可能因一些原因造成性能上的退化或失效，如黏滞阻尼器可能因为漏油的问题造成性能的下降，黏弹性阻尼器的黏弹性材料与钢板之间的脱落，摩擦阻尼器接触面的属性可能随着时间而发生变化。如何从初始设计和再生设计出发，保证减振系统性能退化甚至失效后高层建筑的安全性能，是未来需要面对的重大挑战。

3. 油阻尼器减振系统的性能量化评估

在减振结构及其减振系统的设计和性能评估中，需要准确量化计算减振系统（阻尼器）附加给结构的阻尼和刚度，常以有效阻尼比和有效刚度为量化指标。针对有效阻尼比估计问题，诸多文献提出了不同的近似估计方法，但均需引入一定的简化分析假设，如应变能方法需假设阻尼器和结构的运动为简谐运动，且通常需要借助于结构有限元模型或其模态参数。减振系统的附加刚度亦会影响减振结构的动力学特性，现有减振系统性能研究多围绕有效阻尼展开，在有效阻尼比及有效刚度评估两个方面没有统一的理论基础和框架。

在工程实践中，由于结构的复杂性和施工误差的不确定性，减振结构的设计和性能评估实际是依据可靠度理论展开的，其设计和评估结果通常偏于保守，且存在很强的不确定性。对于实际减振结构，依据结构设计详细（图纸）建立的初始有限元模型通常存在不可忽略的模型误差。这一误差来源于多方面如材料参数、几何尺寸及边界条件的不确定性，以及建模简化假设的引入。这些不可避免地误差导致初始有限元模型不能准确代表结构实体，造成不准确的结构和阻尼器响应预测，影响实际减振结构及其减振系统性能评估的准确性。

针对上述问题，提出了基于监测数据的减振系统性能评估框架，可用于量化评估和验证减振系统附加给结构的有效阻尼比和有效频率，即减振系统对结构模态参数（频率和阻尼比）的影响，为减振结构及减振系统的实际工作性能评估提供精准量化方法，统一减振系统的有效阻尼比及有效刚度评估的理论基础。以图1油阻尼器钢框架结构为例，利用结构及油阻尼器的地震响应监测数据，量化分析油阻尼器减振系统的实际工作性能，例证所提减振系统性能评估框架的可行性和准确性。

该油阻尼器钢框架结构中配备了一套健康监测系统：在结构第 1、4、8 层中布置了双向加速度计，用于采集监测楼层在结构长边和短边方向的地震加速度响应；在结构第 1 层和第 8 层中，沿结构长边和短边方向选取了 4 组油阻尼器，并安装了力和位移传感器，用于采集油阻尼器在地震中的力和位移响应。图 2 分别给出了一组实测的楼层和油阻尼器地震响应数据（结构东西方向），该组数据采集于 2003 年 5 月 26 日发生的一次较大地震（$M_w=7.0$），亦是结构自建成以来的首次大地震。以该组监测数据为例，采用所提方法量化油阻尼器附加给结构的有效阻尼比和有效频率，并利用修正有限元模型验证量化结果的准确性。

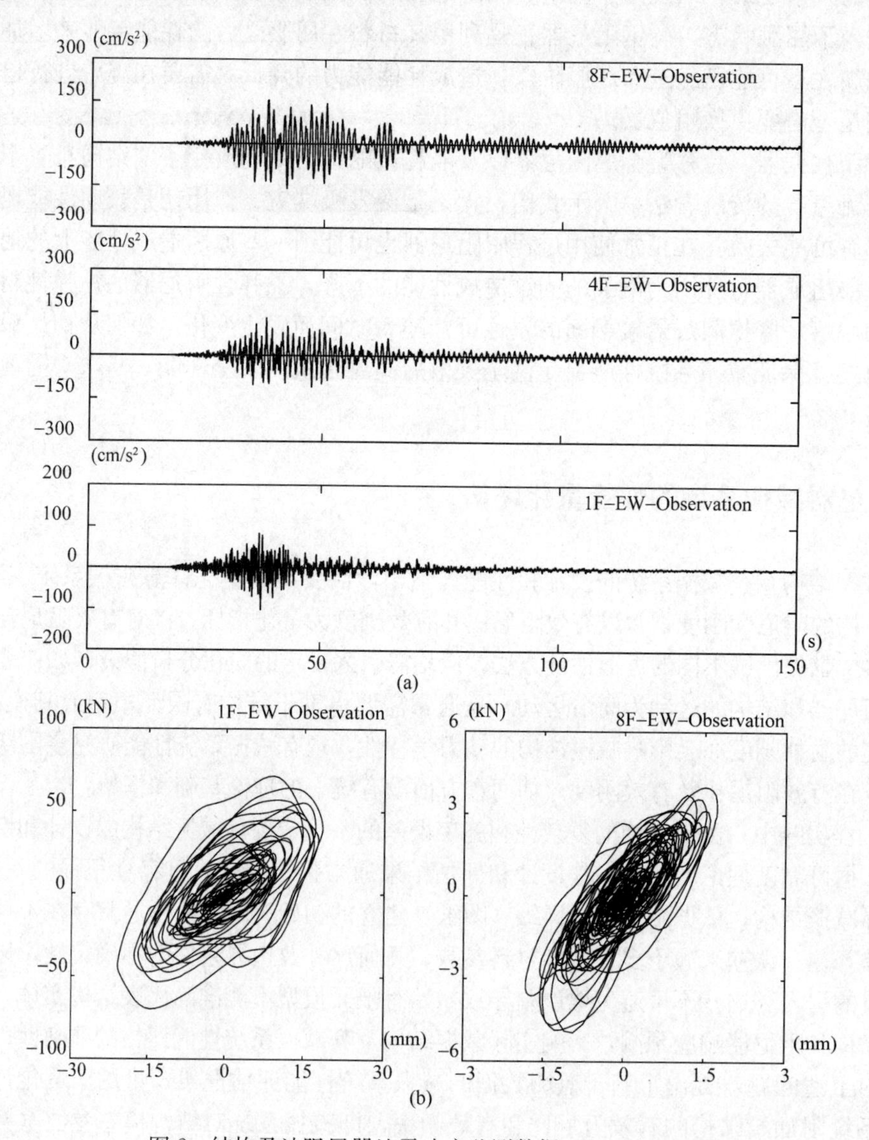

图 2　结构及油阻尼器地震响应监测数据（2013/05/26）

Fig. 2　Earthquake measurements of structure and oil dampers (2013/05/26)

(a) 楼层加速度监测数据；(b) 油阻尼器的阻尼力和位移监测数据

3.1 结构有限元建模及解析修正

3.1.1 有限元模型建模及预测误差

利用结构设计图纸及日本钢结构设计规范，在 SAP2000 结构分析软件建立了结构初始有限元模型，如图 3 所示。为简化结构有限元建模，忽略楼梯间、内隔墙及外墙装饰等非结构构件。梁、柱及阻尼器支持采用框架单元模拟，楼板为混凝土预制楼板，采用壳单元模拟，模拟单元的几何尺寸参数取设计值，材料参数取设计值。

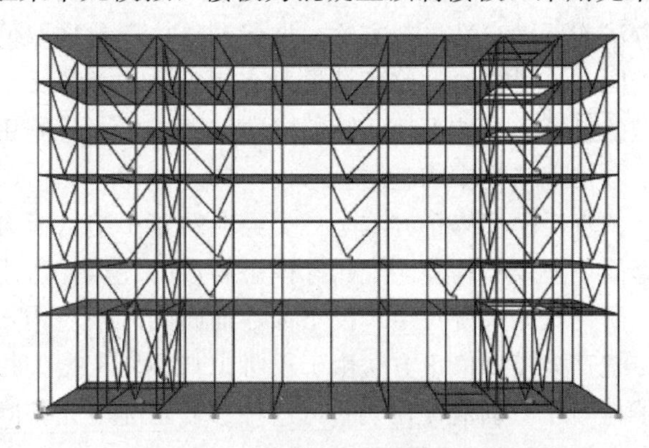

图 3 结构初始有限元模型

Fig. 3 Initial finite element model of the building

考察该初始有限元模型的预测误差，取结构第 1 层实测地震加速度为模型激励，如图 2 所示。假定主体结构的前 3 阶阻尼比均为 0.02，其余高阶阻尼比取 0.1。采用线性 Maxwell 模型模拟油阻尼器的动力学行为，其模型参数取基于实际监测数据的识别值。图 4 给出了结构强震阶段（30～55s）初始有限元模型预测的楼层加速度及油阻尼器阻尼力时程响应。对比实际监测数据，初始有限元模型存在较大预测误差，利用初始有限元模型的模态参数估计油阻尼器对结构阻尼和刚度贡献是不可靠的，有必要修正初始有限元模型。

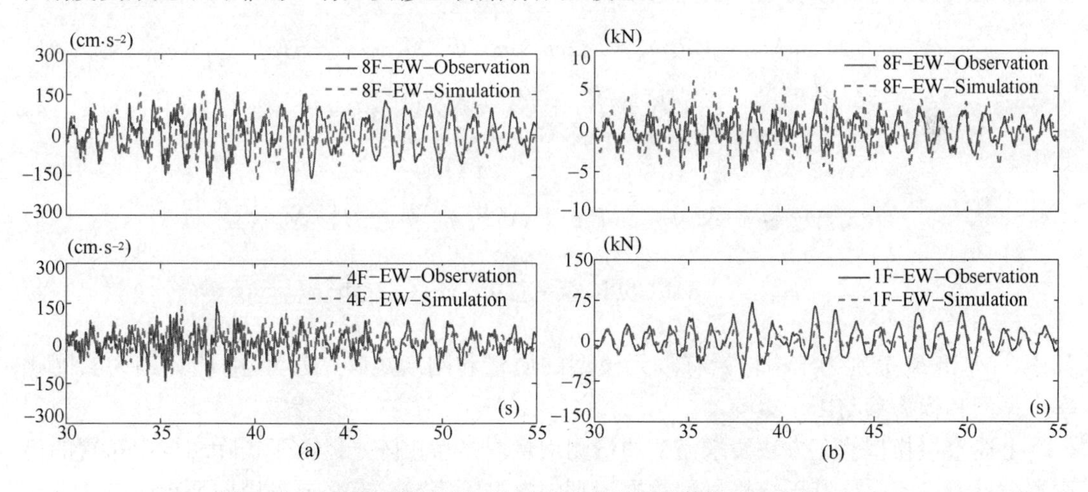

图 4 初始有限元模型预测的响应和实际监测数据（2013/05/26）

Fig. 4 Measured and predicted responses by initial FE model (2013/05/26)

（a）楼层加速度时程响应；（b）油阻尼器力时程响应

3.1.2 结构动力模型的改进直接修正方法

结构振动的实际量测数据，如模态参数，通常认为能更准确反映结构的真实动力特性，常用于修正结构有限元模型，减小模型预测误差。与该过程相关的技术称为模型修正或更新，模型修正技术已在机械和航空航天等领域获得了深入且富有成果的研究。对于复杂的大体积土木工程结构，结构的不确定模型参数较多，而结构量测信息的不完备，其模型修正问题通常被描述为一病态方程，加之结构分析耗时及模型不确定性，相关模型修正技术研究进展较慢。

模型修正实质为一系统识别问题，属反问题的一种。现有的模型修正技术可分为直接修正方法和迭代修正方法两类。迭代修正方法的基本思路是选取特定的模型参数为待修正参数，如结构质量和刚度矩阵元素或材料密度、弹性模量和尺寸参数等，利用确定性优化方法或不确定性更新方法，迭代调整物理参数取值，使修正后的模型预测值与实际量测值趋同。该类方法的修正结果具有明确的物理意义，但严重依赖主观的模型参数选择，不合理的选择会导致不切实际的参数修正结果，且修正过程需要大量结构分析，限制其在复杂土木工程结构中的应用。

直接模型修正方法是一类经典的基于结构模态参数的模型更新技术，该类方法具有计算简单高效和精确匹配目标模态参数的优点，适用于精确刻画结构模态参数的改变，可用于研究减振系统对结构模态参数的影响。当应用于地震激励作用下结构动力模型修正时，传统直接模型修正方法由于仅考虑与频率和振型相关的约束，修正模型预测的模态参与因子与实际识别值存在误差，这种参数匹配不完备问题不可避免影响修正模型的精度，但鲜有文献涉及相关问题研究。

针对上述模态参数匹配不完备问题，在经典的直接模型修正方法基础上，提出了一种改进的直接模型修正方法，通过施加模态正交性、特征方程和误差矩阵最小化约束，利用拉格朗日乘子方法推导了最优质量和刚度矩阵，并假定结构阻尼为经典的比例阻尼模型，修正后的结构质量和刚度矩阵表达为

$$M = M_a + M_a \Phi m_a^{-1}(I - m_a) m_a^{-1} \Phi^T M_a + M_\Delta + M_\Delta^T \tag{1}$$

$$K = K_a + K_\Delta + K_\Delta^T \tag{2}$$

其中，

$$M_\Delta = \frac{1}{c} M_a \Phi m_a^{-1}(P - P_B)(P_a^T m_a^{-1} \Phi^T - I_n^T) M_a \tag{3}$$

$$K_\Delta = \frac{1}{2} M\Phi(\Phi^T K_a \Phi + \Omega)\Phi^T M - K_a \Phi\Phi^T M \tag{4}$$

其中，M_a和K_a分别为结构初始有限元模型的质量和刚度矩阵，Ω、Φ和P分别为实测频率、振型和模态参与因子矩阵。

上述直接模型修正方法要求完整的量测信息或振型矩阵，但由于实际结构中布置的传感器数量通常有限，基于结构量测数据识别的振型不完整，有必要采用振型扩阶方法补充完整振型。图5给出了初始有限元模型的计算振型（初始振型）与扩阶振型的对比，其中，采用的扩阶技术为经典的力残余最小化技术，由计算结果可知第2阶振型相近，但第1阶和第3阶扩阶振型显著不同于初始振型，体现了初始有限元模型的误差。

值得指出的是，上述模型修正仅针对主体结构，待修正模态参数亦与主体结构相对应。实际监测数据仅包含整体结构的模态参数信息，受附加阻尼器对结构的阻尼和刚度影响，主体结构的模态参数仍需进一步确定。忽略阻尼器构件对振型的影响，取整体结构的振型为主体结构的振型，仅考虑阻尼器设备对频率和阻尼比的影响。基于经典的模态应变

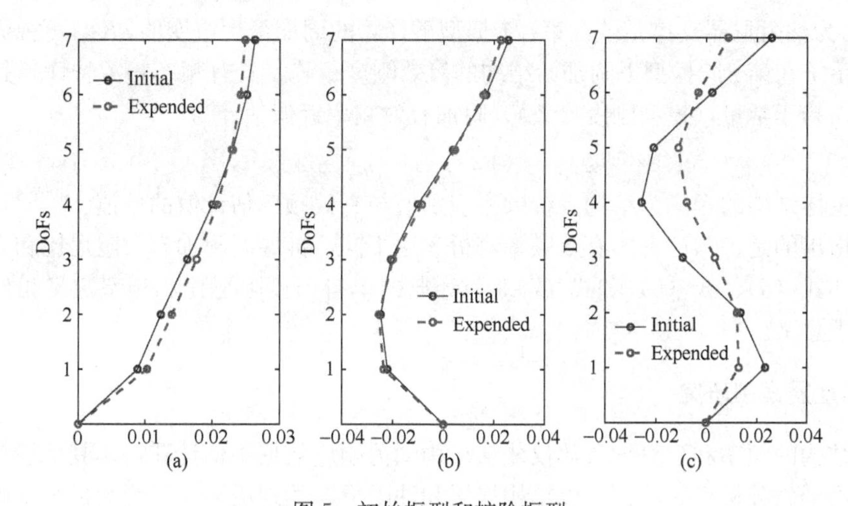

图 5　初始振型和扩阶振型

Fig. 5　Initial and expanded mode shapes

(a) 第 1 阶振型；(b) 第 2 阶振型；(c) 第 3 阶振型

能概念，以下将解析估计油阻尼器设备附加给整体结构的有效阻尼比和有效频率，进而确定主体结构的相应模态参数。

3.2　油阻尼器的附加有效阻尼和有效刚度

取经典 Maxwell 模型模拟油阻尼器及附属支撑系统的动力学行为，模型参数取基于油阻尼器实际监测数据的识别值。为量化油阻尼器减振系统附加给结构的有效阻尼和有效刚度，基于经典的模态应变能概念，提出有效阻尼比和有效频率的解析方法，量化估计油阻尼器减振系统对结构模态参数的影响，进而利用结构系统模态参数的识别值，获取结构主体的模态参数用于模型解析修正。

3.2.1　附加有效阻尼比

假设结构和油阻尼器受简谐激励作用，且结构的变形与振型成比例。基于应变能概念，油阻尼器支撑系统附加给结构第 i 阶模态的有效阻尼比为

$$\zeta_{\mathrm{eq},i}=\frac{1}{4\pi}\frac{E_{\mathrm{D},i}}{E_{\mathrm{S},i}}=\frac{\phi_i^{\mathrm{T}}C_{\mathrm{D},i}\phi_i}{2\omega_{\mathrm{S},i}} \tag{5}$$

其中，$E_{\mathrm{S},i}$ 和 $E_{\mathrm{D},i}$ 分别为在第 i 阶振型下计算的结构周期最大应变能和油阻尼器周期最大耗能，$\omega_{\mathrm{S},i}$ 为结构系统频率识别值，$C_{\mathrm{D},i}$ 为油阻尼器支撑系统在第 i 阶振型下的等效阻尼矩阵，与 Maxwell 模型参数（阻尼系数和刚度系数）相关，亦与油阻尼器的空间布置相关，具有频率相关特性，其元素随频率增大而逐步减小。结构主体的阻尼比为结构系统阻尼比识别值与附加有效阻尼比的差值。

3.2.2　附加有效频率

与有效阻尼比定义统一，基于应变能概念，定义油阻尼器支撑系统的应变能与结构总应变能之比，

$$\eta_{\mathrm{eq},i}=\frac{E_{\mathrm{B},i}}{E_{\mathrm{S},i}}=\frac{\phi_i^{\mathrm{T}}K_{\mathrm{D},i}\phi_i}{\omega_{\mathrm{S},i}^2} \tag{6}$$

其中，$E_{B,i}$为油阻尼器支撑系统在第i个振型下计算的周期最大应变能，$K_{D,i}$可视为为油阻尼器支撑系统在第i阶振型下附加给结构的等效刚度矩阵，具有频率相关特性，其元素随频率增大而趋于常量（模型刚度参数），附加有效频率近似估计为

$$\omega_{eq,i}=(1-\sqrt{(1-\eta_{eq,i})})\omega_{S,i} \tag{7}$$

相应地，主体结构的频率为结构系统频率识别值与有效频率估计值的差值。

值得指出的是，与已有的有效频率解析表达不同，推导的附加有效阻尼比和有效频率解析表达［式（5）、式（6）和式（7）］与经典的单自由度阻尼比和频率定义相似，具有明确的物理意义。

3.3 实际地震案例研究

实际结构的频率和阻尼比参数仅能从结构动力响应数据中估计获得，其估计精度亦仅能通过结构动力分析得以验证。以结构楼层和油阻尼器的地震响应监测数据为基础，识别结构系统的模态参数及油阻尼器模型的参数，采用 2.2 节方法估计油阻尼器附加给结构的有效阻尼比、有效频率以及主体结构的模态参数，并利用 2.1 节模型解析修正方法修正结构初始有限元模型，通过对比修正模型的预测响应与实际监测数据，验证上述参数估计的准确性，说明油阻尼器减振系统量化评估的正确性。

利用式（5）和式（7）分别计算油阻尼器支撑系统附加给结构的有效阻尼比和有效频率，前三阶有效频率（角频率）分别为：0.11、0.40 和 0.14；前三阶有效阻尼比分别为：1.86％、0.64％和 0.06％。主体结构前三阶频率（角频率）为：5.50、17.72 和 33.98；前三阶阻尼比为：1.28％、5.49％和 9.33％。由计算结果可知附加有效阻尼比随模态阶次升高迅速减弱，油阻尼器支撑系统显著提高了该结构的第 1 阶阻尼比，其他附加高阶阻尼比贡献可忽略。有效频率估计值相对较小，是因为结构侧向刚度受结构高阶模态主导，油阻尼器支撑系统的刚度相对较小，仅能影响结构低阶模态信息，并随模态阶次的升高，这种影响势必逐步减弱。

为验证附加有效阻尼比和有效频率的估计精度，利用主体结构频率、阻尼比和扩阶振型修正主体结构的质量、刚度和阻尼矩阵，并利用修正模型预测结构楼层和油阻尼器的地震响应，其中，高阶阻尼比和油阻尼模型参数取值与初始有限元模型保持一致。取预测响应的标准均方差（e_N）衡量模型预测精度，标准均方根误差的取值越接近 1，代表修正模型预测精度越高，说明模态参数（频率、阻尼比和振型）估计值越精确。

图 6 和图 7 分别给出了修正模型预测的强震阶段（30～55s）楼层加速度和油阻尼器阻尼力时程响应。其中，阻尼修正模型忽略了附加有效频率贡献，即结构本体频率等同于整体结构频率；阻尼和刚度修正模型则考虑附加有效阻尼和有效频率的贡献。与图 4 对比可知，阻尼和刚度修正模型的响应可精确拟合实际监测数据，表明了建议方法估计的主体结构的模态参数、附加有效阻尼比和有效频率是可靠且准确的，同时，亦说明了振型扩阶技术及模型修正方法的准确性。

图 6 和图 7 中亦给出了给出了修正模型预测响应的标准均方差。在预测精度方面，修正模型均显著优于初始有限元模型，阻尼和刚度修正模型较阻尼修正模型预测精度更高，说明了附加有效频率估计的准确性，亦表明了为精确预测结构地震响应，油阻尼器支撑系统的附加刚度贡献不可忽略。

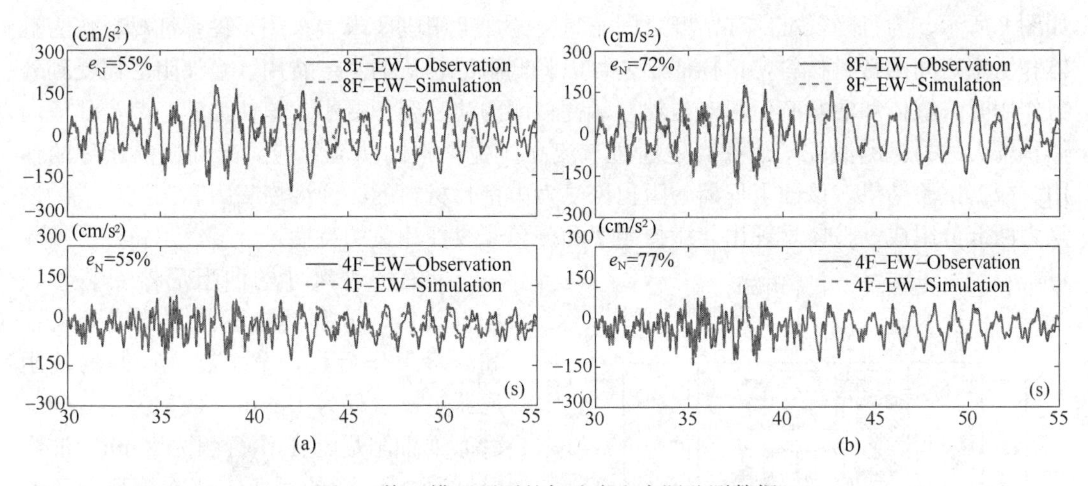

图 6　修正模型预测的加速度和实际监测数据

Fig. 6　Measured and predicted structural responses by updated FE model（2013/05/26）

（a）阻尼修正模型；（b）阻尼和刚度修正模型

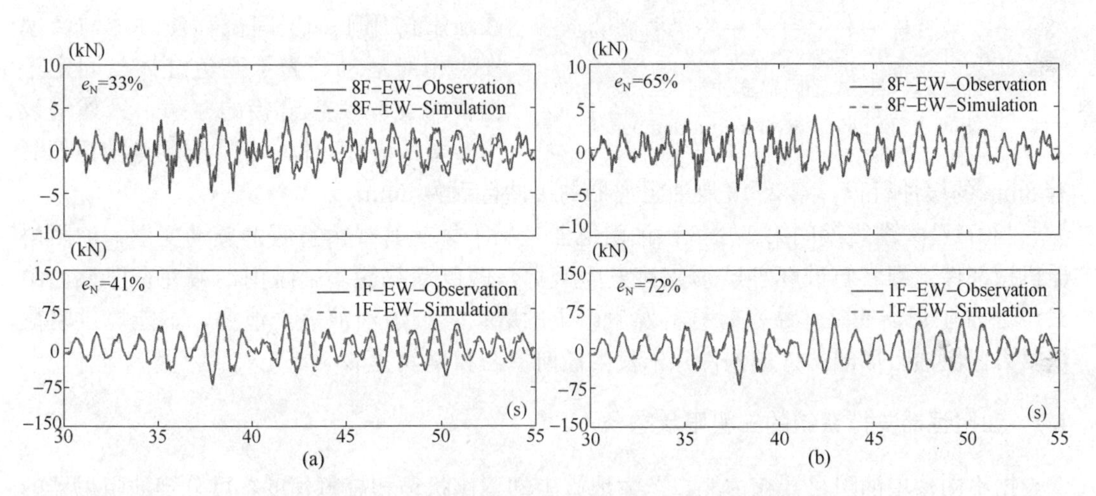

图 7　修正模型预测的阻尼器力和实际监测数据

Fig. 7　Measured and predicted damper responses by updated FE model（2013/05/26）

（a）阻尼修正模型；（b）阻尼和刚度修正模型

实际地震案例验证了油阻尼器性能量化评估和验证方法的有效性和准确性，改进的直接模型修正方法可精确匹配目标模态参数，为精准量化评估油阻尼器对结构模态参数影响提供了可靠支撑，计算结果表明附加有效阻尼比和有效频率估计的高精度，表明改进应变能方法估计的准确性。此外，油阻尼器性能量化评估和验证方法仅利用结构系统的模态参数，计算简单高效，便于在实际工程中应用。

4. 油阻尼器减振系统的破坏和再生

油阻尼器为日本东北工业大学的 Kawamata 和 Funaki 等[4,5]人自主研发的阻尼装置，由位于中部的一个缸体和一对分别位于两侧的活塞组成，油阻尼器的详细构造和设计尺寸如图 8 所示。为了降低缸体和活塞之间密封性的加工精度要求，采用了黏弹性材料作为缸体和活塞之间的密封材料，这样的工艺可以降低阻尼装置的生产费用。当油阻尼器受到外力作用时，缸体和活塞产生相对运动，活塞一侧的油腔容积变小，迫使油液经节流孔流向体积增大的另一侧油腔，油液高速通过节流孔产生很强的湍流效应，起到耗散能量的作用。根据试验结果，该油阻尼器的阻尼恢复力由密封材料的黏弹性恢复力和湍流形成的恢复力两部分组成，整体呈现出与激振频率和振幅非线性相关的特性。

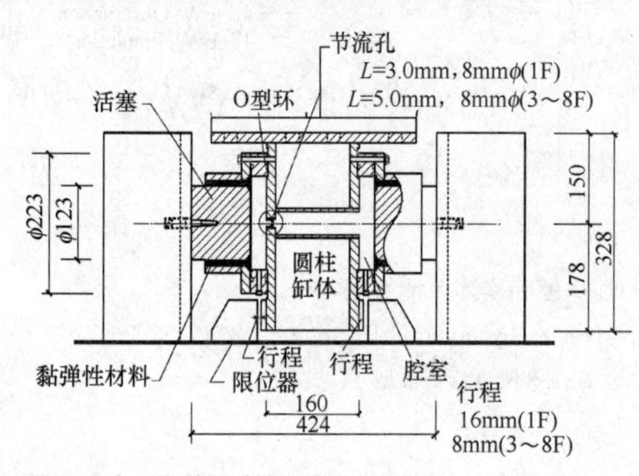

图 8　油阻尼器详细

Fig. 8　Detail of installed oil dampers

共有两种尺寸的油阻尼器安装于该栋行政楼，分别具有不同的节流孔尺寸和活塞工作行程，第 1 层（1F）的油阻尼器工作行程为 16mm，第 3～8 层（3～8F）的油阻尼器工作行程为 8mm。油阻尼器通过 U 形的固定支座将楼板与活塞头连接起来，中间部分的缸体与 V 形斜撑连接，如图 1（c）和图 1（d）所示。在地震作用下，层间位移使缸体与活塞发生相对运动，为了避免缸体与固定于楼板的支座发生直接的碰撞，另外增设了活塞的缓冲行程，第 1 层的油阻尼器具有 8mm 的缓冲行程，第 3～8 层油阻尼器的缓冲行程为 5mm。

“3. 11”大地震期间第 1 层 8 组油阻尼器经历了远超其最大行程的震动变形，油阻尼器机械部件遭到肢解破坏，U 形支座发生破坏，两侧活塞脱落，缸内油液完全泄漏。第 3、4 层油阻尼器的黏弹性密封材料发生严重的磨损，缸内油液完全泄漏，不能提供阻尼恢复力，但油阻尼器的机械构件并未发生任何损伤和塑性变形。

4.1　油阻尼器支撑系统的三极限状态

据不同楼层油阻尼器在“3. 11”大地震中的工作状态和震害程度，可分辨油阻尼器的正常工作状态和破坏状态，定义油阻尼器支撑系统的三极限状态：正常工作极限状态、丧失工作能力极限状态及完全破坏极限状态。

正常工作极限状态是油阻尼器正常工作（提供阻尼耗能）的界限，超越该界限油阻尼器部件（缸体和支座）将发生接触碰撞。若这种接触行为持续发展，将会对油阻尼器造成损伤（密封黏弹性材料磨损），当损伤累积到一定程度，油阻尼器缸内油液发生泄漏，进而破坏油阻尼器工作能力，即丧失工作能力极限状态出现，如结构第 3、4 层油阻尼器的震害。丧失工作能力极限状态代表油阻尼器自身的破坏，随着油阻尼器变形的进一步发展，附属支撑和支座仍能起到“支撑”的作用，完全破坏极限状态则是描述由于过度的变形造成支撑和支座的破坏，具体表现为支撑的屈曲或支座的破坏，此时油阻尼器系统完全丧失工作能力，如第 1 层油阻尼器震害。

考虑到不同楼层油阻尼器的震害与其位移变形直接关联，取位移指标量化油阻尼器的三极限状态。取油阻尼器最大设计行程为正常工作极限状态的位移阈值，丧失工作能力极限状态和完全破坏极限状态较复杂，与密封黏弹性材料特性、支座及支撑有关，极限状态的位移阈值应通过构件试验确定。

经典的 Maxwell 模型常用于描述油阻尼器正常工作状态，但不能刻画油阻尼器出现极限状态后的动力学特征，基于 Hertz 接触理论，提出 Hertz-Maxwell 组合模型模拟油阻尼器的正常工作和极限状态，该组合模型可采用如下数学方程描述

$$F_{\mathrm{d}}=\begin{cases} c_{\mathrm{d}}\operatorname{sgn}(\dot{x}_{\mathrm{d}})\ |\dot{x}_{\mathrm{d}}|^{\alpha}; & x<u_1\ \&\ \max(x)\leqslant u_2 \\ 0; & x<u_1\ \&\ \max(x)>u_2 \\ k_{\mathrm{b}}\ (x-u_1)^{n_{\mathrm{c}}}+c_{\mathrm{b}}\dot{x}; & u_1\leqslant x<u_3\ \&\ \max(x)<u_3 \\ 0; & x\geqslant u_3 \end{cases} \tag{8}$$

其中，x 和 F_{d} 分别为油阻尼器支撑系统两端的位移变形和阻尼力，u_1、u_2 和 u_3 为油阻尼器三极限状态的位移阈值，k_{b}、c_{b} 和 n_{c} 分别为接触刚度系数、阻尼系数和指数参数。图 9 绘出了 Hertz-Maxwell 模型模拟的油阻尼器支撑系统的三极限作状态，其中，阻尼力和位移取为标准化无量纲值。

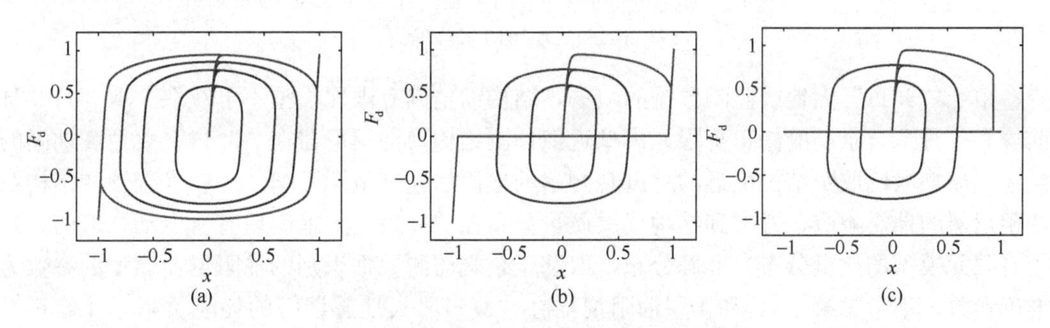

图 9　油阻尼器支撑系统三极限状态

Fig. 9　Three limit states for oil damper brace system

(a) 正常工作极限状态；(b) 丧失工作能力极限；(c) 完全破坏极限状态

4.2　油阻尼器减振系统的破坏重演

3.11 大地震造成结构所在地区电力设施失效，结构健康监测系统未采集到结构及油阻尼器的地震响应数据。结构于“3.11”大地震前后（3 月 9 日和 4 月 7 日）经历了两次较大地震，属东日本大地震序列，为“3.11”大地震的前震和余震。利用监测系统采集的 3 月 9 日和 4 月 7 日地震动数据修正结构有限元模型，修正模型用于预测油阻尼器在 3.11 地震期间的响应，重演油阻尼器的破坏过程[6]。

监测系统于 3 月 9 日采集的结构地基加速度数据最大值为 32cm/s^2（东西方向）和 26cm/s^2（南北方向），结构地震响应相对较小、由基础模态主导。图 10（a）给出了修正有限元模型预测的楼层加速度最大值，对比监测楼层的峰值加速度量测值，修正有限元模型可精确预测结构的地震响应，验证了修正有限元模型的准确性。4 月 7 日发生的地震对结构所在城市（仙台市）的震害烈度与“3.11”地震大致相当，该地震破坏了在“3.11”大地震中存活下来的部分建筑结构，但未对油阻尼器建筑造成进一步的损伤，结构地基的峰值加速度监测值为 176cm/s^2（东西方向）和 289cm/s^2（南北方向），结构地震响应较大、受多个低阶模态影响。考虑到“3.11”大地震破坏了第 1、3 和 4 层油阻尼器，大地

震后破坏的油阻尼器被取出并未重填修补，在结构有限元模型中除去对应的油阻尼器模型，图 10（b）给出了修正有限元模型预测的楼层加速度响应的最大值，对比实际监测数据可知修正有限元模型仍保持较高预测精度。

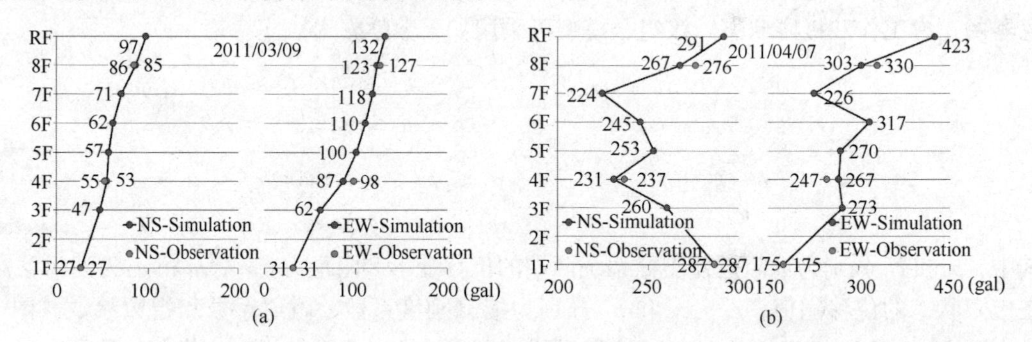

图 10　楼层加速度预测值和量测值

Fig. 10　Measured and predicted acceleration of floors

(a) 2011/03/09；(b) 2011/04/07

由于“3.11”大地震监测数据的缺失，结构的精确有限元模型不可获取。取 3 月 9 日前震和 4 月 7 日余震的修正有限元模型近似预测结构及油阻尼器在“3.11”大地震期间的响应，模型激励取距结构最近的地震观测站的地面加速度量测数据。为较准确预测结构及油阻尼器的地震响应，考虑到结构楼层质量分布在“3.11”大地震后有所变动，取前震修正有限元模型的质量分布为准确分布，取基于余震监测数据识别的油阻尼器模型的参数为准确参数，不考虑第 1、3 和 4 层油阻尼器在“3.11”大地震期间的耗能贡献，在修正有限元模型中除去相应油阻尼器模型。图 11 给出了组合修正有限元模型预测的结构第 4 层楼板的加速度时程响应，结构在“3.11”大地震期间持续了近百秒的强烈震动。

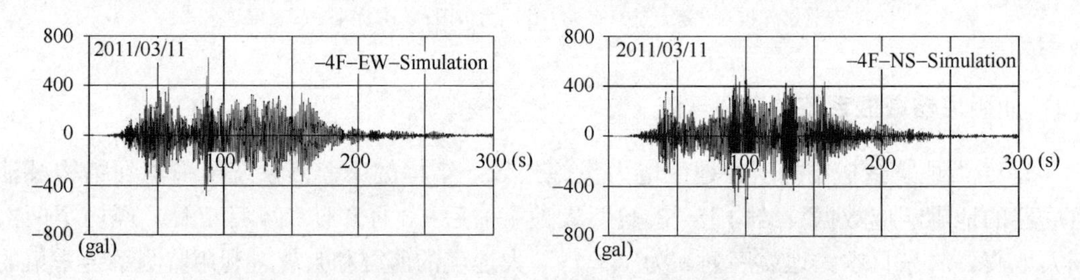

图 11　第 4 层楼板加速度时程响应预测（2011/03/11）

Fig. 11　Predicted acceleration of the 4th floor (2011/03/11)

油阻尼器在地震期间呈现了不同的工作状态和破坏状态，油阻尼器支撑系统的三极限状态（正常工作、丧失工作能力和完全破坏极限状体）的位移阈值需要量化估计。利于油阻尼器的设计信息，正常工作极限位移值取为 16＋8mm（1F）和 8＋5mm（3～8F）。阻尼器丧失工作能力极限状态较难准确量化确定，与黏弹性密封材料属性及变形相关。考虑到结构第 4 层油阻尼器丧失工作能力，但油阻尼器机械构件并未遭到破坏，假设该层油阻尼器的地震响应处于但未超越完全破坏极限状态，即假设第 4 层层间位移（预测）最大值为完全破坏极限位移阈值。

进一步假设油阻尼器的完全破坏与正常工作限位移的比值相同，计算第4层油阻尼器的完全破坏与正常工作极限状态位移比值，假定第1层油阻尼器的极限位移比值等同于第4层油阻尼器极限位移比值，进而可估算第1层油阻尼器完全破坏极限位移阈值(60mm)。图12给出了组合修正有限元模型预测的第1层和第4层油阻尼器的位移响应(层间位移)，其中，红色线代表油阻尼器的正常工作极限状态，超过该极限状态，阻尼器支撑与支座发生接触和碰撞，黑色虚线代表油阻尼器支撑系统的完全破坏极限状态，超过该极限位移值，油阻尼器及支撑将完全退出工作。由图示预测结果可知第1层油阻尼器在90s左右遭到破坏且经历了远超其正常工作极限位移阈值的震动变形，第4层油阻尼器的最大地震位移变形亦超出了其正常工作极限位移阈值，出现油液泄露、丧失继续工作的能力。

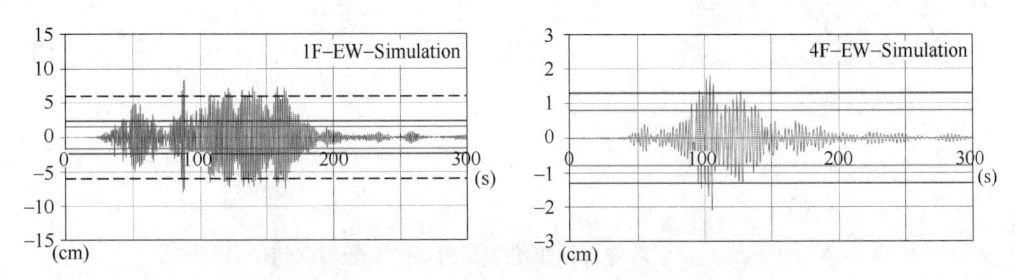

图12　第1、4层油阻尼器位移变形预测（2011/03/11）

Fig. 12　Predicted damper displacement of the 1st and 4th floors (2011/03/11)

值得指出的是油阻尼器破坏主要原因是其工作行程设计值与“3.11”地震的需求变形不匹配，底层油阻尼器的可允许行程（阻尼器的工作行程加上缓冲行程）仅为24mm，与其底部大空间的层高8m相比较而言，仅为0.3%的层间位移，未超过日本的钢结构规范中针对层间位移的弹性极限0.5%。基于结构地震响应的预测分析，其底层层间位移大大超过了0.3%，达到了0.8%的层间位移。

4.3　减振系统的再生设计策略

“3.11”大地震虽破坏了油阻尼器减振系统，但未对结构本体造成直接的损伤，通过震后损伤调查和安全性评估，结构中遭受破坏的油阻尼器被移除，结构重新投入使用。油阻尼器减振系统的破坏，使结构设计者面临如何恢复和再生减振系统的设计问题。如前所述，油阻尼器减振系统的破坏为世界首例，无既有设计案例参考和相关设计规范指导，减振系统的再生设计需要突破常规，寻求新的设计思路[7]。

结构减振系统的再生设计需满足校方提出的四点要求：1）使结构有足够的能力抵抗与“3.11”大地震相当的大地震；2）在预算范围内的高性价比再生设计；3）新增阻尼器的安装简单不需要大规模施工；4）减振性能优于原油阻尼器减振系统。

结构的第1、3、4层油阻尼器均遭到破坏，为控制预算保留原第3层和第4层的油阻尼器设计，在原位置处修复和更换新油阻尼器。结构第1层油阻尼器提供了减振系统的主要阻尼耗能，为获取更优的减振性能，第1层油阻尼器必须被更高性能的阻尼器替换。一个备选经济方案为选择既有的不同性能原型油阻尼器，但通过结构有限元模型时程分析，所有基于原型油阻尼器的再生减振系统均会遭到“3.11”大地震的破坏，油阻尼器的位移变形超过其设计极限值。

结构减振系统再生方案需寻求其他耗能设备，如叠层橡胶垫（隔振支座）。选择了不同材料和截面尺寸的橡胶垫作为备选方案，通过构件测试和有限元模拟验证了备选橡胶垫的耗能和变形能力，并考察备选橡胶垫在“3.11”地震作用下的变形是否超越其设计极限值，通过对比不同备选方案的计算结果，最终选取锡芯叠层橡胶垫替换第1层油阻尼器，锡芯叠层橡胶垫的尺寸及力学特性如图13所示。结构再生减振系统包括第1层锡芯叠层橡胶垫和第3～8层油阻尼器，如图14（a）所示，属混合被动控制减振系统。

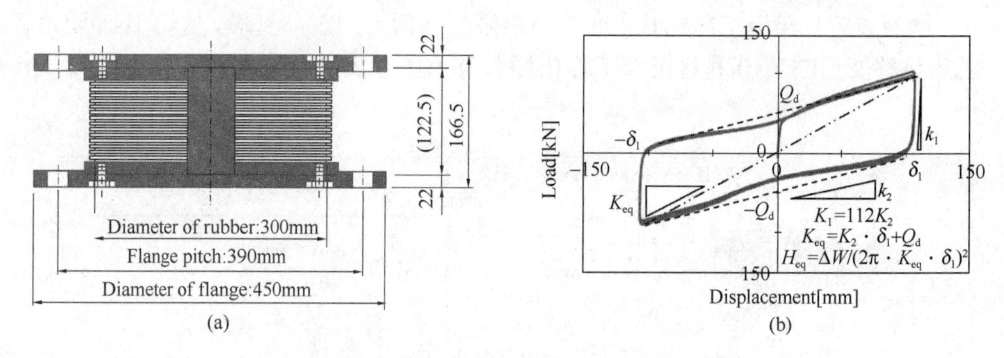

图13　锡芯叠层橡胶垫的几何尺寸及滞回曲线

Fig. 13　Dimension and hysteresis curve of thin-rubber bearing

（a）尺寸参数；（b）力和位移滞回曲线

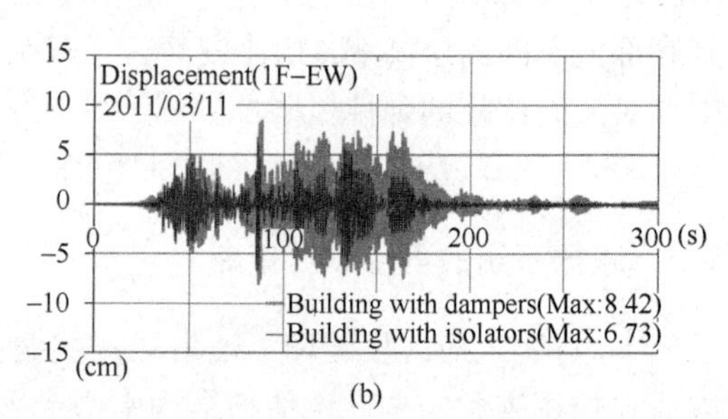

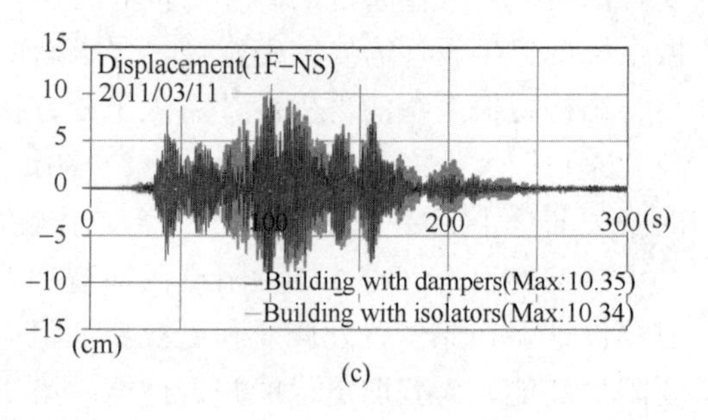

(a)

图14　再生减振系统及有限元模型响应预测对比（2011/03/11）

Fig. 14　Rehabilitated control system and comparations of predicted damper responses（2011/03/11）

（a）再生减振系统；（b）东西方向；（c）南北方向

为对比研究原油阻尼器和再生混合减振系统的性能，利用装备原油阻尼器的和再生的减振系统的有限元模型预测结构在“3.11”大地震中的响应，图 14（b）、（c）给出了有限元模型预测的第 1 层层间位移时程响应，由图示计算结果可知再生减振系统有效抑制了结构位移幅值，其响应幅值显著小于油阻尼器的响应幅值，表明再生减振系统可提供更多阻尼耗能、优于原油阻尼器减振系统，此外，锡芯支座的最大位移变形未超过其设计极限位移值（10cm），满足“3.11”大地震中对再生减振系统的变形需求，再生减振系统可靠性更高。

值得指出的是“3.11”大地震是日本有地震观测记录以来的规模最大的地震，地震的强度及破坏力超出了结构设计的预期，属极罕遇地震。油阻尼器及其他减隔震系统的破坏案例，为减振系统的设计和应用敲响了警钟。减振构件可能在极端荷载条件下破坏，抑或在正常使用过程中出现劣化、性能下降的问题。因此，需要更为全面地、深入地研究消能构件的性能对结构抗震性能的影响，在设计阶段，也需要将消能减震系统与主体结构结合起来进行整体结构的抗震概念设计，目前这方面还缺少深入的研究。

4.4 再生减振系统性能的实际地震验证

从油阻尼器减振系统到再生混合减振系统，结构历经了五个状态的演变，在此期间结构经历了多次强烈地震，利用结构地震响应监测数据量化推断结构性能状态的演变趋势，进而验证再生减振系统的实际工作性能[7]。结构的五个性能状态为：1）初始完整状态，取 2011 年 3 月 9 日地震监测数据推断结构状态；2）“3.11”地震状态，无监测数据可利用；3）油阻尼器减振系统破坏状态，由 2011 年 4 月 7 日地震监测数据推断；4）部分油阻尼器恢复状态，代表地震监测数据日期取为 2012 年 12 月 7 日；5）减振系统再生状态，取 2013 年 8 月 4 日地震监测数据推断结构状态。

结构减振系统的再生工作于 2012 年 2 月 15 日结束，2013 年 8 月 4 日结构所在城市（仙台）发生了一次中等强度的地震（矩震级 6.0），是再生减振系统投入使用后结构经历的第一次较大地震，为检验再生减振系统的实际工作性能提供了绝佳机会，结构健康监测系统量测的结构地基最大加速度为 32cm/s^2（东西方向）和 28cm/s^2（南北方向），利用采集的结构响应数据修正结构有限元模型，图 15 给出了修正有限元模型预测的结构第 4 层楼层加速度响应，由计算结果可知加速度预测值可较精确拟合实际量测值，表明修正有限元模型的准确性。

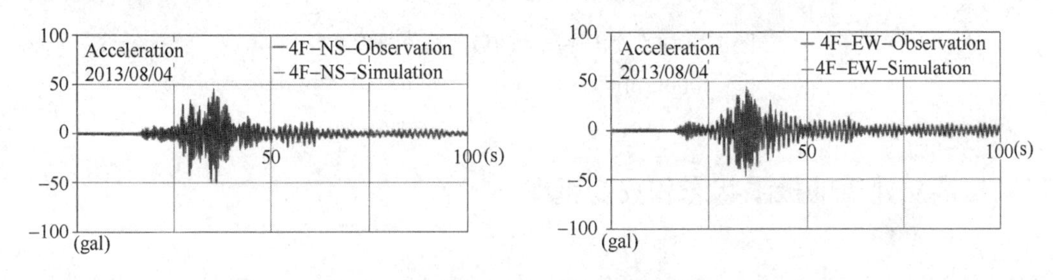

图 15 结构第 4 层楼板加速度时程响应预测（2013/08/04）

Fig. 15 Measured and predicted acceleration of the 4th floor (2013/08/04)

为对比油阻尼器和再生减振系统对结构地震响应影响，即对比结构的初始完整状态和再生减振系统状态差异，图 16 给出结构第 4 层和第 8 层加速度对结构第 1 层（地基层）加速度的传递函数，包括基于实际量测数据计算的传递函数和基于修正有限元模型预测计算的传递函数，对比两个状态的传递函数可知再生减振系统状态具有较小的（基频处）最大幅值放大系数，说明了再生减振系统提供了更多的阻尼耗能。

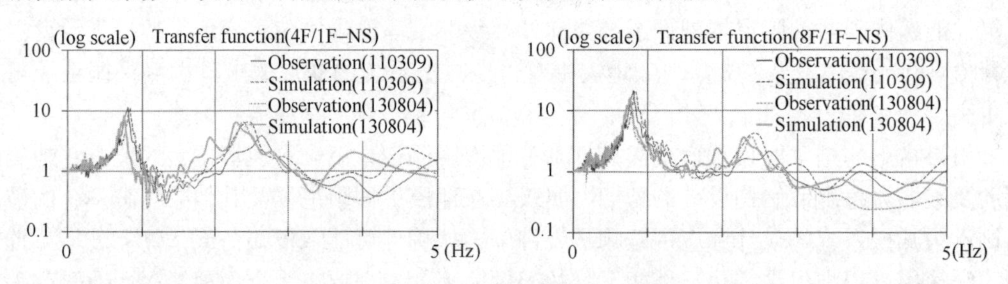

图 16 结构加速度响应的传递函数对比（2011/09/09 和 2013/08/04）

Fig. 16 Comparison of accelerations transfer functions（2011/09/09 vs 2013/08/04）

图 17 给出了加速度传递函数最大幅值放大系数的变化趋势，除“3.11”大地震外，其余四次地震的加速度传递函数由实际监测数据计算获得，选取的五次地震与结构五个性能状态一一对应。由计算结果可知“3.11”大地震后传递函数的最大幅值放大系数显著增大，在 2011/04/07 地震中为最大值，并在随后的 2012/12/07 地震中有所减小，这是因为第 3、4 层油阻尼器恢复工作，提供了额外的阻尼耗能。对比 2011/03/09 地震和 2013/08/04 地震计算结果，后者最大幅值放大系数明显小于前者，两次地震分别对应结构的初始完整状态和再生减振系统状态，在提供附加阻尼耗能方面，再生减振系统优于初始油阻尼器减振系统，再生减振系统更易实现结构最优性态的控制。

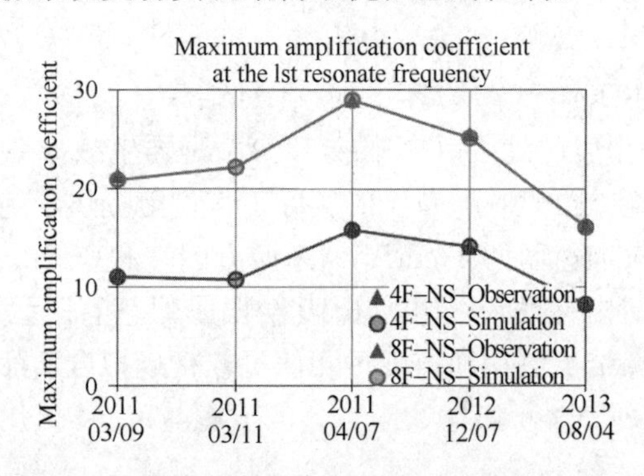

图 17 加速度传递函数的最大幅值放大系数

Fig. 17 Maximum amplification coefficients of acceleration transfer functions

5. 减振高层建筑的极限状态和灾变机理

5.1 减振系统的极限状态

一方面改造加固结构采用了减振器才使结构满足了抗震性能上的要求，另一方面消能减震技术降低新建建筑的抗震需求和位移需求，减少构件的尺寸，减轻了构件的重量，采用减振器的振动控制策略使减振器成为重要的结构构件。由于它们在提高结构抗震性能方面的重要作用，减振器的性能退化和失效会对结构产生不利的影响。

以工程中广泛采用的黏滞阻尼器为例，许多实验研究都证实了黏滞阻尼器具有良好的耗能性能，但通常都是在阻尼器的极限位移（速度）范围内进行测试和性能监测，没有测试阻尼器在极端荷载条件下的性能。Miyamoto 等[8]指出了黏滞阻尼器可能的几种极限状态：承载力极限状态、位移极限状态、混合承载力—位移极限状态，针对黏滞阻尼器建立了可以考虑其极限状态和失效模式的简化模型，并可将该模型应用于消能减震结构的非线性动力分析[9]。用于高层建筑中控制风振的黏滞阻尼器，其变形要求通常很小（几个毫米），但却有很大的阻尼力，这些黏滞阻尼器在高频微幅风振的作用下，可能会发生减振器的疲劳问题[10]，但这一方面的研究非常罕见，黏滞阻尼器的疲劳极限状态也是需要进一步的深入研究。

隔震垫的铅芯在常温环境下经过塑性变形的循环加载可使铅的多晶体发生再结晶，因此具有良好的耗能性能。隔振结构在风荷载作用下，隔震垫会经历微幅而频繁的往复位移，而目前尚不清楚风荷载的作用是否会引起铅芯裂缝的产生，除了隔震垫的位移极限状态，疲劳极限状态也需展开深入的研究。

5.2 减振高层建筑的灾变机理

由于存在减振装置的非线性作用，减振高层建筑的动力响应也是非线性的。在极端动力荷载作用下，减振装置和结构本身都可能进入弹塑性损伤状态，甚至发生局部破坏或整体失效。因此，在减振高层建筑的性能化设计中，需要准确分析不同荷载水准下的结构非线性动力响应，从而对减振装置、结构构件和结构整体的弹塑性损伤和非线性变形等建立可靠的量化评估，这是减振高层结构性能设计与控制的理论基础。

我国现行抗震设计规范基于“大震不倒”的抗震设计实践用于减振高层建筑的设计与分析面临诸多挑战：1）过于粗略的弹塑性变形分析掩盖了实际结构构件与被动减振保护系统在强地震作用下的复杂力学行为；2）未能就结构构件和减振保护系统的实际极限状态给出清晰的界定；3）不能体现减振结构体系在被动减振保护系统突然出现极限破坏后的整体抗震性能。因此建立更加科学、合理的考虑的设计方法已势在必行。

Kasai 等[11,12]在世界最大的振动台 E-Defense 上对一足尺消能减震结构进行了三维模拟振动台试验，测试了安装 5 种不同的阻尼器的消能减震结构的地震响应和抗震性能，这些阻尼器包括了黏滞阻尼器、黏弹性阻尼器、金属阻尼器和摩擦阻尼器。这些试验表明，阻尼器能大量耗散地震输入结构的能量，增加数倍于原有结构的阻尼比，有效地衰减结构的位移、速度和加速度响应，降低结构的层间剪力。但这些消能减震结构试验的研究重点均是在阻尼器正常工作情况下结构的动力响应及抗震性能，目前还尚未有考虑阻尼器极限状态的消能减震结构振动台试验，也无考虑阻尼器极限状态的消能减震结构连续破坏的研究。

目前从减振结构的试验来看，存在的主要问题有：减振器本身的力学性能试验较难真实反应减振器安装在整体结构中的受力条件，高层结构的减振振动台试验较少，足尺模型的振动台试验更少，在缩尺模型中减振器本身的缩尺比例带来的性能误差尚不明确。因此非常有必要针对减振结构的试验开展研究，积累试验数据和经验，为减振高层结构设计提供辅助手段。

6. 基于三个极限状态的研究思路

6.1　三个极限状态的提出

减振结构可明确划分为两部分：本体结构与减振系统，通过调整减振系统的耗能比例，可有效地控制本体结构的塑性变形，实现预先设定的结构的各项性能水准。本体结构功能部分由若干功能子结构组成，保证结构的各种正常使用功能。减振系统功能部分由减振器组成，其可靠度低于本体部分，在遇大震时失效破坏，但其破坏不影响结构的主要功能，引起的损失相对较小，易于迅速修复以恢复结构的正常使用功能。在各种正常使用荷载及小震等非灾害荷载作用下，结构主要功能部分和减振系统共同发挥作用，本体结构处于弹性状态，结构的各项性能满足正常使用功能，在大震等灾害荷载作用下，结构的减振系统开始发挥分灾作用，部分减振器显著进入塑性消能阶段甚至失效破坏，以保证本体结构部分屈服程度不大甚至于不屈服，从而达到所要求的建筑物的各项性能指标，维护整个结构体系的各种正常使用功能。

根据以上的抗震思想，极端灾害下减振高层建筑具有三个极限状态：高层建筑正常运行时的极限状态（有减振系统耦合）；减振系统在不同工况下的极限状态；减振系统失效后的建筑本体结构的极限状态（无减振系统耦合），如图 18 所示。

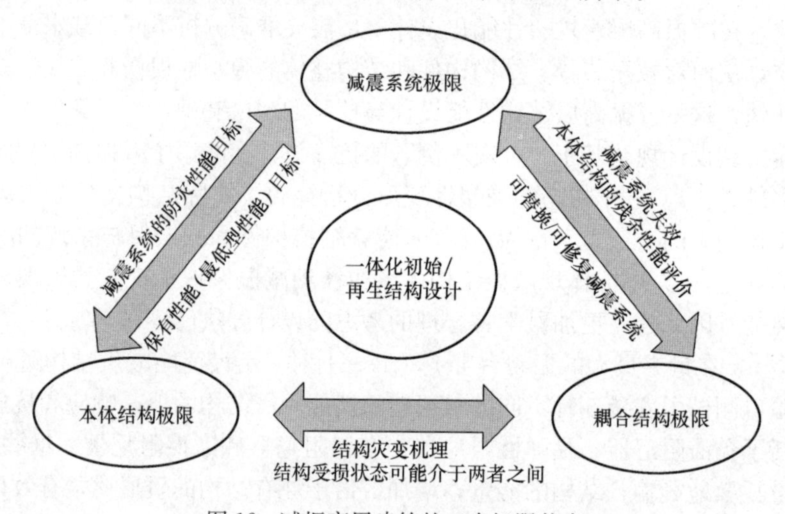

图 18　减振高层建筑的三个极限状态

Fig. 18　Three limit states of vibration-controlled, high-rise buildings

根据前述减振系统的灾害案例，表明减振系统具有工作极限状态，在极端情况下若减振系统完全失效，无法起到任何作用，本体结构具有的极限性能应能满足最低设防目标的要求，以保证结构体系“大震不倒”。在正常情况下，通过调整、设定减振系统的抗灾性能目标，可使减振系统耦合的整体结构具有比本体结构更好的抗灾性能，在极端灾害作用下，本体结构与减振系统也能协同工作，进入预先设定的理想极限状态。但在实际结构中，由于减振系统和本体结构在正常使用过程中也可能发生性能退化或经历一些突发事件，以及灾害造成的损伤的不确定性，使减振系统和本体结构的损伤模式和失效路径偏离了预先设定的模式，未能达到原先设定的性能目标。实际应用中，减振结构的灾变机理应介于本体结构的极限状态和耦合结构的极限状态之间。另外，在灾害发生后，减振系统或本体结构可能生部分损伤，应对灾后的结构进行残余性能评估；也应该将减振系统具有可替换性、可修复性的思想贯穿于减振结构设计、使用阶段及再生阶段的全过程，形成减振结构的一体化初始/再生设计方法。

6.2 极端灾害下减振高层建筑全寿命性能演化过程

高层建筑的性能是指其在抵御灾变时能够表现出的最大保有性能，图 19 中纵轴所示的建筑本体结构和减振系统性能的耦合。建筑本体结构性能是指减振建筑去除减振系统后建筑本体结构的保有性能（圆点填充部分）；减振系统性能是指建筑结构安装的减振系统的自有性能（斜线下方空白部分）；一个完整的减振建筑具有的保有性能包括了建筑本体结构性能和减振系统性能，同时二者间还存在一部分非线性耦合的性能重叠（斜线填充部分）。图中的三种填充方式对应了减振高层建筑的三个极限状态，也就是建筑本体结构极限、减振系统极限及它们耦合时的极限。

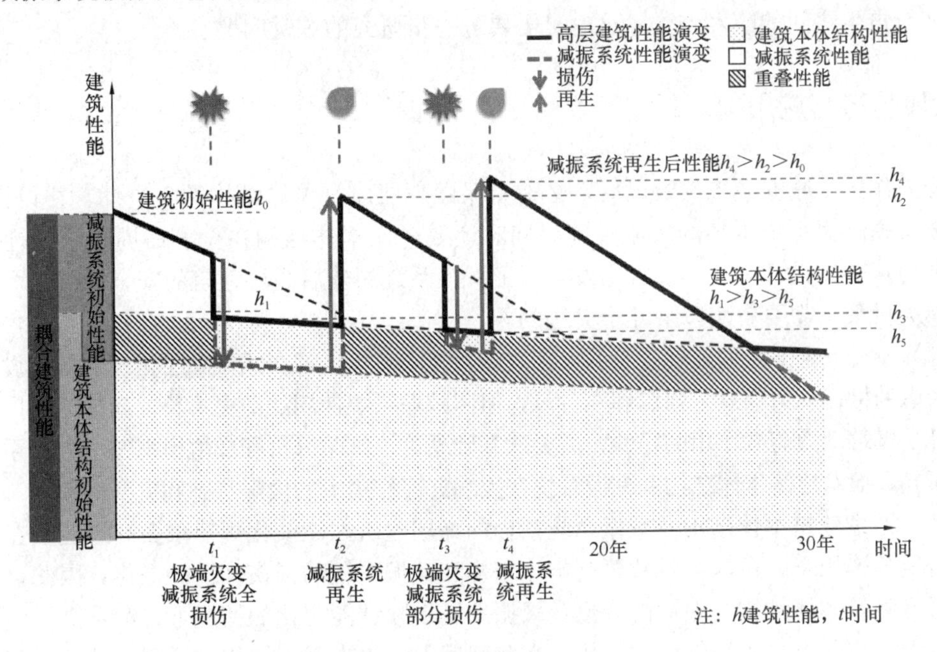

图 19　极端灾害下减振高层建筑全寿命性能演化

Fig. 19　Life-cycle performance evolving of vibration-controlled building

图 19 中粗实线代表减振高层建筑性能随时间变化的过程（简化为直线，忽略非线性演化的过程），粗虚线（包括与粗实线重叠的部分）表示减振系统性能随时间变化的过程。减振建筑在使用过程中，日常的劣化和可能的突发灾变，均会使其保有性能随时间不断变化，性能发展的整体趋势是不断降低的（粗实线）。图中细虚线（包括与粗实线重叠部分）表示了减振系统为减振建筑所提供的附加性能的演化过程，当遇到突发灾变或达到使用寿命时，减振系统有可能完全失效，这时就无法为建筑提供任何附加的性能保障。图中粗实

线和粗虚线的变化对应了建筑结构的性能演变。

建筑初始性能 h_0 为新建建筑本体结构性能 h_1 与新装减振系统性能的耦合。极端灾害时，减振系统首先起到抵御作用，假设图中 t_1 时刻发生的灾害超出了减振系统的设计极限，减振系统将完全失效，如图中向下的红色箭头所示。由于减振系统与建筑本体结构之间存在部分重叠性能，此时的建筑保有性能将跌落至 h_3，即建筑本体结构性能水平。在不进行减振系统再生的情况下，随着时间的推移，建筑保有性能将继续缓慢下降，如图中 t_1 和 t_2 间的粗实线所示。如对减振系统再生，如图中向上的箭头所示，建筑的保有性能将提高。理想的再生设计应使再生后的建筑保有性能 h_2 在 t_2 时刻超过其初始性能 h_0，即 $h_2>h_0$。图中假设 t_3 时刻再次发生的灾害使减振系统部分失效，建筑保有性能将再次跌落至建筑本体结构性能水平 h_5。当在 t_4 时刻对减振系统再次再生时，建筑保有性能将提高到 h_4 水平，理想情况下再生后的保有性能 h_4 应超过之前保有性能 h_2，即 $h_4>h_2$。从性能控制角度来说，每次建筑保有性能因遭遇灾害下降后，由于建筑本体结构保有性能始终在逐渐劣化，总有 $h_5<h_3<h_1$，而理想的减振建筑再生设计应使得建筑每次再生后的保有性能均超过之前水平，也就是使 $h_4>h_2>h_0$。图中的向上箭头、向下箭头对应了灾变和再生时的建筑性能的突变。如何有效地控制损伤发生后建筑结构的性能、设定再生性能目标以及制定灾变与再生之间的应对方法，这也是需要进一步研究的关键问题。

7. 发展趋势与应用前景

强震作用下减振系统失效以后，必须要保证高层建筑本体结构的安全性，在设计中，不仅要考虑减振系统本身的极限性能，同时还要考虑本体结构在全寿命周期内的性能退化。如何从初始设计和再生设计出发，保证减振系统失效后高层建筑的安全性能是今后面临的重大挑战，需要解决以下几个关键问题：

（1）极端灾害下减振高层建筑的灾害效应及极限破坏机理在减振高层建筑的结构设计中主要考虑的极端动力荷载是强震和强台风，但因减振系统使用的历史不长，尚缺乏在广泛范围内经受极端动力荷载作用的检验，需要从基础理论、试验分析和数值模拟三个方面系统地展开研究，揭示建筑本体结构和减振系统在极端灾害作用下的极限状态和灾变机理。

（2）考虑建筑本体结构及减振系统耦合的高层建筑性能演化规律减振高层建筑在强震或强台风的作用下，可能会导致建筑本体结构和减振系统的损伤和性能退化，因此，需要研究考虑建筑本体结构极限性能变化及减振系统极限性能变化的情况下，耦合高层建筑的破坏极限、评估方法和性能演化规律。在此基础上，提出基于结构灾后现状的减振控制方法，以此对灾后建筑本体和减振系统进行再生设计与修复，从而实现对减振高层建筑耦合性能的提高。

（3）高层建筑抗灾性能控制理论要实现灾后减振高层建筑性能的再生和提高，就必须防患于未然，考虑减振高层建筑的初始一体化设计。因此在设计阶段，既要考虑极端灾害对本体结构和减振系统的损伤作用，也需要考虑建筑寿命期内减振系统和材料的常时或突发的性能退化，使得减振系统具有可替换性、可修复性和可提高性，同时作为分灾系统可承担极端灾害给整体结构带来的损伤，甚至失效破坏，但不影响本体结构的主要功能。以此建立本体结构的最低设防目标，确立分灾系统的抗灾性能目标，确保整体结构具有鲁棒性。

减振高层建筑的应用历史只有30多年，许多结构都没有经历过强震、强台风的考验，更缺乏减振器的灾害资料，对其减振器的极限性能的研究开展亦不充分，再加上结构体系本身的复杂性，人们对其在极端灾害作用下的动力损伤演化过程和破坏机理等方面的认识还不够深入。尽管减振技术的应用日益广泛，但其抗灾问题并未真正解决，抗灾研究还落后于发展的需要。因此，研究减振器的极限性能，研究考虑减振器的极限性能的减振结构在极端动力作用下的连续破坏机理和极限性能，进而发展控制结构损伤的基于性能的分析与设计方法，可保障减振结构的安全，有效控制灾害造成的经济损失，为结构防灾和城市防灾提供科学依据和技术支持，是城市、社会、环境可持续发展的重要内容。

参考文献

[1] Xue Songtao. Study on a real 8F steel building with oil damper damaged during the 2011 Great East Japan Earthquake [C]. 15th World Conference on Earthquake Engineering, Lisbon, Portugal, 2012.

[2] Kasai K. Performance of seismic protection technologies during the 2011 Tohoku-Oki Earthquake[J]. Earthquake Spectra, 2013, 29(s1): S265-S293.

[3] JSSI. Report of Response-Controlled Buildings[R]. Tokyo, Japan: Japan Society of Seismic Isolation (JSSI) Investigation Committee, 2012.

[4] Kawamata S. Passive control of building frames by means of liquid dampers sealed by viscoelastic material [C]. 12th World Conference on Earthquake Engineering, Auckland, New Zealand, 2000.

[5] Funaki N. Vibration response of a three-storied full-scale test building passively controlled by liquid dampers sealed by viscoelastic material[C]. 16th International Conference on structural Mechanics, Washington DC, USA, 2001.

[6] Xue Songtao. Performance Study of an Eight-story Steel Building Equipped with Oil Dampers Damaged During the 2011 Great East Japan Earthquake Part 1: Structural Identification and Damage Reasoning [J]. Journal of Asian Architecture and Building Engineering. 2015, 14(1): 181-189.

[7] Xue Songtao. Performance Study of an 8-story Steel Building Equipped with Oil Damper Damaged During the 2011 Great East Japan Earthquake Part 2: Novel Retrofit Strategy[J]. Journal of Asian Architecture and Building Engineering. 2016, 15(2): 303-313.

[8] Miyamoto H K. Limit states and failure mechanisms of viscous dampers and the implications for large earthquakes[J]. Earthquake Engineering & Structural Dynamics, 2010, 39(11): 1279-1297.

[9] Miyamoto H K. Collapse risk of tall steel moment frame buildings with viscous dampers subjected to large earthquakes part I: Damper limit states and failure modes of 10-storey archetypes[J]. The Structural Design of Tall and Special Buildings, 2010, 19(4): 421-438.

[10] Yoshida M. Evaluation of fatigue damage to a damper induced by a typhoon[J]. Journal of Wind Engineering and Industrial Aerodynamics, 1998, 74: 955-965.

[11] Kasai K. Damping identification of a full-scale passively controlled five story steel building structure [J]. Earthquake Engineering & Structural Dynamics, 2013, 42(2): 277-295.

[12] Kasai K. Full scale shake table tests of 5-story steel building with various dampers[C]. In 7th International Conference on Urban Earthquake Engineering (7CUEE) & 5th International Conference on Earthquake Engineering (5ICEE), Toronto, 2010.

8 基于减震和隔震技术的既有 RC 框架建筑抗震韧性提升

解琳琳，钟勃健，李爱群，杨参天，王心宇
（北京建筑大学，北京）

摘　要：既有 RC 框架建筑是现代城市的重要组成部分，其能否在震后快速恢复功能对于城市防灾至关重要。为了评估并提升城市的抗震韧性，应首先针对既有框架结构建筑开展韧性评价与提升相关研究。本文以既有 RC 框架建筑为基本研究对象，基于《建筑抗震韧性评价标准》评价了该建筑的抗震韧性水准，明确了经济损失、修复时间和人员伤亡情况。在此基础上，基于消能减震技术和隔震技术对该类建筑进行了抗震韧性提升，研究了两种技术对抗震韧性水准的影响规律，本文的研究成果可为既有 RC 框架建筑的地震韧性评价和提升提供重要参考。

关键词：既有 RC 框架建筑，抗震韧性评价，消能减震，隔震

Resilience-based retrofitting of existing urban RC-frame buildings using energy dissipation and seismic isolation

Xie Linlin，Zhong Bojian，Li Aiqun，Yang Cantian，Wang Xinyu
（Beijing University of Civil Engineering and Architecture，Beijing）

Abstract：The improvement of seismic resilience of existing reinforced-concrete（RC）frame buildings，which is essential for the seismic resilience of a city，has become a critical issue. To evaluate and improve the seismic resilience of the city，the seismic resilience evaluation and improvement of the existing frame structure building should first be studied. An existing RC frame building is herein selected as the research object. Based on the Standard for Seismic Resilience Assessment of Buildings，the seismic resilience of this building was assessed and the corresponding restoration cost，repair time and casualties were also evaluated. Furthermore，the seismic resilience was improved using the energy dissipation technology and seismic isolation technology，respectively. The influence of the two technologies on the seismic resilience of this building were compared. The research outcome can provide an important reference for the assessment and improvement of the seismic resilience of existing RC frame buildings.

Keywords：existing RC frame buildings，seismic resilience assessment，energy dissipation，seismic isolation

---

基金项目：北京市自然科学基金项目（8192008）；中国地震局工程力学研究所基本科研业务费专项资助项目（2019D17）。

通讯作者：李爱群（1962-），男，湖南耒阳人，教授，博士，博导，主要从事工程防灾减灾研究（E-mail：liaiqun@bucea.edu.cn）。

1. 引言

随着城市建设的不断发展，我国各类城市中建设了大量钢筋混凝土框架结构建筑，包括医院、学校、办公楼、政府大楼等。这些既有建筑是现代城市的重要组成部分。为了评估并提升城市的抗震韧性，应首先针对城市中广泛存在的这些既有框架结构建筑开展韧性评价与提升相关研究。

建筑抗震韧性评价方法是近年来结构工程与地震工程界的研究热点，大量学者在这一领域开展了相关研究[1-13]。2012 年，美国联邦应急管理署（FEMA）颁布了下一代基于性能的抗震设计方法：《FEMA P-58 建筑抗震性能评估》[14,15]，FEMA P-58 以建筑物在遭受地震灾害后的人员伤亡、经济损失和恢复时间衡量建筑的抗震韧性，并提供了辅助软件工具 PACT。奥雅纳工程咨询有限公司（Arup）提出了基于韧性的抗震设计方法[16]，以系统失去功能的时间作为评价指标，对建筑进行抗震韧性评价。然而，上述评价方法无法直接适用于我国建筑物的抗震韧性评估，这是由于各国建筑构件的设计方法和工程做法存在差异，国外学者建立的数据库并未包含我国建筑构件的易损性数据。

针对这一问题，由清华大学负责，住房和城乡建设部标准定额研究所、中国建筑科学研究院有限公司和中国地震局工程力学研究所等单位共同参与起草、编制了我国第一部关于建筑抗震韧性的规范，即《建筑抗震韧性评价标准》GB/T 38591—2020[17]（以下简称《评价标准》）。该标准包含了我国典型建筑构件的易损性数据库和后果函数。目前，基于《评价标准》对我国既有框架建筑进行抗震韧性评价和提升的相关研究相对较少。基于此背景，对典型既有框架建筑进行抗震韧性评价，根据评价结果进行相应的抗震韧性提升，用以保障既有建筑在震后实现损伤低、恢复快等功能具有重要的现实意义。

已有研究和建筑震害经验均表明：减隔震技术是提升建筑韧性的重要手段[18-24]，该技术能有效提升建筑的抗震性能，保证主体结构低损伤甚至无损伤，并有效降低非结构构件、重要设备的损伤，为建筑“服务功能不间断”目标提供有效的技术保障。

因此，本文针对既有 RC 框架建筑案例，基于《评价标准》，首先评价了既有 RC 框架建筑的抗震韧性等级，明确了经济损失、修复时间和人员伤亡情况。在此基础上，基于消能减震技术和隔震技术设计了该建筑的加固方案，评价了加固后建筑的抗震韧性，对比研究了两种加固技术对抗震韧性水准的影响规律，本文的研究成果可为既有 RC 框架建筑的地震韧性评价和提升提供重要参考。

2. 建筑抗震韧性评价方法简介

《评价标准》集成了震害调查经验、结构性能化设计、损伤控制理论以及韧性评价方法等研究成果，制订出建筑在给定地震水准下的损伤状态、修复费用、修复时间和人员损伤的评估方法，并规定了建筑的韧性等级标准。本研究以《评价标准》为基准展开相关研究。基于评价标准的抗震韧性评价流程（图 1），其主要内容如下：

在建筑进行韧性评价前，先判断建筑是否可修复。当某一建筑的多次弹塑性时程分析结果的层残余变形的最大平均值大于层残余变形限值时，则判定该建筑不可修复，不进行韧性评级，评价标准建议混凝土公共建筑的层残余位移角限值宜取 0.5%；若小于限值，则按以下流程进行评估：

（1）集成建筑信息，包括建筑的楼层高度、楼层面积等基本信息以及结构构件和非结构构件的数量、种类和尺寸等信息；

（2）建立结构分析模型，分别在设防地震和罕遇地震下进行弹塑性时程分析，提取最大层间位移角和楼面加速度峰值等工程需求参数；

（3）确定构件的损伤状态，根据提取的工程需求参数结合构件的易损性数据库确定构件的损伤状态；

（4）建筑损伤评估，根据构件的损伤状态，评估在给定地震水准下建筑的修复费用、修复时间以及人员伤亡；

（5）判定建筑的抗震韧性等级，根据建筑的损伤情况，判定建筑的韧性等级。

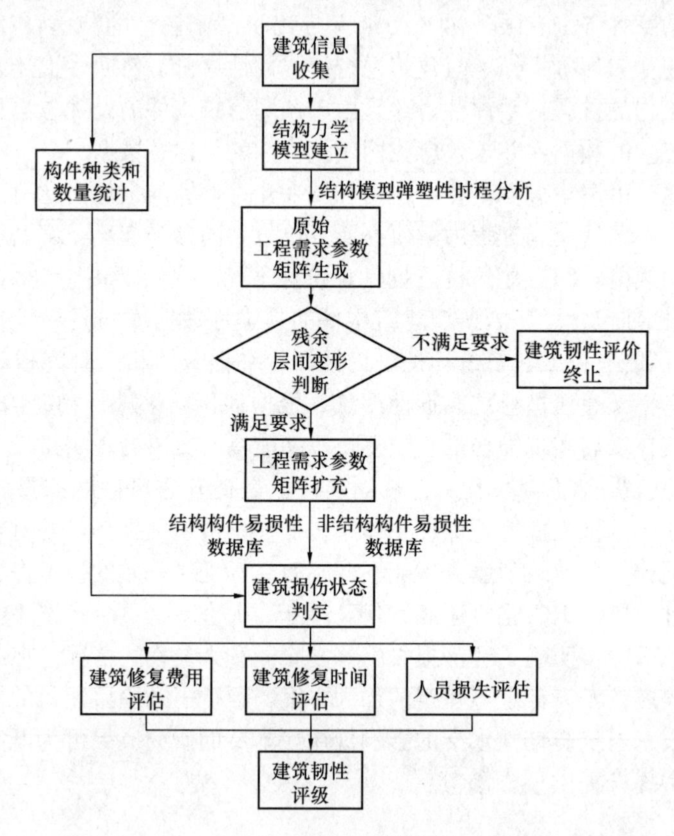

图1　抗震韧性计算流程

建筑抗震韧性评价共有三个评价指标，分别为建筑修复费用、修复时间和人员损伤。在评估给定地震水准下建筑的修复费用、修复时间以及人员损伤指标时，其值应采用蒙特卡洛方法计算得到的具有 84%保证率的拟合值。根据 84%保证率的拟合值判断三个指标的等级，每个指标对应三个等级，由一星至三星抗震韧性等级提高，并取三个指标最低等级作为该建筑的抗震韧性等级。

建筑修复费用指标的等级如表 1 所示，其评价指标为震后建筑修复费用与现行建造成本的比例。评价需要分别在设防地震和罕遇地震下进行。设防地震下，若建筑的修复费用小于10%，满足一星标准，即可进行更高等级的评价；若罕遇地震下建筑修复费用介于5%与10%之间，则判定该建筑的修复费用等级为二星，若大于10%则为一星。

建筑修复费用指标的等级 **表1**

等级	地震水准	建筑修复费用指标
三星	罕遇地震	≤5%
二星	罕遇地震	(5%，10%)
一星	设防地震	(0，10%)

建筑修复时间指标的等级如表2所示，按照一定的修复次序，取所有楼层的最长修复时间作为评价指标。设防地震下，若建筑的修复时间小于30d，满足一星标准，即可进行更高等级的评价；若罕遇地震下建筑修复时间介于7d与30d之间，则判定该建筑的修复时间等级为二星，若大于30d则为一星。

建筑修复时间指标的等级 **表2**

等级	地震水准	建筑修复时间指标
三星	罕遇地震	≤7d
二星	罕遇地震	(7d，30d)
一星	设防地震	(7d，30d)

人员损伤指标的等级如表3所示，其评价指标为房屋建筑中伤、亡人数占全部人数的比例。设防地震下，若人员受伤率小于千分之一，且人员死亡率小于万分之一，则满足一星标准，即可进行更高等级的评价；若罕遇地震下人员受伤率小于千分之一，且人员死亡率小于万分之一，则判定人员损失等级为二星，若受伤率或者死亡率任何一个不满足，则为一星。

人员损伤指标的等级 **表3**

等级	地震水准	人员损失指标
三星	罕遇地震	γ_H≤10～4，且γ_D≤10～5
二星	罕遇地震	γ_H≤10～3，且γ_D≤10～4
一星	设防地震	γ_H≤10～3，且γ_D≤10～4

3. 既有RC框架建筑抗震韧性评价

本文选取了一座既有RC框架结构办公楼作为案例建筑。该建筑总面积为3952m^2，共4层，无地下室，建筑高度14.7m，结构高宽比为0.45。抗震设防烈度为8度(0.20g)，场地类别为Ⅱ类。结构三维示意图和平面布置图分别如图2和图3所示。

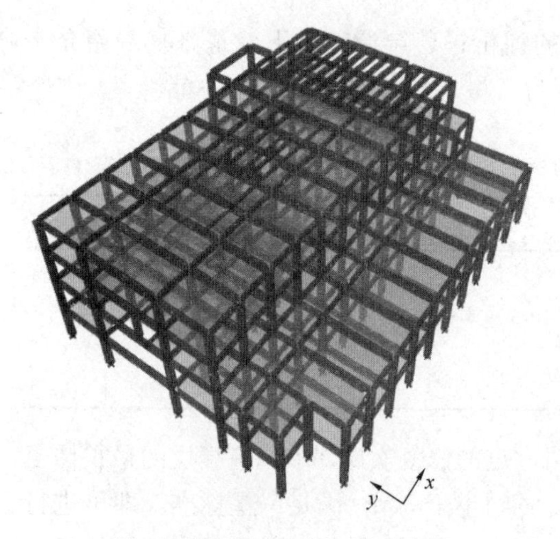

图 2　案例建筑结构三维示意图

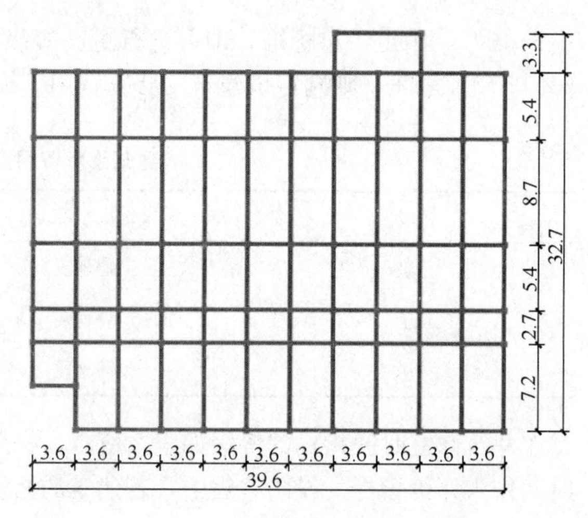

图 3　案例建筑结构平面布置图

3.1　案例建筑构件信息

本研究结合实地考察和竣工图纸，统计了建筑抗震韧性评价所需的构件信息，包括结构构件和建筑非结构构件的种类、数量、材料、几何尺寸等。案例建筑的主要结构构件信息见表 4。案例建筑的主要非结构构件信息见表 5。

案例建筑结构构件类别及数量　　**表 4**

构件类型	构件尺寸（mm×mm）	编码	起始楼层	终止楼层	X 向	Y 向	无方向
框架柱	500×500	C. F. KZ. 001	1	2	0	0	75
	500×500	C. F. KZ. 002	3	3	0	0	52
	500×500	C. F. KZ. 003	4	4	0	0	46
框架梁	250×500	C. F. KL. 001	1	2	66	0	0
	300×800	C. F. KL. 002	1	4	1	0	0
	300×800	C. F. KL. 003	1	2	0	26	0
	300×800	C. F. KL. 004	1	2	0	22	0
	300×500	C. F. KL. 005	1	2	0	19	0
	250×500	C. F. KL. 001	3	3	45	0	0
	250×500	C. F. KL. 006	3	3	11	0	0
	300×800	C. F. KL. 007	3	3	0	26	0
	300×800	C. F. KL. 008	3	3	0	9	0
	250×500	C. F. KL. 001	3	3	0	6	0
	250×500	C. F. KL. 001	4	4	38	0	0
	300×800	C. F. KL. 004	4	4	13	0	0
	250×500	C. F. KL. 006	4	4	10	0	0
	300×800	C. F. KL. 009	4	4	0	22	0
	300×800	C. F. KL. 010	4	4	0	8	0
	300×500	C. F. KL. 005	4	4	0	4	0

主要非结构构件易损性类别　　表 5

非结构构件类型	构件易损性编号	起始楼层	终止楼层	有无方向性	EDP 类型
玻璃幕墙	BE. F. 01. 01	1	4	有方向	MIDR
填充墙	BI. P. 01. 01	1	4	有方向	MIDR
隔墙饰面	BI. D. 01. 01	1	4	有方向	MIDR
吊顶	BI. C. 01. 01	1	4	无方向	MAFA
冷水管	MEP. P. 01. 01	1	4	无方向	MAFA
热水管 1	MEP. P. 02. 01	1	4	无方向	MAFA
热水管 2	MEP. P. 02. 05	1	4	无方向	MAFA
污水管	MEP. P. 03. 01	1	4	无方向	MAFA
喷头立管	MEP. M. 06. 01	1	4	无方向	MAFA
支管及风口	MEP. M. 07. 01	1	4	无方向	MAFA
VAV 箱带卷盘	MEP. M. 08. 01	1	4	无方向	MAFA
消防喷淋水管	MEP. P. 05. 01	1	4	无方向	MAFA
楼梯	BI. S. 02. 01	1	4	无方向	MIDR
电梯	MEP. L. 01. 01	1	4	无方向	MAFA
冷水机组	MEP. M. 02. 01	1	4	无方向	MAFA
冷却塔	MEP. M. 03. 01	1	4	无方向	MAFA
空气处理机组	MEP. M. 10. 01	1	4	无方向	MAFA
电机控制箱	MEP. E. 02. 01	1	4	无方向	MAFA

3. 2　结构地震响应分析

3. 2. 1　分析模型建立与地震动选取

为提取抗震韧性评价所需的工程需求参数，根据标准要求，采用商用有限元软件 Perform-3D 建立了案例结构的弹塑性分析模型。梁、柱构件采用纤维截面模型模拟，混凝土材料通过五折线形式定义材料本构，钢筋通过三折线定义。采用该模型进行模态分析得到的结构前三阶周期分别为 0. 597s、0. 508s 和 0. 471s，与 PKPM 弹性分析模型得到的周期误差不超过 2. 5%，说明该模型具有可靠性和合理性，可用于建筑地震响应分析和工程需求参数提取。

根据《评价标准》，采用时程分析法进行地震响应分析时，应选用不少于 11 条的实际强震记录（大于 2/3）和人工模拟的加速度时程曲线。本研究共选取 8 条天然地震波和 3 条人工波，天然波均选自美国太平洋地震研究中心（PEER）数据库，人工波采用 SIMQKE _ GR 软件生成。11 条地震波的归一化波形如图 4 所示，加速度反应谱与规范谱对比如图 5 所示，符合规范相关要求。

3. 2. 2　结构弹塑性时程分析

采用上述弹塑性分析模型和 11 条地震动时程记录开展案例建筑在设防地震和罕遇地震作用下的弹塑性时程分析，其中地震动采用双向输入，主方向与次方向的比例取为 1 ∶ 0. 85。限于篇幅，本文仅展示结构弱轴（*X* 向）方向数据。

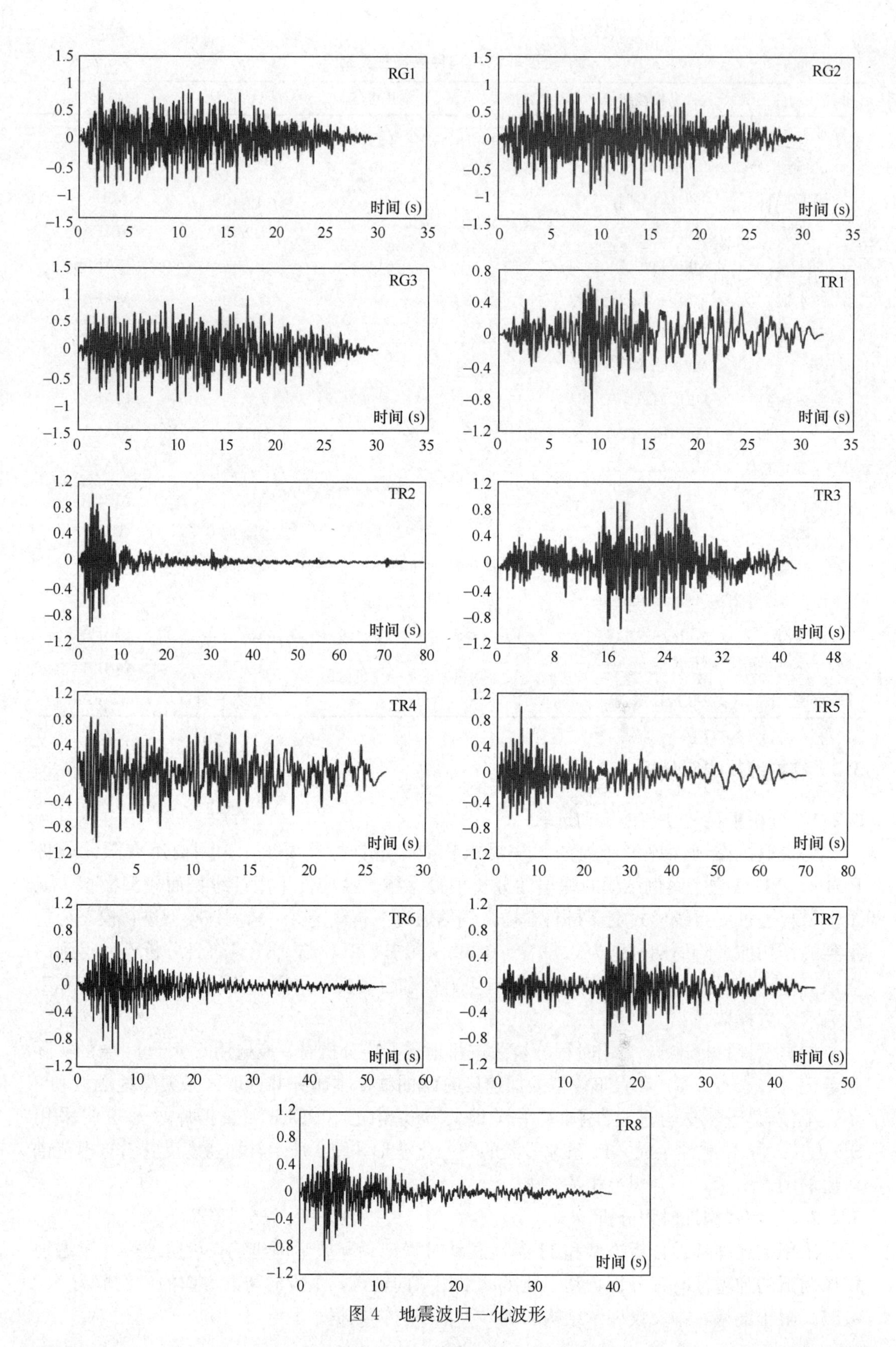
RG1
时间 (s)
RG2
时间 (s)
RG3
时间 (s)
TR1
时间 (s)
TR2
时间 (s)
TR3
时间 (s)
TR4
时间 (s)
TR5
时间 (s)
TR6
时间 (s)
TR7
时间 (s)
TR8
时间 (s)

图 4　地震波归一化波形

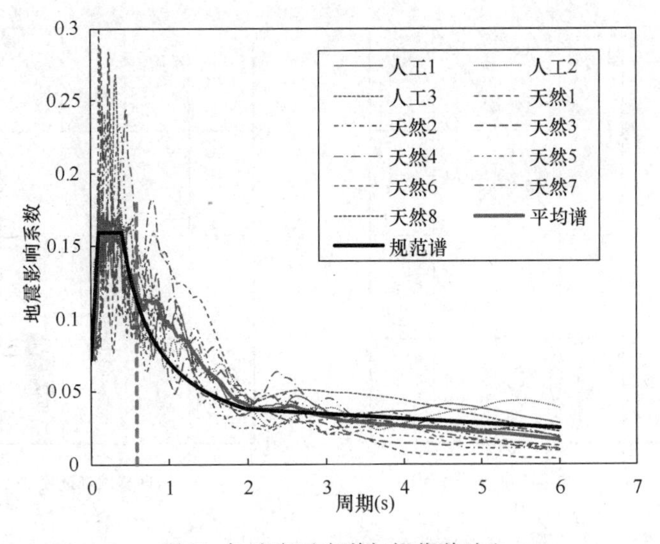

图 5 加速度反应谱与规范谱对比

案例结构在设防地震和罕遇地震作用下最大层间位移角（MIDR）如图 6 所示。结构最大层间位移角出现在 2 层，在设防地震和罕遇地震作用下，平均值分别为 0.4%和 0.82%，包络值分别为 0.54%和 1.06%。

案例结构在设防地震和罕遇地震作用下楼面加速度峰值（MAFA）如图 7 所示。结构最大楼面加速度出现在 4 层，在设防地震和罕遇地震作用下，相对地面输入加速度峰值的比值平均值分别为 1.99 和 1.50，包络值分别为 2.33 和 1.75。

同时值得注意的是，根据《评价标准》，残余层间变形决定建筑是否可以修复。经计算，案例结构的残余层间变形均小于《评价标准》建议的残余变形限值 0.5%，因此该建筑具有震后可修复性，可对其开展抗震韧性评级。

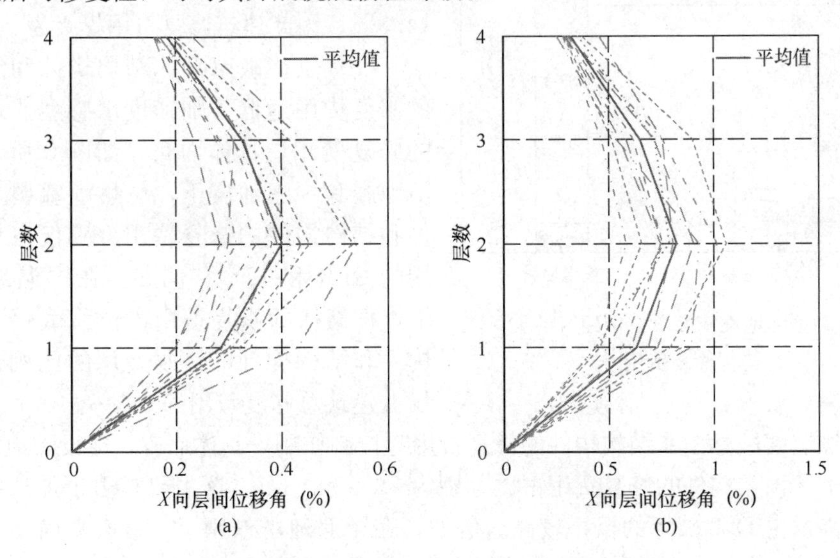

图 6 案例结构最大层间位移角（MIDR）分布

（a）设防地震作用下（X 向）；（b）罕遇地震作用下（X 向）

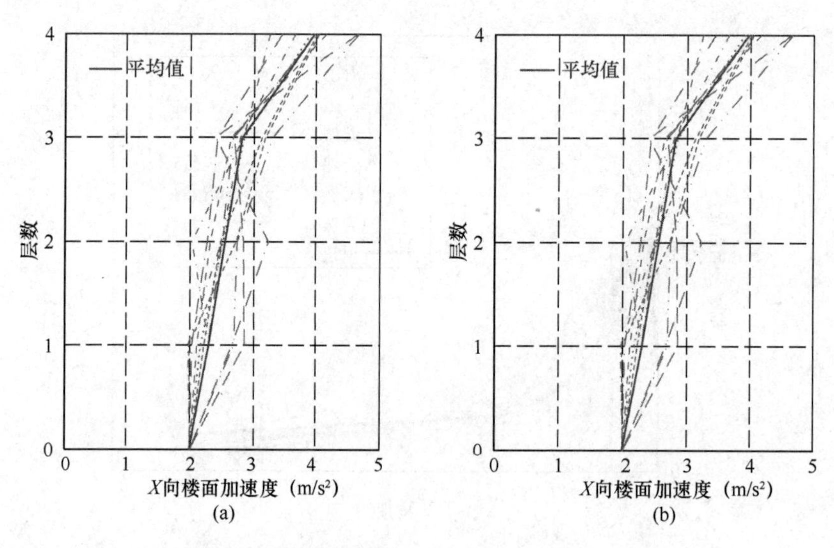

图 7　案例结构楼面加速度峰值（MAFA）分布

（a）设防地震作用下（X 向）；（b）罕遇地震作用下（X 向）

3.3　抗震韧性评价

3.3.1　修复费用评级

案例建筑在设防地震和罕遇地震作用下的修复费用如图 8 所示，各类构建修复费用占比如图 9 所示。由图可知：

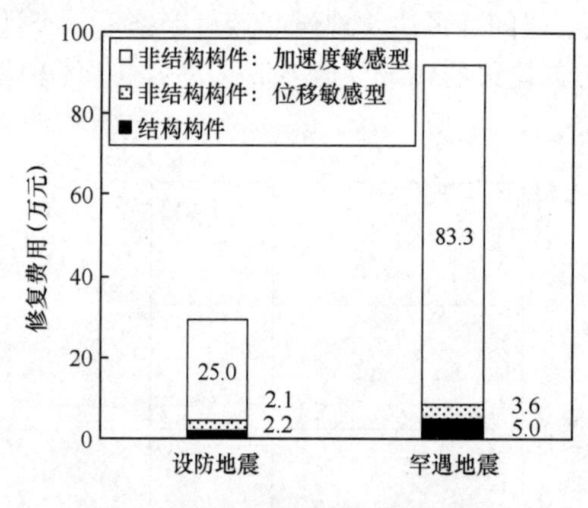

图 8　不同地震水准下案例建筑各类构件修复费用

（1）根据定额计算案例建筑的建造成本为 640.46 万元，该建筑在设防地震和罕遇地震下的修复费用指标 κ 分别为 4.6% 和 14.3%，因此建筑修复费用等级为一星。

（2）位移敏感型非结构构件和结构构件的修复费用远低于加速度敏感型非结构构件的修复费用。具体而言，如图 8 所示，在设防地震和罕遇地震下，位移敏感型非结构构件和结构构件的修复费用分别仅占总修复费用的 21% 和 17%。例如，在罕遇地震作用下，框架柱的修复费用占位移敏感型非结构构件和结构构件的修复费用的比例最高，但仅占建筑总修复费用的 6.3%。

（3）加速度敏感型非结构构件的修复费用起控制作用。具体而言，在设防地震和罕遇地震作用下，该类构件的修复费用占比分别为 79% 和 83%。这主要是由于加速度敏感型非结构构件发生 DS3 破坏的概率较高。例如，在罕遇地震作用下，最大楼面加速度达到 0.61g，此时曳引电梯发生 DS3 破坏的概率为 80%。同时，相比于位移敏感型非结构构件和结构构件，加速度敏感型非结构构件的修复费用相对较高。例如，每台曳引电梯的修复费用高达 20 万元。

（4）为提升案例建筑的修复费用评级，应控制该建筑的最大楼面加速度，减少加速度敏感型非结构构件的损伤。

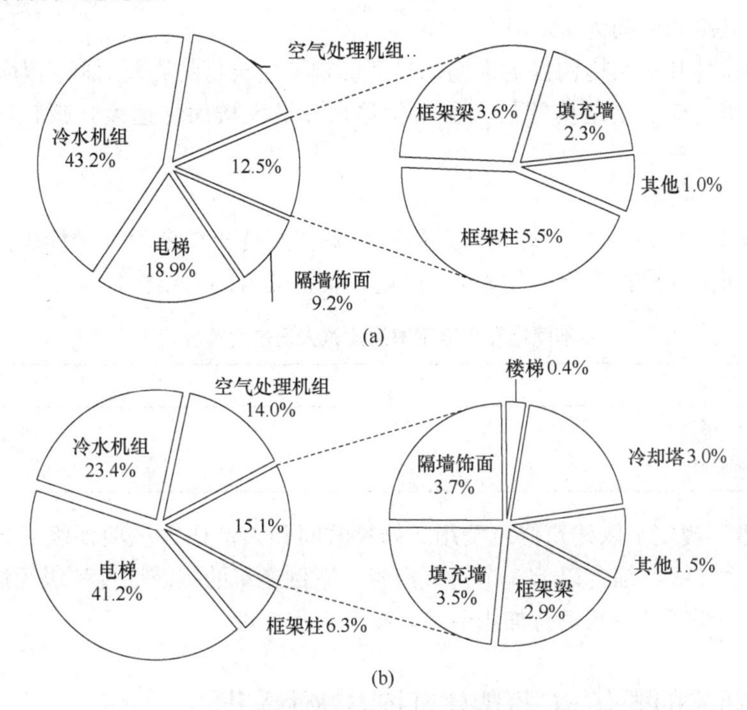

图 9　不同地震水准下案例建筑各类构件修复费用占比

（a）设防地震；（b）罕遇地震

3.3.2　修复时间评级

根据《评价标准》，建筑修复工作按开工时间先后分为两个阶段：（1）结构构件与楼梯修复和（2）非结构构件修复。第一阶段修复工作结束后方可开展第二阶段修复工作。《评价标准》要求计算修复时间应充分考虑修复工作的先后次序和并行作业，取主要修复工作的最长时间组合作为建筑修复时间评价指标。案例建筑在设防地震和罕遇地震作用下的修复时间如图 10 所示。由图可知：

（1）设防地震和罕遇地震作用下，案例结构的修复时间分别为 37.9d 和 52.6d，均超过了 30d。因此案例建筑的建筑修复时间指标的等级无法达到一星标准。

（2）设防地震和罕遇地震作用下，第一阶段的修复时间比第二阶段的修复时间长。结构构件和曳引电梯的修复分别对第一和第二阶段的修复时间起控制作用。值得注意的是，结构构件的修复时间比非结构构件的修复时间长，这主要是因为结构构件的数量远大于曳引电梯的数量。例如，在罕遇地震作用下，建筑的二层层间位移角最大，导致该楼层梁柱构件发生 DS2 破坏，所需修复时间分别为 6.2 和 5.6 人・d。同时，该

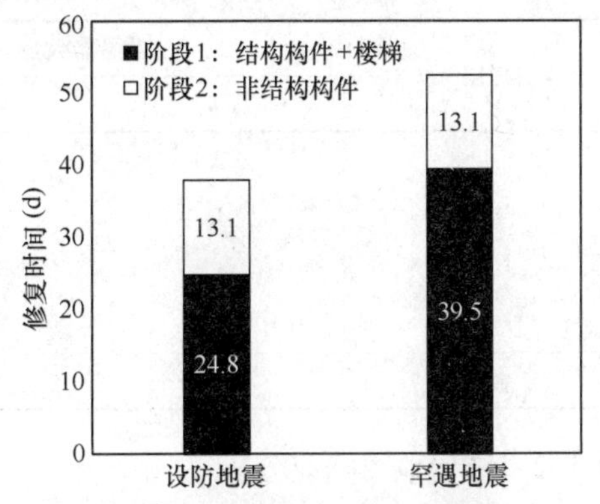

图 10　不同地震水准下案例建筑各类构件修复时间

层的框架梁和框架柱数量分别为67个和75个，且每$100m^2$需要2名工人维修非结构构件。因此，罕遇地震下二层的结构构件维修时间为39.5d。与此相对，建筑中仅设置了一部曳引电梯，其维修时间为13.1d。

（3）结构构件和非结构构件的损伤均显著影响了建筑物的修复时间。为提高案例建筑的修复时间评级，应控制结构的最大层间位移角和最大楼面加速度，减轻这两类构件的损伤。

3.3.3 人员伤亡评级

《评价标准》中，人员伤亡评级由人员受伤率和人员死亡率两个指标确定。不同地震水准下案例建筑的人员伤亡指标见表6。因此，案例建筑的人员伤亡评级为一星。

不同地震水准下案例建筑人员伤亡情况 **表6**

地震水准	人员受伤率 γ_H	人员死亡率 γ_D
设防地震	2.76×10^{-5}	0.00
罕遇地震	3.61×10^{-3}	6.29×10^{-4}

《评价标准》规定，取建筑修复费用、修复时间和人员伤亡三项评级的最低等级作为建筑的抗震韧性评级，综合以上3个小节内容，案例建筑的抗震韧性评级无法达到一星标准。因此，有必要对该建筑进行加固改造，提升抗震韧性。

4. 基于减震技术的既有RC框架建筑抗震韧性提升

4.1 减震加固方案设计及结构响应分析

4.1.1 消能减震设计方案

根据研究团队的设计经验，综合考虑建筑使用功能，为保证该消能减震方案在罕遇地震作用下仍有较高的减震率，在建筑1～4层的外围沿结构X、Y向均匀布置了34个黏滞阻尼器，阻尼器分布情况如图11所示。黏滞阻尼器的设计参数和楼层布置数量详见表7所示。

阻尼器布置方案及设计参数 **表7**

楼层	阻尼器数量		阻尼器设计参数	
	X向	Y向	阻尼系数C/ kN·$(s/m)^{\alpha}$	阻尼指数α
1	5	5	1000	0.3
2	4	4	1000	0.3
3	4	4	1000	0.3
4	4	4	1000	0.3
合计	17	17		

4.1.2 结构地震响应分析

在第2节建立的Perform-3D弹塑性分析模型基础上，布置黏滞阻尼器，建立案例建筑减震加固后的弹塑性分析模型。其中黏滞阻尼器单元由Fluid Damper（黏滞阻尼器组

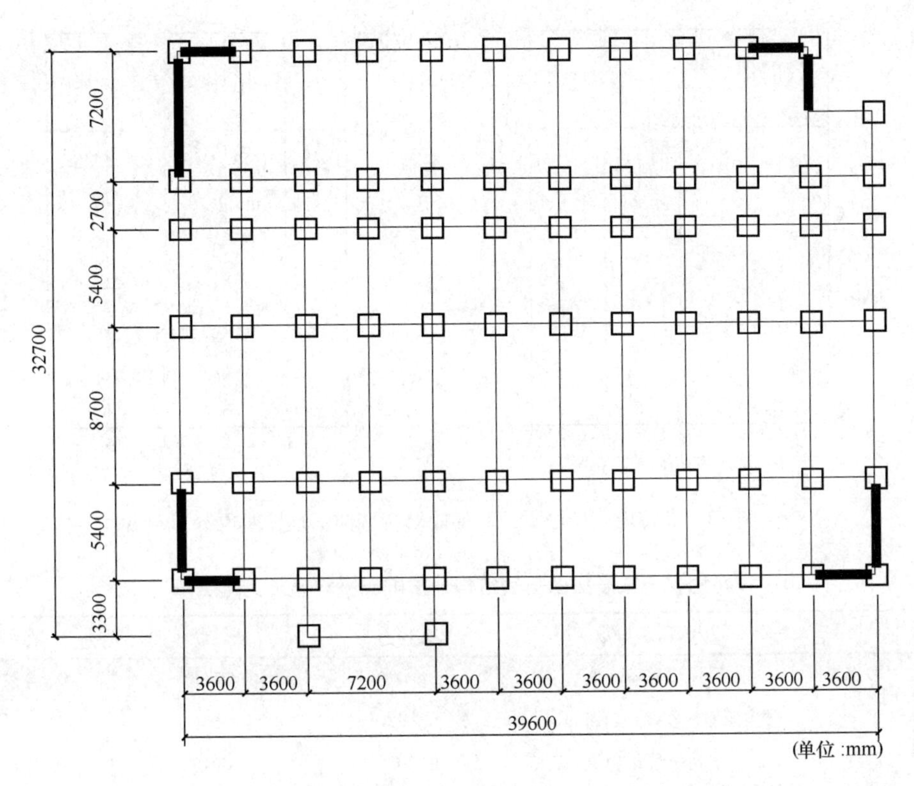

图 11　阻尼器布置图

件）和 Linear Elastic Bar（线弹性杆组件）串联组成，黏滞阻尼器组件采用多折线段来模拟轴力与轴向变形率之间的关系。采用 2.2 节所选取的 11 条地震波对减震加固后的模型进行设防地震、罕遇地震下的弹塑性时程分析，地震动采用双向输入。

设防地震作用下，减震加固后的层间位移角和楼面加速度峰值分布情况如图 12、图 13 所示；结构响应峰值平均值及其减震率如表 8 所示。

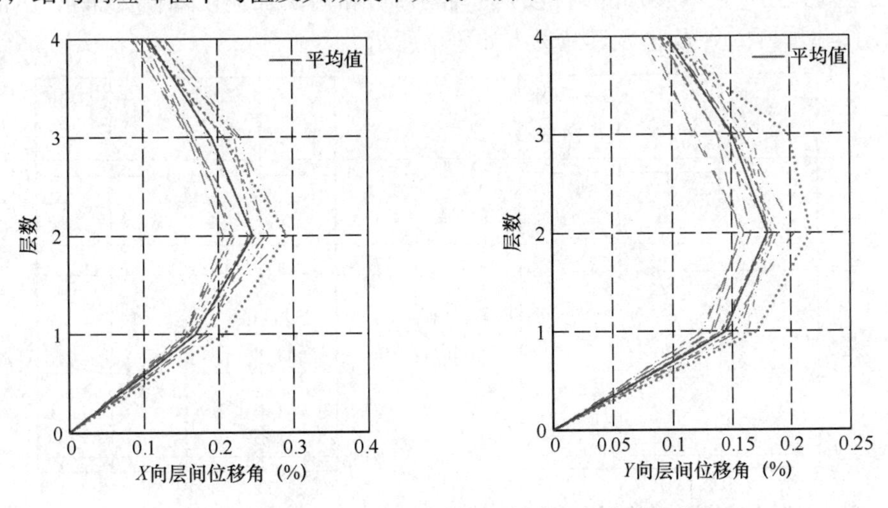

图 12　设防地震作用下减震加固结构层间位移角分布

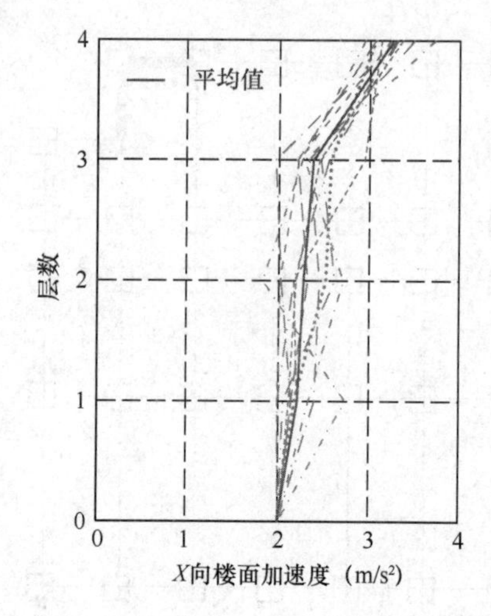

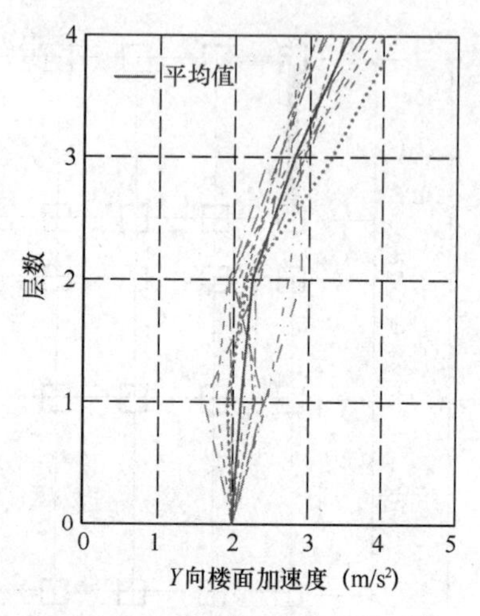

图 13　设防地震作用下减震加固结构楼面加速度峰值分布

设防地震下减震加固前后结构地震响应平均值及其减震率　　表 8

方向	工程需求参数	减震前	减震后	减震率
X 向	层间位移角峰值	0.0040	0.0025	38.4%
	楼面加速度峰值（m/s^2）	3.97	3.27	17.6%
Y 向	层间位移角峰值	0.0026	0.0018	30.7%
	楼面加速度峰值（m/s^2）	4.95	3.53	28.7%

从上述图表中可看出，设防地震下减震加固结构的层间位移角和楼面加速度明显降低。以结构 X 向为例，层间位移角峰值平均值为 0.0025，较减震前减少 38.4%；楼面加速度平均值峰值减少 17.6%，减震后值为 3.27m/s^2。可见，在结构中设置黏滞阻尼器明显提升了案例建筑在设防地震下的抗震性能。

罕遇地震作用下，减震加固结构的层间位移角和楼面加速度峰值分布情况如图 14、图 15 所示；结构响应平均值及其减震率如表 9 所示。

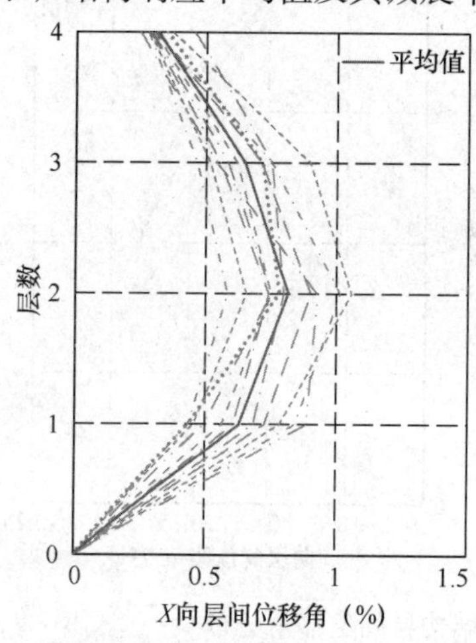

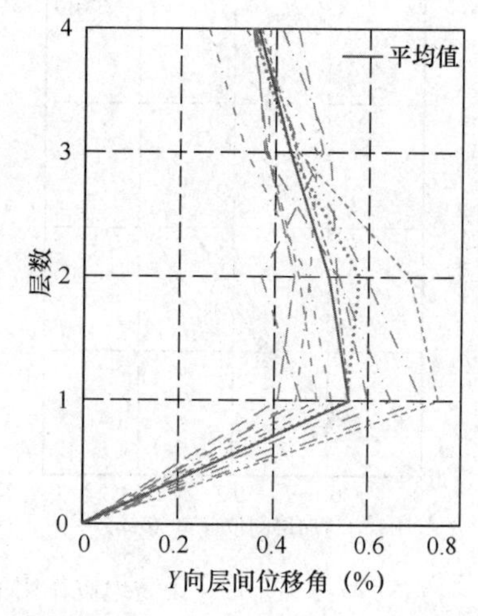

图 14　罕遇地震作用下层间位移角分布

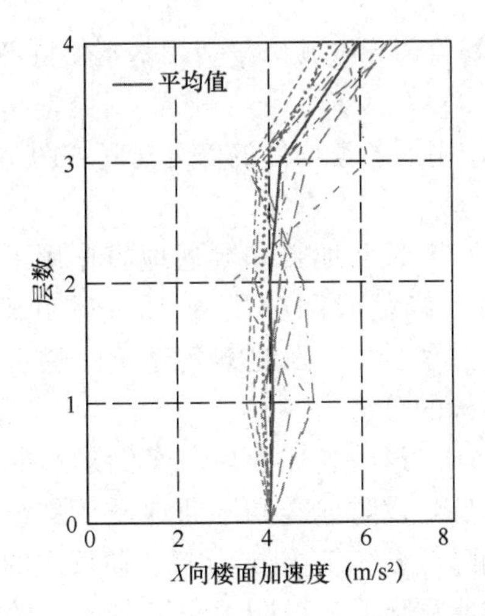

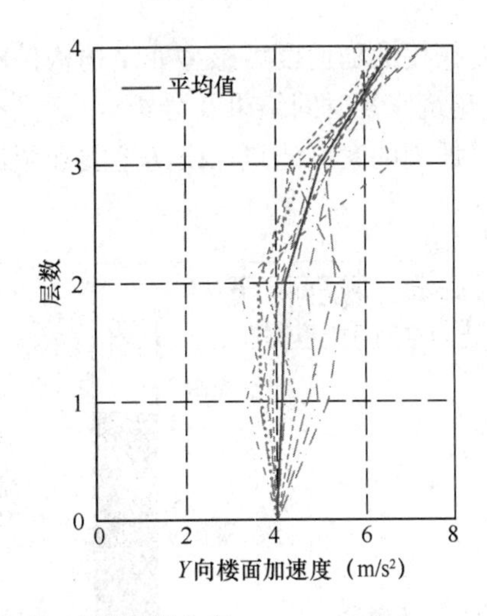

图 15　罕遇地震作用下楼面加速度峰值分布

罕遇地震下减震加固前后结构地震响应平均值及其减震率　　表 9

方向	工程需求参数	减震前	减震后	减震率
X 向	层间位移角峰值	0.0082	0.0062	24.0%
	楼面加速度峰值（m/s²）	5.99	4.83	19.4%
Y 向	层间位移角峰值	0.0056	0.0038	32.0%
	楼面加速度峰值（m/s²）	6.68	5.50	17.7%

从上述图表中可看出，与设防地震作用下结果类似，罕遇地震下减震加固结构的层间位移角和楼面加速度明显降低。以结构 X 向为例，减震结构 X 向的峰值加速度较原结构减小 24.0%；最大层间位移角减小 19.4%。这表明在结构中设置黏滞阻尼器同样明显提升了案例建筑在罕遇地震作用下的抗震性能。

4.2　减震加固案例建筑抗震韧性评价

4.2.1　建筑修复费用结果及分析

减震加固的案例建筑在设防地震和罕遇地震作用下各类构件的修复费用如图 16 所示，由图可知：

(1) 该建筑在设防地震和罕遇地震下的修复费用指标 κ 分别为 2.5% 和 13.7%，因此建筑修复费用等级为一星。

(2) 设防地震和罕遇地震下，因结构层间位移角减震率较高，加固后层间位移角低于多种结构构件和位移敏感型非结构构件的损伤阈值，降低其损伤数量、损伤程度，进而引起修复费用明显降低。

(3) 设防地震下，加固后楼面加速度值低于电梯的损伤阈值，进而引起加速度敏感型非结构构件的修复费用大幅降低。罕遇地震下，加固后楼面加速度值仍超过对修复费

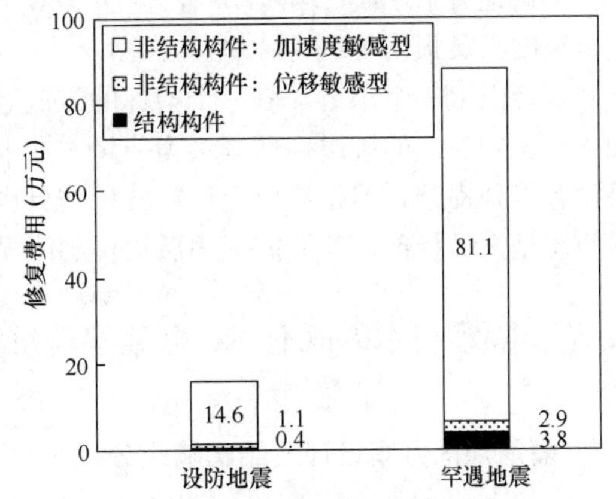

图 16　不同地震水准下减震加固案例建筑各类构件修复费用

用其控制作用的加速度敏感型非结构构件的损伤阈值，导致其修复费用基本保持不变。

4.2.2 建筑修复时间结果及分析

减震加固的案例建筑在设防地震和罕遇地震作用下各类构件的修复时间如图 17 所示。由图可知：

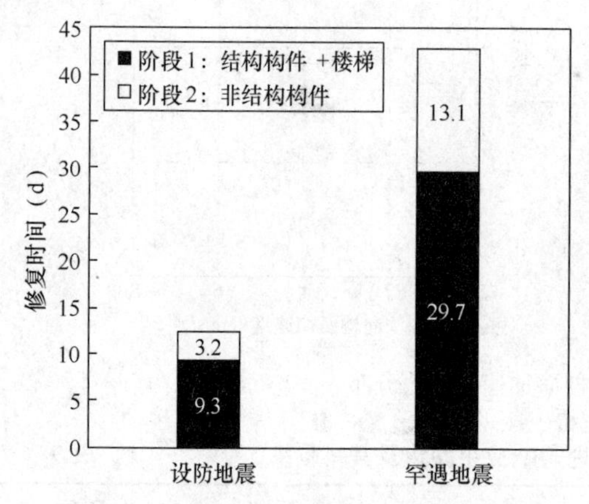

图 17 不同地震水准下减震加固案例建筑各类构件修复时间

（1）设防地震和罕遇地震作用下，案例结构的修复时间分别为 12.46d 和 56.41d，均超过了 30d。因此案例建筑的建筑修复时间指标的等级为一星。

（2）经减震加固后，案例建筑的修复时间降低，设防地震和罕遇地震下总修复时间分别降低 67.1%和 18.6%，阶段 1 的修复时间均明显降低，设防地震下阶段 2 的修复时间降低效果更为显著，罕遇地震下阶段 2 的修复时间基本保持不变。这是因为阶段 2 的修复时间由电梯修复起控制作用，在减震加固后，电梯仍有较高概率发生 DS3 破坏。

4.2.3 人员损伤结果及分析

减震加固前后，结构 X 向设防地震和罕遇地震下 84%保证率对应的人员损伤对比如表 10 所示。

减震加固前后不同地震水准下人员损伤对比 **表 10**

地震水准	人员伤亡指标	减震前	减震后
设防地震	人员受伤率 γ_H	2.76×10^{-5}	2.55×10^{-6}
	人员死亡率 γ_D	0.00	0.00
罕遇地震	人员受伤率 γ_H	3.61×10^{-3}	1.50×10^{-3}
	人员死亡率 γ_D	6.29×10^{-4}	2.57×10^{-4}

由表 10 可知，减震加固后案例建筑在设防地震和罕遇地震下的人员受伤率和人员死亡率均明显降低。减震后，设防地震下 $\gamma_H\leqslant1.0\times10^{-3}$，且 $\gamma_D\leqslant1.0\times10^{-4}$，满足一星标准；罕遇地震下 $\gamma_H>1.0\times10^{-3}$，且 $\gamma_D>1.0\times10^{-4}$，不满足二星要求，即减震加固后案例建筑的人员损伤等级为一星。

综合以上三个小节，取三个指标的最低等级作为减震加固后的抗震韧性等级，即减震加固后案例建筑的抗震韧性等级为一星。在本例消能减震方案下，案例建筑的抗震韧性等级没有得到提升，但是从以上计算分析结果也可以看出无论是建筑的修复费用、建筑修复时间还是人员损伤，经减震加固后均得到明显优化。

5. 基于隔震技术的既有 RC 框架建筑抗震韧性提升

5.1 隔震加固方案设计及结构响应分析

5.1.1 隔震设计方案

隔震支座的规格、数量和分布经反复计算确定，共计使用 75 个支座。包括 39 个铅芯隔震支座（均为 LRB500）和 36 个天然橡胶隔震支座（26 个 LNR500 和 10 个 LNR400），隔震支座平面布置如图 18 所示，其力学性能参数见表 11。隔震支座长期面压为 8.4MPa，满足丙类建筑不超过 15MPa 的要求，隔震结构屈重比为 4.18%。

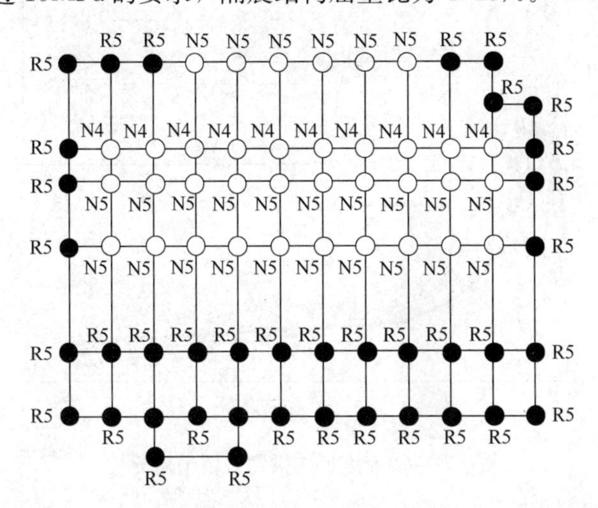

图 18 案例建筑隔震层布置图

隔震支座参数表 **表 11**

型号	符号	100%等效水平刚度（kN/m）	屈服后刚度（kN/m）	屈服力（kN）	橡胶层总厚度（mm）
LRB500	R5	1150	600	73	98
LNR500	N5	620	—	—	98
LNR400	N4	490	—	—	78

5.1.2 结构地震响应分析

时程分析采用 ETABS 软件，其中上部结构为弹性模型，隔震层为弹塑性模型。隔震支座采用 Rubber Isolator 单元与 Gap 单元并联组合模拟。隔震加固前后的前三阶周期对比如表 12 所示，隔震后结构的周期明显增大，一阶周期从 0.594s 延长至 2.064s。

隔震加固前后周期对比 **表 12**

周期（s）	T_1	T_2	T_3
非隔震结构	0.594	0.497	0.451
隔震结构	2.064	2.051	1.777

3.2.1 节所选取的 11 条地震波，其加速度反应谱与规范谱对比如图 19 所示。在 0.594s（非隔震结构周期）和 2.064s（隔震结构周期）的加速度反应谱值与规范反应谱值最大误差均不超过 35%，平均误差均不超过 20%，符合规范要求，在此仍采用上述地震波对隔震结构进行时程分析。

隔震与非隔震结构楼层剪力平均值比值见表 13，可以看出减震系数的平均值为 0.30（小于 0.40），满足规范要求。罕遇地震下隔震层最大位移为 146.2mm，小于隔震支座有

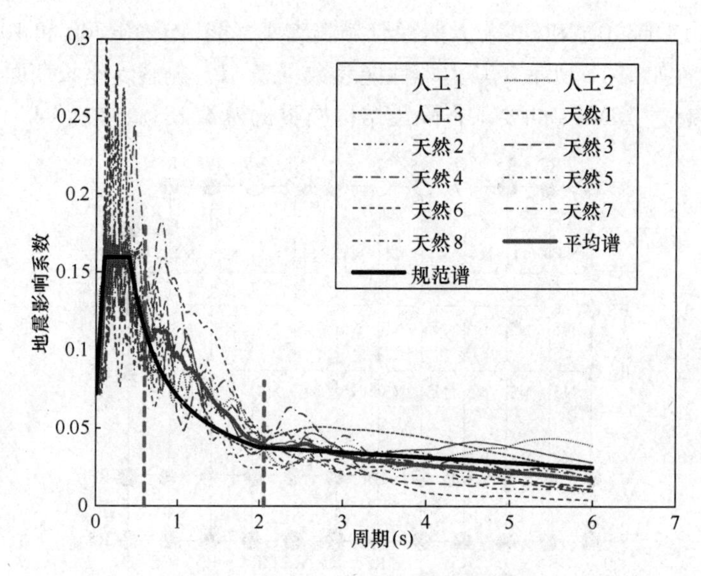

图19　加速度反应谱与规范谱对比

效直径的0.55倍和支座厚度的3倍的较小值（220mm），所选支座位移能力满足罕遇地震作用下的水平变形要求。隔震支座的极大面压为12.14MPa，远未超过30MPa限值，极小面压为0.14MPa，支座不产生拉应力。

隔震结构与非隔震结构楼层剪力比值　　　表13

楼层	*X*向平均值	*Y*向平均值
4	0.14	0.10
3	0.18	0.13
2	0.24	0.18
1	0.30	0.23

设防地震作用下，隔震加固后的层间位移角和楼面加速度峰值分布情况如图20、图21所示；结构响应峰值平均值及其减震率如表14所示，隔震结构0层楼面加速度为隔震层楼面加速度。

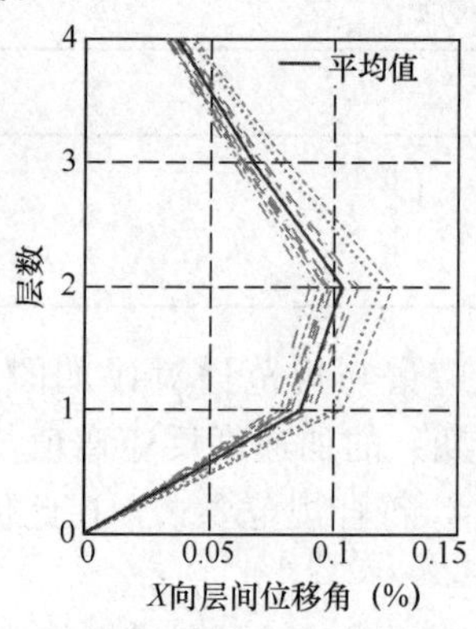

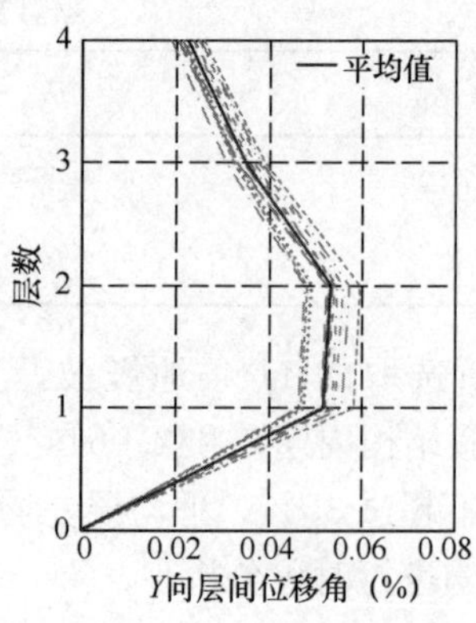

图20　设防地震作用下层间位移角分布

设防地震下隔震加固前后结构地震响应平均值及其减震率　　表 14

方向	工程需求参数	减震前	减震后	减震率
X 向	层间位移角峰值	0.0040	0.0010	75.0%
	楼面加速度峰值（m/s^2）	3.97	1.55	61.0%
Y 向	层间位移角峰值	0.0026	0.0005	80.8%
	楼面加速度峰值（m/s^2）	4.95	1.28	74.1%

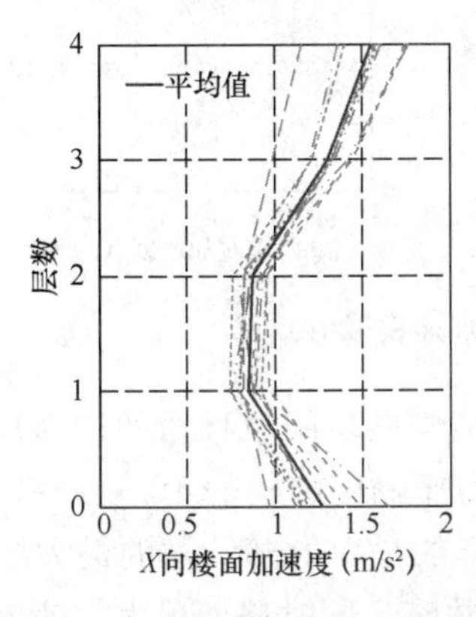

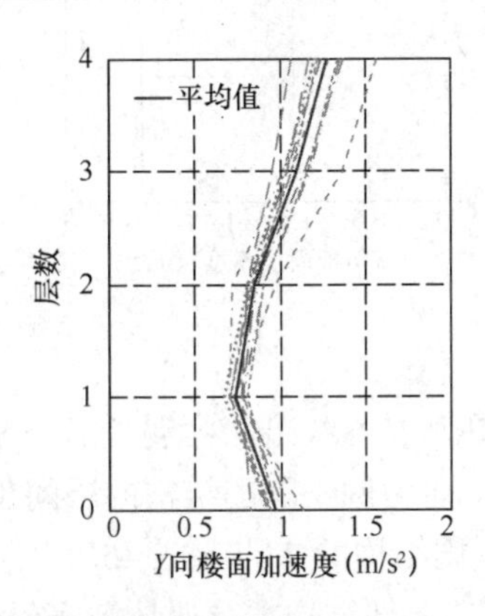

图 21　设防地震作用下楼面加速度峰值分布

从上述图表中可看出，设防地震下隔震加固结构的层间位移角和楼面加速度显著降低。以结构 X 向为例，层间位移角峰值仅为非隔震结构的 25.0%，楼面加速度峰值为非隔震结构的 39.0%。可见，隔震加固可以显著提升案例建筑在设防地震下的抗震性能。罕遇地震作用下，隔震加固后的层间位移角和楼面加速度峰值分布情况如图 22、图 23 所示；结构响应峰值平均值及其减震率如表 15 所示。

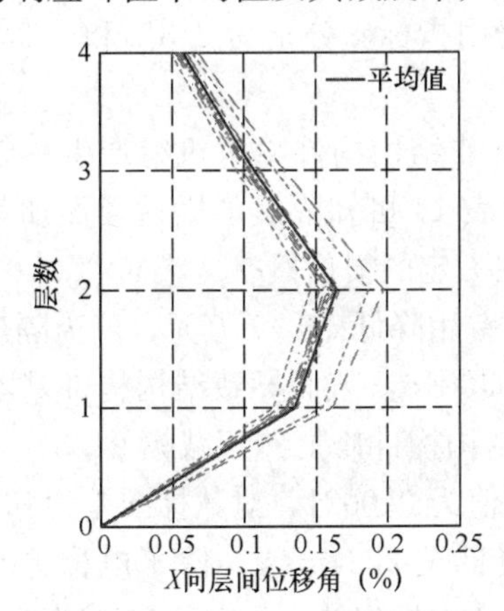

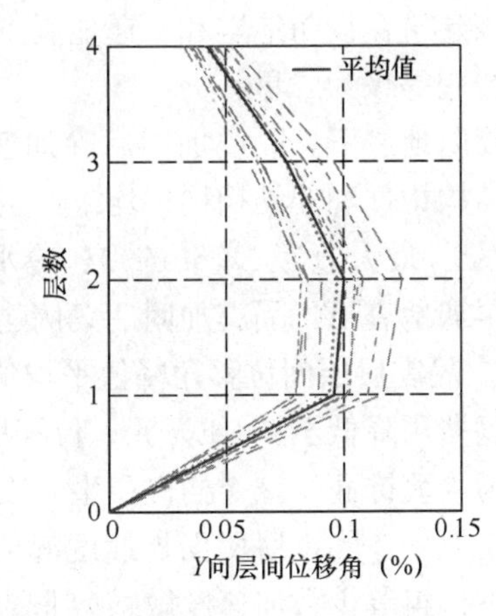

图 22　罕遇地震作用下层间位移角分布

罕遇地震下隔震加固前后结构地震响应平均值及其减震率　　表 15

方向	工程需求参数	减震前	减震后	减震率
X 向	层间位移角峰值	0.0082	0.0016	80.5%
	楼面加速度峰值（m/s^2）	5.99	2.34	60.9%
Y 向	层间位移角峰值	0.0056	0.0010	82.1%
	楼面加速度峰值（m/s^2）	6.68	1.91	71.4%

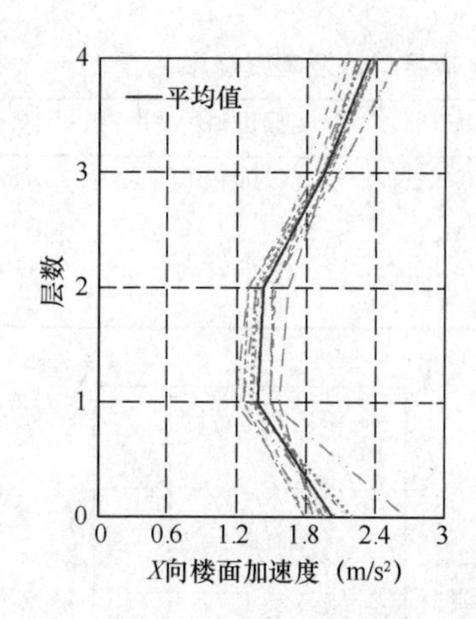

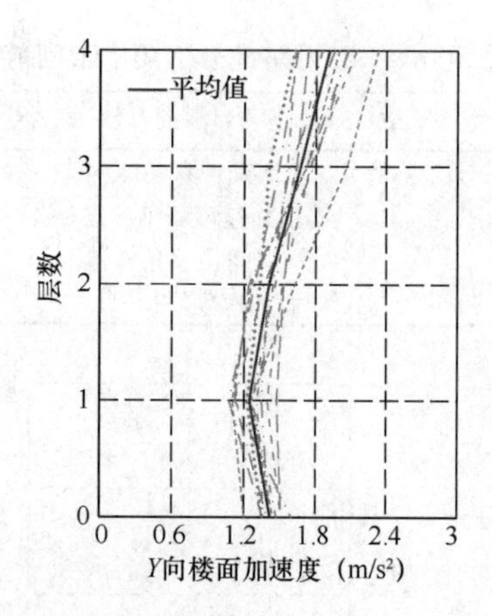

图 23　罕遇地震作用下楼面加速度峰值分布

从上述图表中可看出，罕遇地震下隔震加固结构的层间位移角和楼面加速度显著降低。以结构 X 向为例，隔震结构的层间位移角仅为非隔震结构的 16.7%～21.5%，楼层加速度为非隔震结构的 33.9%～46.1%。值得注意的是，上部结构的最大层间位移角平均值为 1/625，上部结构在罕遇地震下仍处于弹性状态，同时楼面最大加速度平均值也降至 2.34m/s^2，极大提升了该既有建筑的抗震性能。

5.2　隔震结构抗震韧性评估结果及分析

5.2.1　建筑修复费用结果及分析

隔震加固的案例建筑在设防地震和罕遇地震作用下各类构件的修复费用如图 16 所示，由图可知：

（1）该建筑在设防地震和罕遇地震下的修复费用指标 κ 分别为 2.3%和 7.2%，因此建筑修复费用等级为二星。

（2）设防地震下，隔震加固后除加速度敏感型非结构构件冷水机组产生一定的损伤外，其余结构构件和非结构构件均未产生损伤。由表 14 可知隔震加固后楼面加速度峰值为 1.55m/s^2（最大值为 1.76m/s^2），冷水机组存在一定的损伤概率。

（3）罕遇地震下，隔震加固后案例建筑总修复费用降低 45.47 万元，仅为隔震加固前的 50.5%。隔震后层间位移角峰值平均值降低至 0.0016，从而导致结构构件基本未产生损伤，修复费用降低至 0.04 万元；位移敏感型非结构构件修复费用减幅 94.8%，产生损伤的构件为隔墙饰面，该类构件产生 3 级损伤的阈值为 0.0021，对数标准差为 0.6，存在一定的损伤概率。隔震后楼面加速度峰值平均值降低至 2.34m/s^2，除了电梯、冷水机组和空气处理机组等少数加速度敏感型非结构构件仍产生损伤外，绝大部分加速度敏感型非结构构件损伤等级和数量大幅降低，修复费用减幅 44.5%。

不同地震水准下隔震加固案例建筑各类构件修复费用　　表 16

构件类型	设防地震	罕遇地震
结构构件	0.005	0.040
非结构构件：位移敏感型	0.127	0.185
非结构构件：加速度敏感型	14.584	46.136
合计	14.716	46.361

5.2.2 建筑修复时间结果及分析

隔震加固的案例建筑在设防地震和罕遇地震作用下各类构件的修复时间如图 24 所示。由图可知：

(1) 设防地震和罕遇地震作用下，案例结构的修复时间分别为 3.3d 和 15.1d，因此隔震加固的案例建筑的建筑修复时间指标的等级为二星。

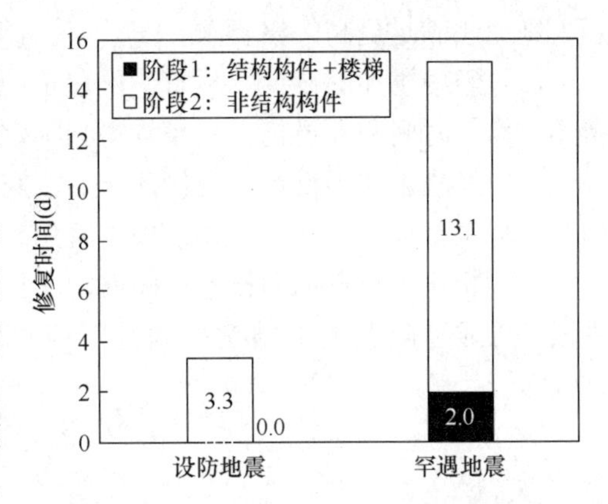

图 24 不同地震水准下隔震加固案例建筑各类构件修复时间

(2) 隔震加固的案例建筑的修复时间由非结构构件的修复控制。这一规律与加固前案例结构的规律相反，这是由于隔震加固后，结构的最大层间位移角 MIDR 得到了有效控制，结构构件基本未产生损伤。罕遇地震下，隔震后楼面加速度虽然也得到了一定程度的控制，但电梯、冷水机组和空气处理机组等少数加速度敏感型非结构构件仍发生了损伤，导致了较长的修复时间。

5.2.3 人员损伤结果及分析

隔震加固前后，案例建筑在设防地震和罕遇地震下 84%保证率对应的人员损伤对比如表 17 所示。

隔震加固前后不同地震水准下人员损伤对比　　表 17

地震水准	人员伤亡指标	隔震前	隔震后
设防地震	人员受伤率 γ_H	2.76×10^{-5}	0.00
	人员死亡率 γ_D	0.00	0.00
罕遇地震	人员受伤率 γ_H	3.61×10^{-3}	0.00
	人员死亡率 γ_D	6.29×10^{-4}	0.00

由表 17 可知，隔震加固后案例建筑在设防地震和罕遇地震下的人员受伤率和人员死亡率均显著降低。隔震后，设防地震下 $\gamma_H\leqslant1.0\times10^{-3}$，且 $\gamma_D\leqslant1.0\times10^{-4}$，满足一星标准；罕遇地震下 $\gamma_H\leqslant1.0\times10^{-4}$，且 $\gamma_D\leqslant1.0\times10^{-5}$，满足三星要求，即隔震加固后案例建筑的人员损伤等级为三星。

综合以上三个小节，取三个指标的最低等级作为隔震加固后的抗震韧性等级，即隔震加固后案例建筑的抗震韧性等级为二星。在本例隔震加固方案下，案例建筑的抗震韧性等级得到明显的提升，建筑的修复时间和人员损伤等级均从一星提升至三星，建筑修复费用等级从一星提升至二星。因此，经隔震加固后，该建筑在震后可更快恢复正常使用功能。

6. 结论

本文以既有 RC 框架建筑为基本研究对象，基于《建筑抗震韧性评价标准》评价了该类建筑的抗震韧性水准，明确了经济损失、修复时间和人员伤亡情况。在此基础上，基于消能减震技术和隔震技术对该类建筑进行了抗震韧性提升，研究了两种技术对抗震韧性水准的影响规律，得出以下主要结论：

（1）既有 RC 框架案例建筑的抗震韧性等级能满足设防地震下的 1 星标准，但罕遇地震下则无法满足更高等级要求。

（2）经减震加固设计，尽管案例建筑的抗震韧性水准并未得到提升，但加固后的修复费用、修复时间和人员伤亡概率均得到一定程度的降低。

（3）经隔震加固设计，案例建筑的抗震韧性等级得到明显的提升，其中建筑的人员损伤等级从 1 星提升至 3 星，建筑修复费用和修复时间等级均从 1 星提升至 2 星。

（4）相比于减震加固技术，隔震加固技术是实现既有框架建筑抗震韧性提升的更有效手段，可显著降低上部结构的地震响应，减轻各类构件的损伤，提升既有建筑的抗震韧性。

参考文献

[1] M. Bruneau , S. E. Chang, R. T. Eguchi, et al. A Framework to Quantitatively Assess and Enhance the Seismic Resilience of Communities[J]. Earthquake Spectra, 2003, 19(4): 733-752.

[2] G. P. Cimellaro, A. M. Reinhorn, M. Bruneau. Framework for analytical quantification of disaster resilience[J]. Engineering Structures, 2010, 32(11): 3639-3649.

[3] S. B. Manyena. The Concept of Resilience Revisited[J]. Disasters, 2006, 30(4): 434-450.

[4] S. E. Chang, M. Shinozuka. Measuring Improvements in the Disaster Resilience of Communities[J]. Earthquake Spectra, 2004, 20(3): 739-755.

[5] G. P. Cimellaro, A. M. Reinhorn, M. Bruneau. Seismic resilience of a hospital system[J]. Structure and Infrastructure Engineering, 2010, 6(1-2): 127-144.

[6] S. Marasco, A. Zamani Noori, G. P. Cimellaro. Cascading Hazard Analysis of a Hospital Building[J]. Journal of Structural Engineering, 2017, 143(9): 1-15.

[7] Y. Dong, D. M. Frangopol. Performance-based seismic assessment of conventional and base-isolated steel buildings including environmental impact and resilience[J]. Earthquake Engineering & Structural Dynamics, 2016, 45(5): 739-756.

[8] 宁晓晴，戴君武．地震可恢复性与非结构系统性态抗震研究略述[J]．地震工程与工程振动，2017，37(03)：85-92.

[9] 谢礼立．城市防震减灾能力的定义及评估方法[J]．地震工程与工程振动，2006(3)：1-10.

[10] C. Renschler, A. Frazier, L. Arendt, et al. A Framework for Defining and Measuring Resilience at the Community Scale: The PEOPLES Resilience Framework[M]. Buffalo: MCEER, 2010.

[11] C. Xiong, X. Lu, X. Lin, et al. Parameter Determination and Damage Assessment for THA-Based Regional Seismic Damage Prediction of Multi-Story Buildings[J]. Journal of Earthquake Engineering, 2016, (3): 461-485.

[12] X. Lu, B. Han, M. Hori, et al. A coarse-grained parallel approach for seismic damage simulations of urban areas based on refined models and GPU/CPU cooperative computing[J]. Advances in Engineering Software, 2014, 70: 90-103.

[13] 方东平，李在上，李楠，等．城市韧性——基于“三度空间下系统的系统”的思考[J]．土木工程学报，2017，50(7)：1-7.

[14] FEMA P-58-1. Seismic performance assessment of buildings: Volume 1-Methodology. Washington,

DC：Federal Emergency Management Agency，2012.

[15] FEMA P-58-2. Seismic performance assessment of buildings：Volume 2-Implementation guide. Washington，DC：Federal Emergency Management Agency，2012.

[16] I. Almufti，M. Willford. REDi rating system[R/OL]. London，UK：ARUP(2013-10) [2018-04-14]. https：//www. Arup. com/publications/research/section/redi-rating-system.

[17] GB/T 18591-2020. 建筑抗震韧性评价标准[S]. 北京：中国建筑工业出版社，2020.

[18] Satish Nagarajaiah，Xiaohong Sun. Response of base-isolated USC hospital building in Northridge earthquake[J]. Journal of Structural Engineering，2000，126(10)：1177-1186.

[19] 濑川丰，项琳斐. 石卷红十字医院与 3·11 震灾[J]. 世界建筑，2012(4)：110-111.

[20] 周云，吴从晓，张崇凌，等. 芦山县人民医院门诊综合楼隔震结构分析与设计[J]. 建筑结构，2013(24)：23-27.

[21] 李爱群，解琳琳，曾德民，等. 高烈度区高层隔震结构研究新进展与应用[J]. 建筑，2018，000(021)：73-75.

[22] 杨参天，解琳琳，李爱群，等. 适用于高层隔震结构的地震动强度指标研究[J]. 工程力学，2018，v. 35(08)：31-39.

[23] Yang Cantian，Xie Linlin，Li Aiqun，et al. Ground motion intensity measures for seismically isolated RC tall buildings[J]. Soil Dynamics and Earthquake Engineering，2019，125，105727.

[24] Li Aiqun，Yang Cantian，Xie Linlin，et al. Research on the rational yield ratio of isolation system and its application to the design of seismically Isolated RC frame-core tube tall buildings[J]. Applied Sciences，2017，7(11).

9 建筑摩擦摆隔震支座及其工程应用

郁银泉，邓　烜，雷远德
（中国建筑标准设计研究院有限公司，北京）

摘　要：本文对摩擦摆支座的产品力学性能特征、摩擦摆支座隔震建筑的动力反应特征、试验研究、支座连接构造等多个方面进行了研究和分析。对比了摩擦摆支座的动摩擦系数参数化模型和试验结果，检验了支座竖向压缩变形与蠕变，分析了采用摩擦摆支座隔震建筑在减震效果、扭转控制、隔震层本构等方面的动力反应特征，通过振动台试验和有限元分析验证了结构有限元模拟的精确性，研究了支座连接螺栓受拉的可能性。结果表明：建筑摩擦摆支座的力学性能能够进行很好的参数化表达；摩擦摆隔震结构能够取得和叠层橡胶隔震支座相近的隔震效果，且不需要经过专门的隔震层刚心调整就能有效抑制隔震层的地震扭转反应；摩擦摆支座隔震结构能够通过简单的有限元模拟得到足够的结构计算精度；摩擦摆支座的连接螺栓不会承受竖向拉力的作用，相关预埋件长度可以大大缩短，降低施工难度。
关键词：摩擦摆支座；建筑隔震；试验研究；有限元分析；连接构造

Friction pendulum isolation bearing for buildings and its engineering application

Yu Yinquan，Deng Xuan，Lei Yuande
（China Institute of Building Standard Design and Research Co. Limited，Beijing）

Abstract：The mechanical properties of the product，the dynamic response characteristics of the vibration isolation building，the experimental research and the connection of the friction pendulum bearings are studied and analyzed. The kinetic friction coefficient of friction pendulum bearings parametric model and test results are compared. the bearing vertical compression deformation and creep of friction pendulum bearing is tested. Isolation building is analyzed in damping effect，torsional control and isolation layer constitutive dynamic response characteristics. It is verified by shaking table test and finite element analysis of structural precision of the finite element simulation. It is studied the support the possibility of connecting bolts in tension. The results show that the mechanical properties of building friction pendulum can be well parameterized. The friction pendulum structure can achieve the isolation effect similar to that of the laminated rubber bearing，and the seismic torsion response of the isolation layer can be effectively suppressed without special adjustment of the stiffness of the isolation layer. The vibration isolation structure with friction pendulum bearings can obtain sufficient precision by simple finite element simulation. The connecting bolts of the friction pendulum bearing will not bear the vertical tension，and the length of the embedded parts can be greatly shortened，thus reducing the construction difficulty.
Keywords：friction pendulum bearing；building isolation；experiment study；finite element analysis；connection

1. 引言

隔震技术作为目前世界上最有效的建筑防震技术之一在多个国家得到了广泛应用。隔震技术的原理为在建筑基础、底部或下部结构与上部结构之间设置由隔震支座和阻尼装置等部件组成具有整体复位功能的隔震层，以延长整个结构体系的自振周期，减少输入上部结构的水平地震作用，达到预期防震要求[1]。一般来说，采用隔震技术之后上部结构的自振周期延长 2～3 倍以上来避开水平地震作用的卓越周期，从而取得较好的隔震效果。隔震支座作为隔震建筑中最重要的力学功能构件，应当具有较大的竖向承载力和竖向刚度、较小的水平刚度和较大的水平变形能力。建筑隔震技术发展与橡胶支座的发明、改进与应用具有密切的关系，我国现有建筑隔震设计相关技术标准中也主要对橡胶隔震支座进行了相关支座参数的规定[1,2]。

摩擦摆隔震支座作为一种支座主体为钢材的刚性滑动隔震支座，自 20 世纪 80 年代美国加州大学 Zayas 等人提出以来在工程中开始了广泛应用[3]，主要包括建筑、桥梁、天然气储罐等，相关产品技术也与橡胶支座一起被列入常用的隔震支座种类中[4-7]。在我国桥梁工程领域，摩擦摆支座作为一种重要的隔震支座类型在大量重要工程得以应用[8,9]，并形成了相关标准[10,11]。该支座在自复位能力、隔震周期、抗扭能力、耐久性等方面具有显著优势，近年来在国内建筑领域内得到了广泛关注，并编制了相应的产品标准[12]。建筑摩擦摆隔震支座与桥梁摩擦摆隔震支座最显著的差异在于使用年限和使用条件，建筑摩擦摆隔震支座一般要求支座具有不低于结构相同的使用年限（一般为 50 年），支座正常使用时仅有温度变形导致的微量位移，而桥梁支座由于磨耗等要求一般使用年限为 30 年；桥梁支座在使用过程中由于车辆、风等作用下长期处于运动状态，支座允许采用润滑剂且方便进行维护和更换，建筑摩擦摆隔震支座常年仅在温度作用下有少量变形且不便于维护，因此不允许采用油脂类进行润滑，且对磨耗的要求较低。本文通过摩擦摆支座产品力学性能、摩擦摆支座隔震建筑动力反应特征、试验研究、支座连接构造等多个方面简要介绍摩擦摆支座隔震建筑的特点。

2. 建筑摩擦摆隔震支座构造及力学性能

2.1 摩擦摆隔震支座构造

在国家标准《建筑摩擦摆隔震支座》中规定了两种最为常见的摩擦摆隔震支座构造Ⅰ型和Ⅱ型，即单摆型和双摆型，具体构造见图 1。其中双摆型支座在一些特殊结构应用过程中会出现一定转角需求，例如大跨度连廊、沉降差较大的大跨结构、大跨度钢结构屋盖等需要支座具有水平变形能力的同时还应具备一定的转动能力，因此需要在中间球冠板中进行一定的改进使之能够适应。在文献［13］研究证明，变摩擦系数摩擦摆隔震支座能够取得最优的隔震效果，保证建筑在中震水平下具有很低的减震系数、保证建筑能够安全运行，大震水准下通过更高的摩擦系数消耗大量的地震能量、提高减震效果，实现方式为多级摆。

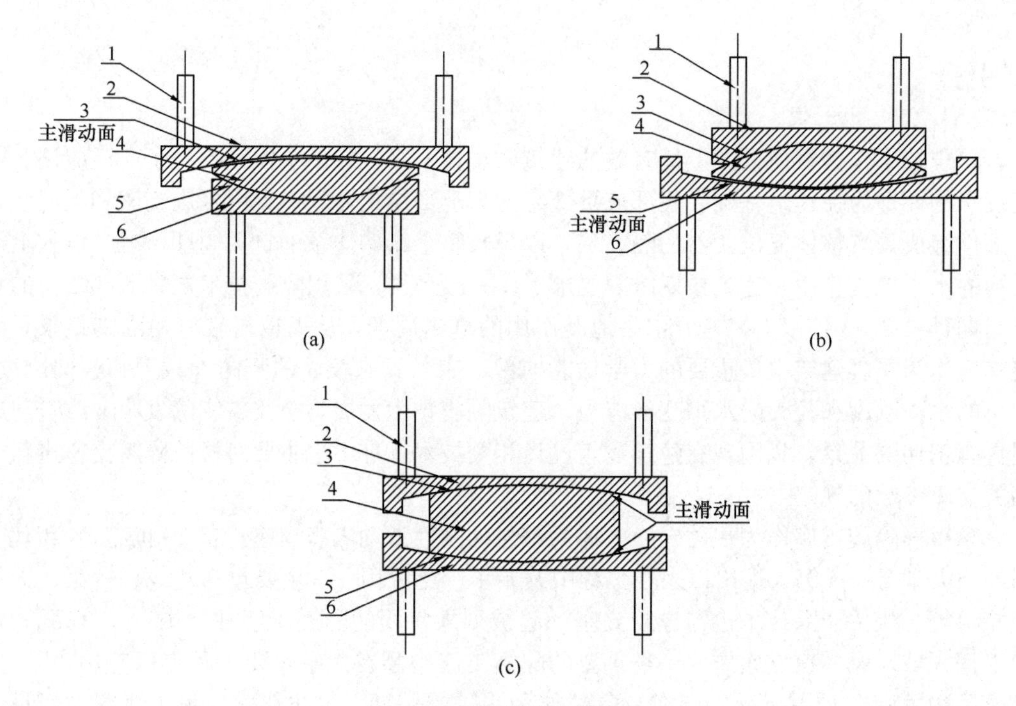

图 1 常见摩擦摆支座构造

1—上下锚固装置；2—上座板；3—上滑动摩擦面；4—球冠体；5—下滑动摩擦面；6—下座板

(a) Ⅰa 型；(b) Ⅰb 型；(c) Ⅱ型

2.2 摩擦摆隔震支座力学性能指标

取一个常规摩擦摆隔震支座，示意图如图 2 所示。其中，R 为支座等效曲率半径；R_1 为支座下滑动摩擦面曲率半径；R_2 为支座上滑动摩擦面曲率半径；d 为球冠体中间高度。

摩擦摆隔震支座的水平变形与水平恢复力关系为三角函数的非线性关系，为简化模型参数忽略二阶小量的影响，可采用双线性本构模型进行简化，如图 3 所示。支座的初始刚度、等效刚度和水平恢复力分别按式 (1)、式 (2)、式 (3)、式 (4) 计算。

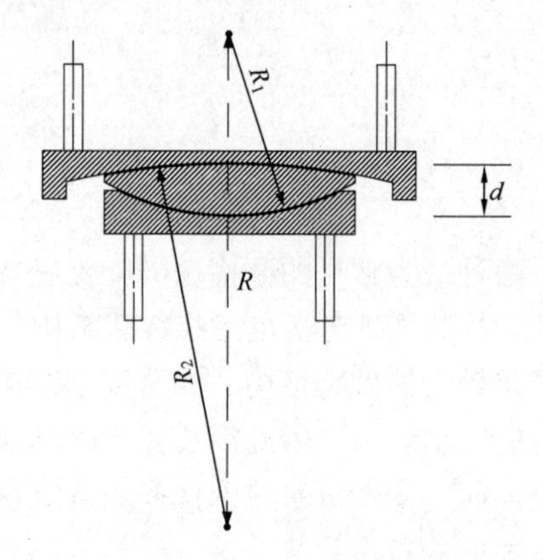

图 2 摩擦摆支座示意图

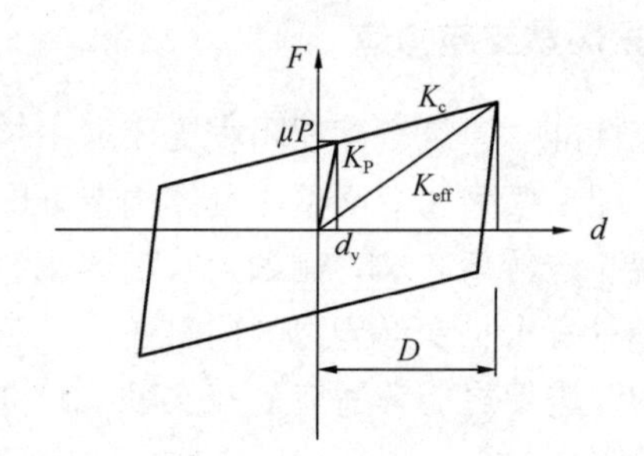

图 3 摩擦摆隔震支座的荷载-位移滞回曲线

$$K_{\mathrm{p}}=\frac{\mu P}{d_{\mathrm{y}}} \tag{1}$$

$$K_{\mathrm{eff}}=\left(\frac{1}{R}+\frac{\mu}{D}\right)\cdot P \tag{2}$$

$$F=\frac{P}{R}D+\mu P\,\mathrm{sgn}(D) \tag{3}$$

$$\mathrm{sgn}(D)=\begin{cases}1 & D\geqslant 0\\ -1 & D<0\end{cases} \tag{4}$$

其中，K_{p}为支座初始刚度；μ为动摩擦系数；P为支座所受竖向荷载；d_{y}为屈服位移，一般取2.5mm；K_{eff}为等效刚度；D为支座水平位移；K_{c}为屈服后刚度。

2.3 隔震支座力学性能试验

2.3.1 动摩擦系数测定

建筑摩擦摆隔震支座的摩擦副由PTFE等高分子摩擦材料和不锈钢板等金属对磨面组成。其摩擦行为与古典库仑摩擦定律有明显的差异，经典库仑理论中认为摩擦力与接触面积以及相对滑动速度无关，而PTFE等黏弹性材料的摩擦行为在现代摩擦理论中认为是与受压面积及相对速度相关的，摩擦系数可采用式（5）进行参数化表达。其中μ_{fast}为快速动摩擦系数（在文献［12］中为动摩擦系数上限），μ_{slow}为慢速动摩擦系数（在文献［12］中为动摩擦系数下限），α为比率参数（一般取值在15～40s/m之间），v为滑动相对速度，σ为压应力（MPa），β为压应力相关性系数。

$$\mu(v,\sigma)=[\mu_{\mathrm{fast}}-(\mu_{\mathrm{fast}}-\mu_{\mathrm{slow}})ge^{-\alpha v}]g\sigma^{-\beta} \tag{5}$$

以下为一个典型的建筑摩擦摆隔震支座动摩擦系数的测定试验。其中支座的力学性能参数见表1，支座的设计竖向荷载（重力荷载代表值工况下）为3000kN，对应压应力25MPa。试验装置采用典型的隔震支座压剪机，测试压应力范围为10～35MPa，每5MPa为一档，在每个压应力下测定速度由慢变快的摩擦系数变化。试验测得的支座力-位移滞回曲线见图4，动摩擦系数统计结果见表2。

支座力学性能参数表　　表1

设计竖向荷载	动摩擦系数		比率参数	滑动半径
kN	慢	快	s/m	m
3000	0.01	0.03	20	4.2

支座在不同速度和压应力作用下的摩擦系数　　表2

速度（mm/s）＼压应力(MPa)	10	15	20	25	30	35
4	0.016	0.015	0.013	0.011	0.010	0.010
15	0.021	0.019	0.017	0.015	0.015	0.015
50	0.030	0.028	0.026	0.024	0.024	0.024
100	0.035	0.034	0.031	0.028	0.028	0.027
150	0.038	0.036	0.032	0.030	0.029	0.027

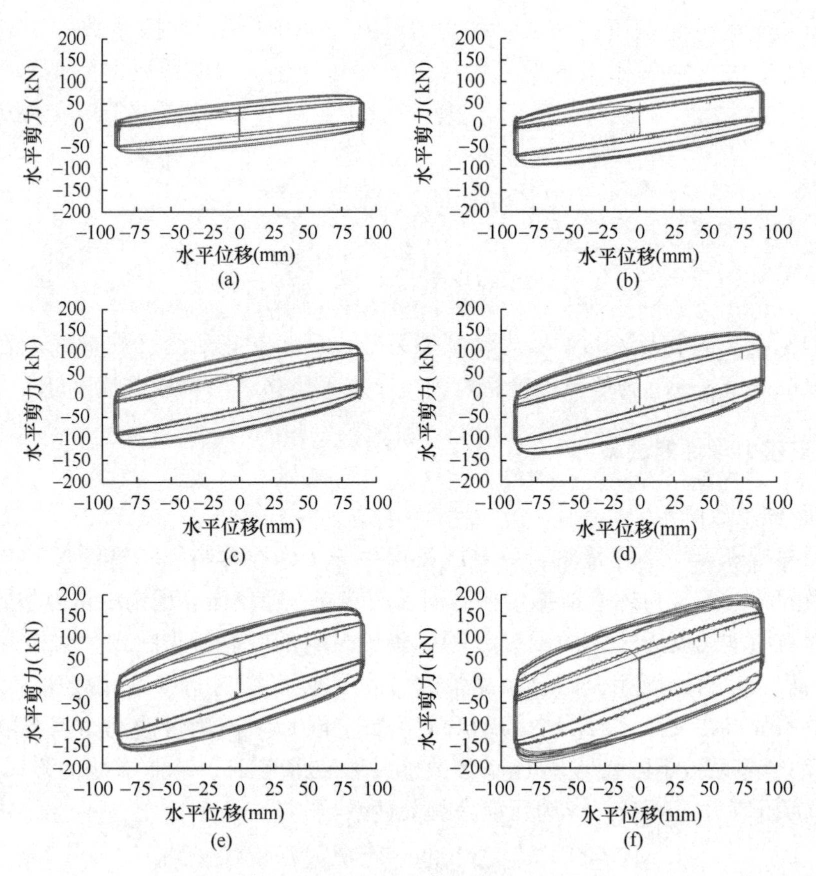

图 4　摩擦摆支座摩擦系数实验测试结果

(a) 10MPa 工况；(b) 15MPa 工况；(c) 20MPa 工况；(d) 25MPa 工况；
(e) 30MPa 工况；(f) 35MPa 工况

通过对表 2 的摩擦系数结果进行公式（5）的拟合，取 $\alpha=20\text{s/m}$，$\beta=0.30$，得到图 5 的拟合结果，其中最大误差 -16.6%，平均误差 -7.72%，最大速度下的平均误差 -3.48%。可见公式和试验结果具有很好的拟合精度。

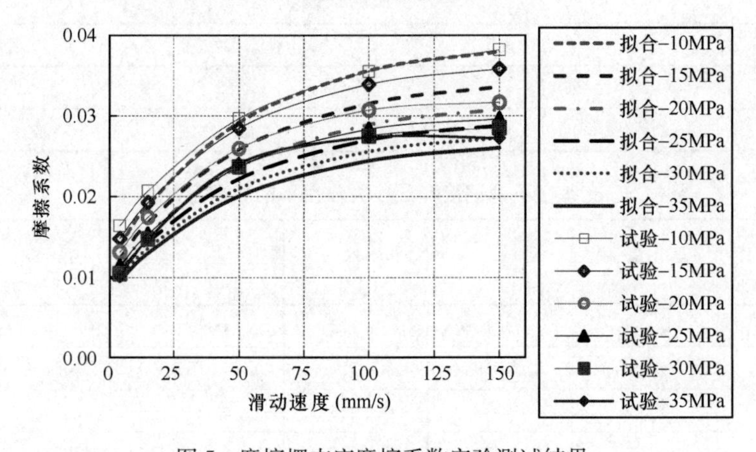

图 5　摩擦摆支座摩擦系数实验测试结果

通过试验结果可以看出当压应力控制在 15～30MPa 时，摩擦系数的相对变化较小（±20%左右），且在地震作用下支座的竖向力不断变化，摩擦系数也呈现动态变化。因此，为简化模型参数，在现有的通用有限元软件中忽略了摩擦系数的压应力相关性，动摩擦系数 μ 表达式见式（6），采用此类近似也具有足够的工程计算精度。

$$\mu = \mu_{\text{fast}} - (\mu_{\text{fast}} - \mu_{\text{slow}}) g e^{-\alpha v} \tag{6}$$

2.3.2 竖向压缩变形及蠕变

由摩擦摆隔震支座的构造上可以看出，摩擦摆支座的竖向压缩变形主要取决于各组件之间的组装间隙和摩擦材料的压缩变形组成，钢组件（上下座板及球冠板）的变形量忽略不计。其中，组装间隙在第一次压缩之后认为是可以忽略不计的，支座的实际竖向弹性刚度主要取决于摩擦材料的变形。摩擦材料受到周边钢圈的约束，支座的竖向压缩刚度随着压应力的增大是逐渐增大的，同时蠕变由于受到控制而趋于稳定。

在标准［12］中规定，支座竖向变形应小于支座总高度的 1%和 2mm 中的较大值。对表 1 中的支座进行竖向压缩变形试验，竖向压力为 3000kN，结果见图 6。可见支座竖向变形均值为 1.89mm，在完成预压之后的支座竖向变形均值在 1.60mm 左右。

PTFE 材料本身具有一定的蠕变性，但在周边钢圈的约束下可以大大降低蠕变性，保证在设计使用年限内蠕变变形量控制在很小的范围内。根据标准［12］中的试验方法，对上述摩擦材料进行了 30MPa 下 240h 的蠕变试验，根据材料的特性认为蠕变应变对数值与时间对数呈线性关系，整理结果见图 7。可见当时间延长时，竖向蠕变的规律较为明显，且呈现蠕变速率变慢的趋势，按照趋势线的规律进行计算，摩擦摆支座的蠕变变形在 50 年时约为 0.81mm，100 年时约为 0.82mm。摩擦隔震支座 PTFE 材料露出高度在 3～4mm 之间，在计算蠕变变形范围内能够有效保证摩擦材料具有足够的厚度以保证支座的正常工作。

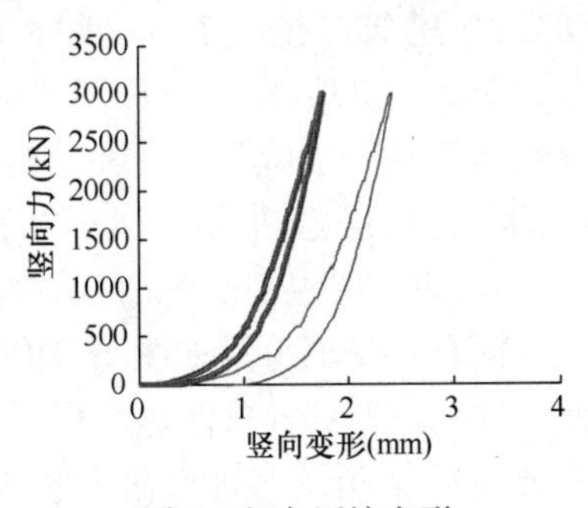

图 6 竖向压缩变形

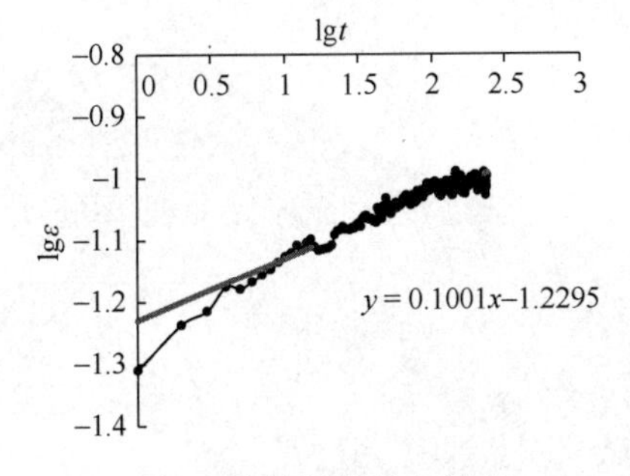

图 7 摩擦材料竖向蠕变

2.3.3 摩擦摆支座的磨耗

摩擦摆隔震支座的摩擦材料与上下座板的接触为大面积的面接触，与一般摩擦学上的普通线接触具有明显的差异。在摩擦学中关于摩擦材料的磨损试验机原理见图 8，摩擦材料与对磨钢环在初始时为线接触，当磨损增大时，接触面逐渐增大形成圆柱面，最终通过测量磨损面的宽度来评价材料的磨损程度。

摩擦摆隔震支座摩擦材料与上下座板的接触类型始终为面接触，即便在地震工况下这种摩擦形态也能够得到较好的保持，与常规摩擦学中的对磨情况有显著差异。因此，在标准［12］的摩擦材料的磨耗试验设计中采用的是简单的双剪试验，通过验证地震工况下可能发生的位移情况下摩擦材料的磨损不超过限值，保证支座在地震作用下正常发挥其隔震功能，试验装置见图9。

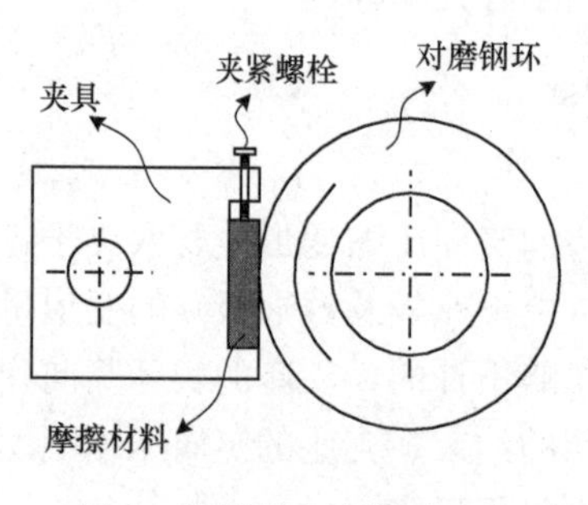

图8　摩擦学中磨损试验机原理图

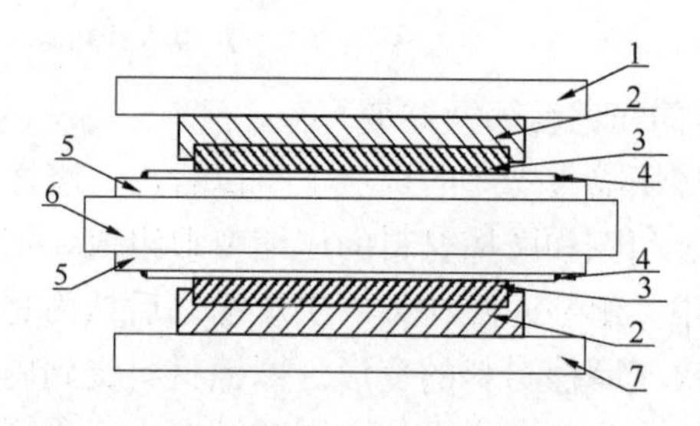

图9　摩擦摆支座摩擦材料磨耗试验装置

1—试验机上承压板；2—嵌放摩擦材料钢板；3—摩擦材料；4—不锈钢板；5—焊接不锈钢板用的基层钢板；6—水平力加载装置；7—试验机下承压板

3. 摩擦摆支座隔震建筑的动力特性

3.1　工程概况

为便于说明，本文选取了一个实际工程结构进行有限元分析计算。该工程实例为一幼儿园教学楼，结构等效宽度为19.65m，高度为16.4m（含隔震层），高宽比为0.83，属于乙类建筑。抗震设防烈度为8度（0.30g），场地土特征周期为0.40s。结构类型为钢筋混凝土框架结构，地上4层，隔震层层高2m，1～3层层高3.7m，第4层层高3.3m，结构三维模型和隔震层平面分别如图10和图11所示。采用ETABS软件建立了隔震结构和非隔震结构的三维有限元模型。梁、柱构件采用空间杆系单元，楼板采用壳单元。隔震模型中，增设隔震支座（摩擦摆隔震支座采用连接单元Isolator2[14]）。

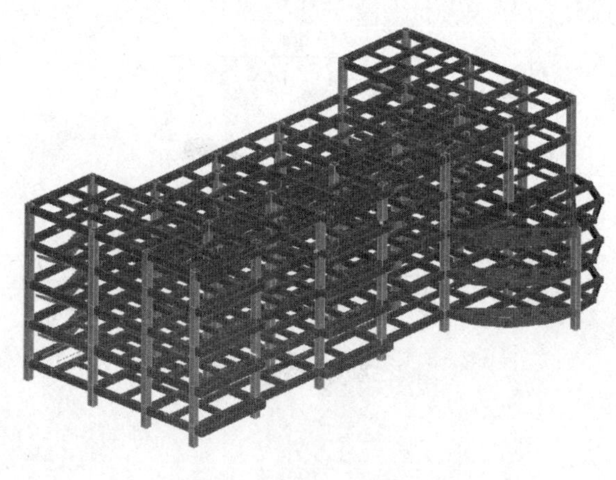
图10　结构三维模型图

隔震结构的动力特性会随着隔震支座水平变形的变化而不断变化，这里隔震支座的等效水平刚度取支座中震计算下的隔震层位移对应的等效刚度。在Etabs中利用RITZ向量法计算出非隔震结构和隔震体系前六阶振型周期如表3所示。可以看出，隔震体系的周期较原结构增大很多，基本周期由原来的0.83s延长至2.51s。

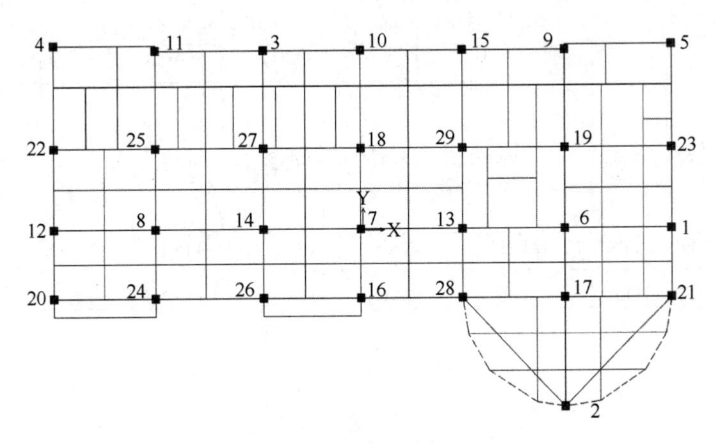

图 11 隔震层平面图

非隔震结构和隔震结构前 6 阶的周期（s） **表 3**

振型	非隔震结构（基础固定）	隔震结构（中震位移）
1	0.83	2.51
2	0.82	2.48
3	0.77	2.40
4	0.27	0.41
5	0.27	0.41
6	0.26	0.39

3.2 三种不同隔震支座布置方案

根据重力荷载代表值作用下的柱底反力，为隔震层选取了 FPS1、FPS2 和 FPS3 三种摩擦摆支座，具体的摩擦摆支座参数如表 4 所示。支座在长期荷载作用条件下的最大面压为 24.59MPa（26 支座），低于限值 25MPa，支座具备一定的安全储备。该工程隔震支座的平面布置图如图 12 所示。

摩擦摆支座设计参数 **表 4**

型号		FPS1	FPS2	FPS3
竖向刚度	（kN/mm）	1330	2000	2670
等效水平刚度	（kN/m）	1867	2800	2293
摩擦系数	（慢）	0.02	0.02	0.01
摩擦系数	（快）	0.05	0.05	0.02
比率参数	（s/m）	40	40	20
滑动半径	（m）	3	3	3
设计竖向荷载	（kN）	2000	3000	4000
数量	（个）	12	12	5

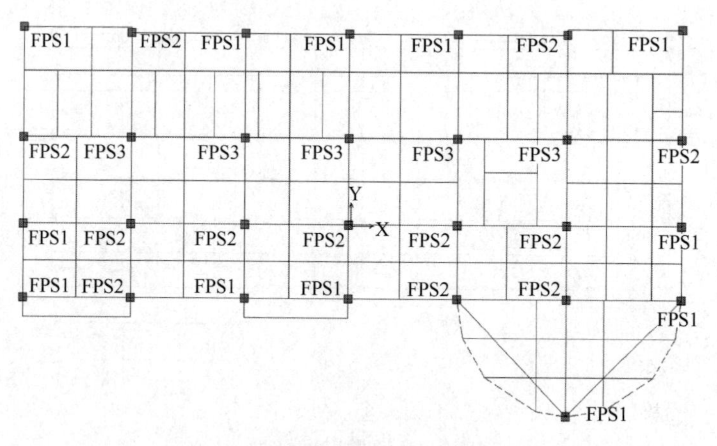

图 12 摩擦摆隔震支座布置图

重力荷载代表值下各支座长期压应力 **表 5**

编号	支座类型	1.0 恒载+0.5 活载（kN）	面压（MPa）
1	FPS1	1610	20.13
2	FPS1	1937	24.21
3	FPS1	1776	22.20
4	FPS1	1449	18.11
5	FPS1	1495	18.69
6	FPS2	2789	22.13
7	FPS2	2432	19.30
8	FPS2	2569	20.39
9	FPS2	2336	18.54
10	FPS1	1732	21.65
11	FPS2	2303	18.28
12	FPS1	1533	19.16
13	FPS2	2524	20.03
14	FPS2	2449	19.44
15	FPS1	1789	22.36
16	FPS1	1950	24.38
17	FPS2	2846	22.59
18	FPS3	3048	18.36
19	FPS3	3597	21.67
20	FPS1	1362	17.03
21	FPS1	1921	24.01
22	FPS2	2120	16.83
23	FPS2	2157	17.12
24	FPS2	2056	16.32

续表

编号	支座类型	1.0恒载+0.5活载（kN）	面压（MPa）
25	FPS3	3378	20.35
26	FPS1	1967	24.59
27	FPS3	3139	18.91
28	FPS2	2485	19.72
29	FPS3	3288	19.81

在本框架结构的隔震层设计中，将摩擦摆支座设计定为方案A。此外，为对比结构在不同支座布置方案下的动力特性，还设计了全橡胶支座隔震层（方案B），包含铅芯橡胶支座（LRB600）和普通橡胶支座（LNR600和LNR700），并进行隔震层的响应分析对比。

根据式（3），摩擦摆支座隔震方案可以根据重力荷载代表值下的支座竖向力计算其隔震层水平回复力曲线，具体见图13。但是在地震荷载作用下，每个支座除受到重力作用，还受到竖向地震动和由于水平地震作用引起弯矩而导致的轴向力重分配，每个支座的竖向力是实时变化的。图13所示的隔震层本构可以认为是该过程中的隔震层本构“均值”。

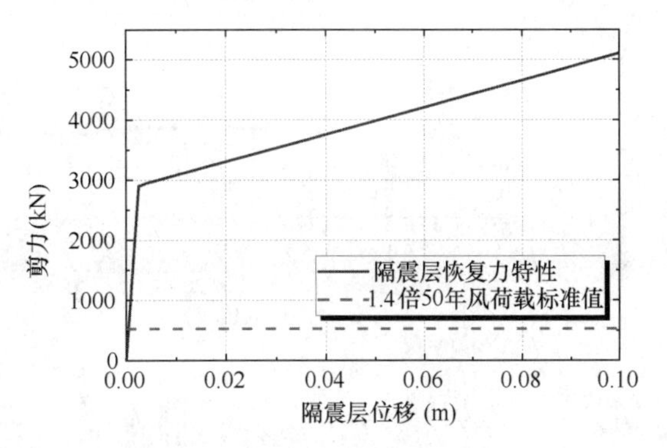

图13　隔震层水平回复力曲线

为探索摩擦摆支座水平恢复力模型的适用性，尤其是对结构峰值反应的影响，本文针对隔震层又设计了第三种支座方案C。在方案C中，以29个不同的铅芯橡胶支座代替方案A中的摩擦摆支座，根据每个摩擦摆的在其重力荷载代表值下的相关参数（水平有效刚度、非线性刚度、屈服力、屈服后刚度比）用铅芯橡胶支座的属性参数进行替代，得到一个方案A的“稳态隔震层本构方案”。

3.3　三种方案的隔震层反应与减震效果对比

针对以上三种隔震层支座设计方案，根据《建筑抗震设计规范》GB 50011—2010（简称抗规）的相关规定，采用了5条天然波（TR1～TR5）和2条人造波（RZ1和RZ2），对框架结构进行了设防地震和罕遇地震条件下的非线性时程分析，其中梁柱单元为线性单元，支座单元为非线性单元。7条地震动的时程曲线如图14所示。其中三种隔震方案的减震效果相近，具体结果见表6。地震动平均谱与设计反应谱对比如图15所示。

三种支座设计方案（A、B、C）在设防地震（中震）和罕遇地震（大震）作用条件下的响应分析结果如表7和表8所示，包括隔震层的水平位移最大值和层间剪力。由表可知，摩擦摆支座隔震层（方案A）的水平位移响应明显要小于橡胶支座隔震层（方案B），前者位移要比后者分别低30%（中震）和17%（大震）。在中震条件下，摩擦摆支座隔震层的层间剪力明显低于橡胶支座，两者间差距约为16%；在大震条件下，两者间差距不明显，但前者要略低于后者。方案C的水平位移和层间剪力与方案A相差较小，这也与两者间基本参数一致相对应。

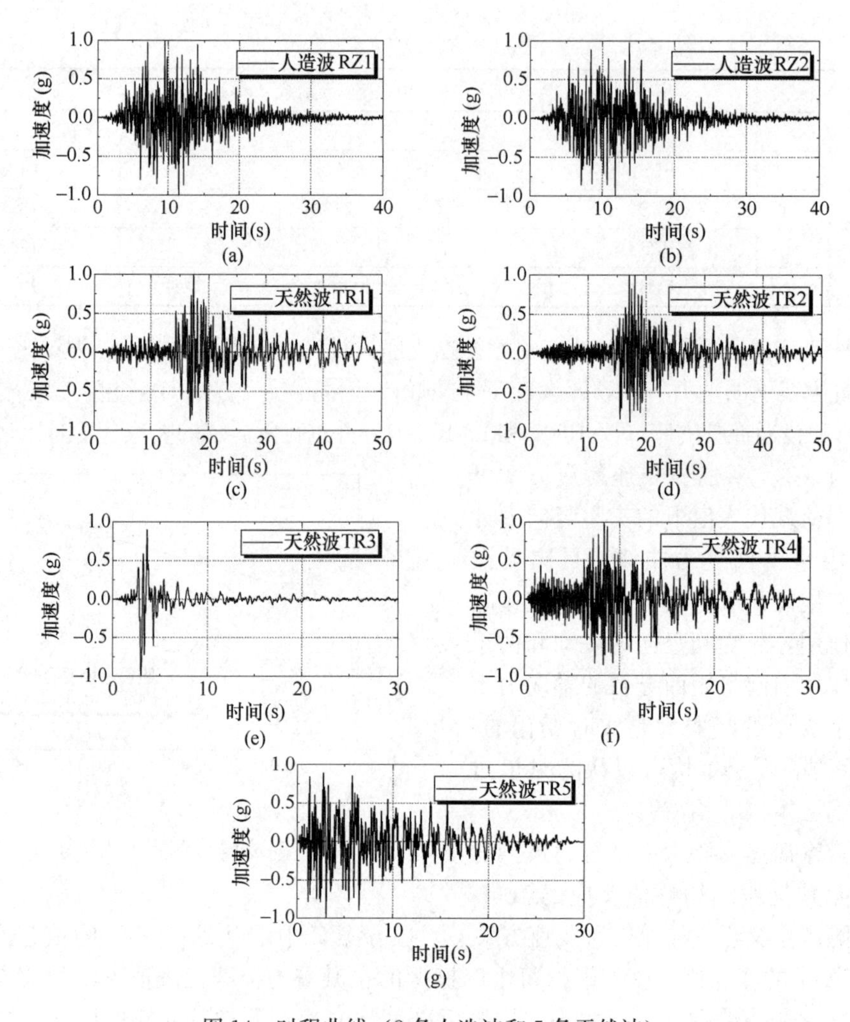

图 14　时程曲线（2 条人造波和 5 条天然波）

(a) RZ1；(b) RZ2；(c) TR1；(d) TR2；(e) TR3；(f) TR4；(g) TR5

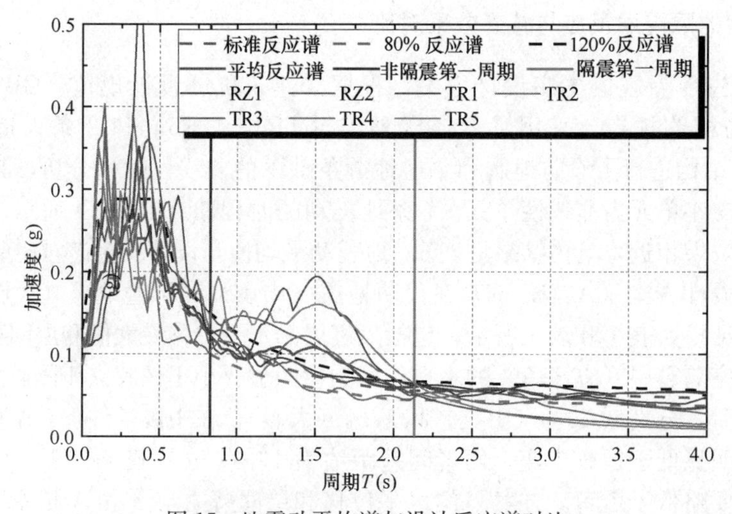

图 15　地震动平均谱与设计反应谱对比

不同支座方案结构减震系数 **表 6**

方案	RZ1	RZ2	TR1	TR2	TR3	TR4	TR5	均值
A	0.25	0.26	0.25	0.25	0.26	0.26	0.22	0.25
B	0.32	0.33	0.36	0.25	0.27	0.31	0.24	0.30
C	0.21	0.25	0.35	0.27	0.24	0.39	0.26	0.27

隔震层水平位移最大值（mm） **表 7**

方案	A		B		C	
作用	中震	大震	中震	大震	中震	大震
RZ1	82.6	229.5	193.2	272.2	82.6	206.5
RZ2	70.6	144.7	160.3	206	70.6	136.6
TR1	127.7	228.1	152.6	233.8	127.7	225.1
TR2	58	146.7	82.3	175.3	58	132.7
TR3	81.9	145.2	137.3	185.7	81.9	143.1
TR4	134.6	312.3	100.5	405.2	134.6	288.6
TR5	108.1	308.7	105.6	339.6	108.1	253.9
均值	95	216	133	260	95	198

隔震层层间剪力（kN） **表 8**

编号	A		B		C	
作用	中震	大震	中震	大震	中震	大震
RZ1	5148	7314	6514	8903	4409	7356
RZ2	4655	6431	5878	7112	4401	5809
TR1	3968	9926	5669	7865	5535	7765
TR2	3771	6652	3768	6281	4125	5713
TR3	5066	6807	5253	6564	4709	5955
TR4	3522	11701	4258	12499	5237	9146
TR5	4053	9898	4398	10677	4736	8354
均值	4312	8390	5105	8557	4736	7157

3.4 对上部结构扭转反应的控制对比

由于摩擦摆支座的水平回复力和竖向作用呈正相关关系，因此隔震层可以呈现和上部质量分布一致的刚度分布，能够有效抑制上部结构的扭转效应。而且随着地震作用下实时

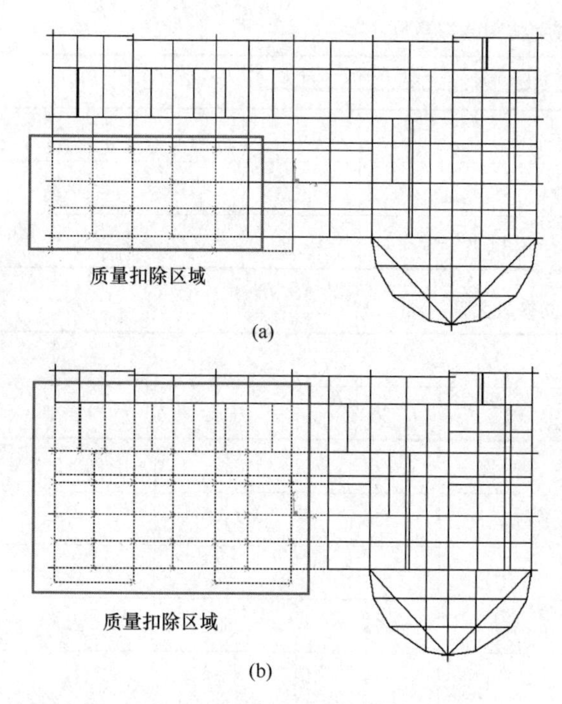

图 16 框架结构质量扣除示意图
(a) 扣除 1/6 质量；(b) 扣除 1/3 质量

的竖向力变化，隔震层反应能够与之实现实时的调整变化，可以近似认为隔震层的刚度与上部结构的竖向力分布是实时对应的，这与普通橡胶隔震支座方案将会有较为明显的差异。为此，对不同质刚偏心率的框架结构，进行罕遇地震作用下的扭转响应分析。在 ETABS 软件中对原有结构进行两次质量扣除，第一次扣除约 1/6 质量，第二次扣除约 1/3 质量，扣除部分如图 16 所示，同时保持隔震支座布置方案保持不变。

原始结构与扣除质量后的结构总质量和隔震层质刚偏心率见表 9。相对于原始结构，扣除质量结构在 X 方向的质刚偏心率均大幅提高，而 Y 向的偏心率提高较不明显。采用与 3.3 节中相同的 7 条地震动对原始结构和扣除质量的结构进行时程分析，分析结果如表 10 和表 11 所示，包括结构的楼层位移比均值和层间位移角均值。由表 10 可知，对于相同的结构，罕遇地震作用下摩擦摆支座（方案 A）的楼层位移比最小，橡胶支座（方案 B）的楼层位移比最大，方案 C 的位移比居中且略高于方案 A。横向比较可以看出，结构的楼层位移比随着质刚偏心率的增大而增大，但摩擦摆支座的位移比增长幅度明显要低于橡胶支座；前者增幅仅为 10%左右，而后者增幅可达 40%，其最大值为 1.49。

对于结构的层间位移角，三种方案间的差距较小，并且结构质刚偏心率的变化对各层位移角的影响也较不明显。

原始结构与扣除质量结构相关参数 **表 9**

		原始结构	扣除 1/6 质量结构	扣除 1/3 质量结构
总质量（t）		6738.1	5619.9	4438.4
质量占比		100%	83.4%	65.8%
质刚偏心率	X 向	0.033 (0.025)	0.205 (0.186)	0.445 (0.410)
	Y 向	0.004 (0.002)	0.064 (0.055)	0.061 (0.051)

注：括号内数值为方案 A 和 C 隔震层的质刚偏心率，括号外数值为方案 B 隔震层的质刚偏心率。

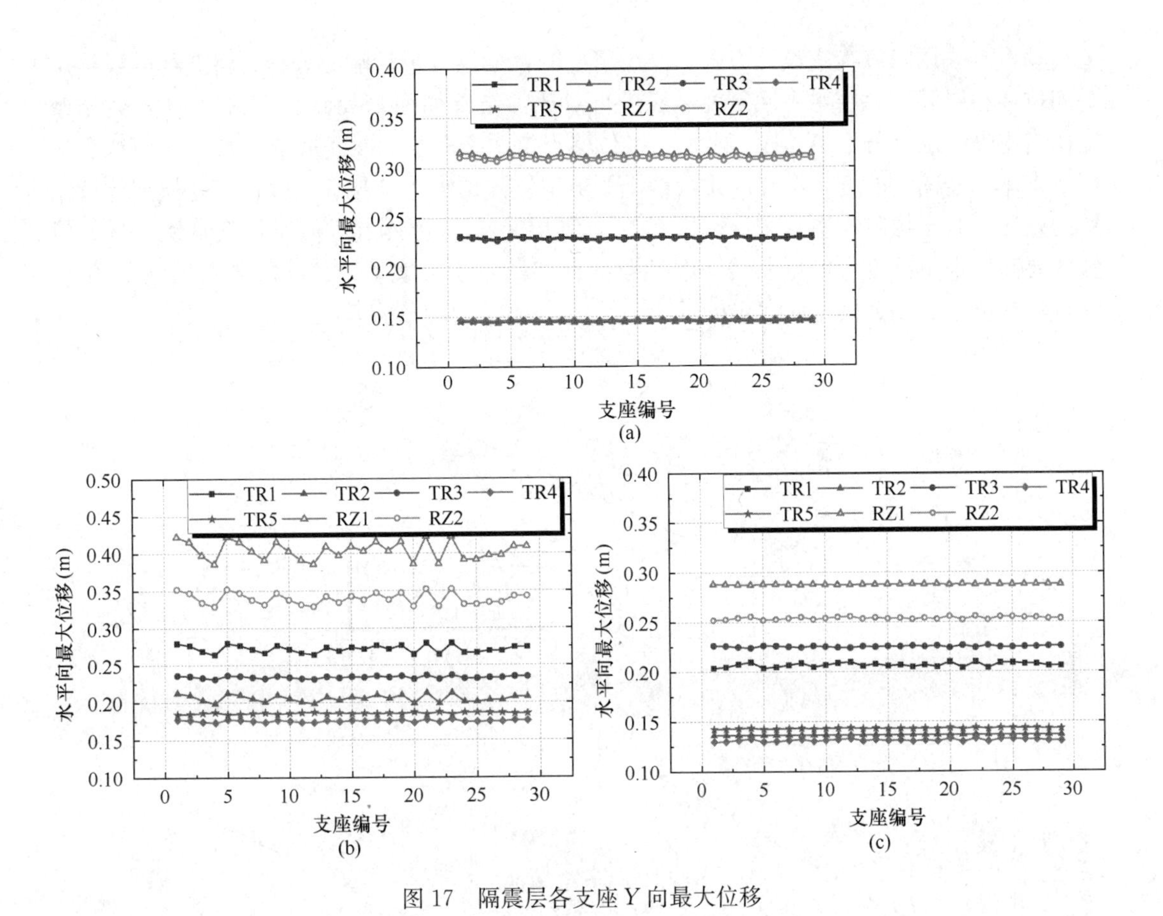

图 17　隔震层各支座 Y 向最大位移

(a) 方案 A；(b) 方案 B；(c) 方案 C

罕遇地震作用下框架结构楼层位移比　　**表 10**

	原始结构		扣除 1/6 质量结构		扣除 1/3 质量结构	
	首层	顶层	首层	顶层	首层	顶层
方案 A	1.032	1.024	1.071	1.071	1.134	1.142
方案 B	1.072	1.062	1.240	1.227	1.485	1.448
方案 C	1.038	1.034	1.120	1.093	1.264	1.260

罕遇地震作用下框架结构层间位移角　　**表 11**

	原始结构		扣除 1/6 质量结构		扣除 1/3 质量结构	
	首层	顶层	首层	顶层	首层	顶层
方案 A	1/478	1/906	1/515	1/817	1/570	1/999
方案 B	1/442	1/1150	1/463	1/1105	1/472	1/1223
方案 C	1/506	1/975	1/509	1/806	1/546	1/892

3.5　隔震层水平回复力模型校核

本节将对 3.2 节中提及的隔震层水平回复力模型进行检验。同样采用 3.3 节中的 7 条地震动，对采用隔震支座设计方案 A、B 和 C 的框架结构进行响应分析，将获得的隔震层平均位移和层间剪力与水平回复力曲线进行对比，结果如图 18 所示。由图可知，对于摩擦摆支座隔震层（方案 A），其层间剪力-位移点绕水平回复力曲线散落分布，说明地震作用下支座处竖向力的变化会在一定程度上影响隔震层的剪力与位移，并造成与模型计算结果的差异。对于橡胶支座隔震层（方案 B），其层间剪力-位移点近似于一条直线，且直线斜率要高于水平回复力模型曲线非线性段的斜率。对于方案 C，其隔震层的层间剪力-位移点正好位于回复力模型曲线上。

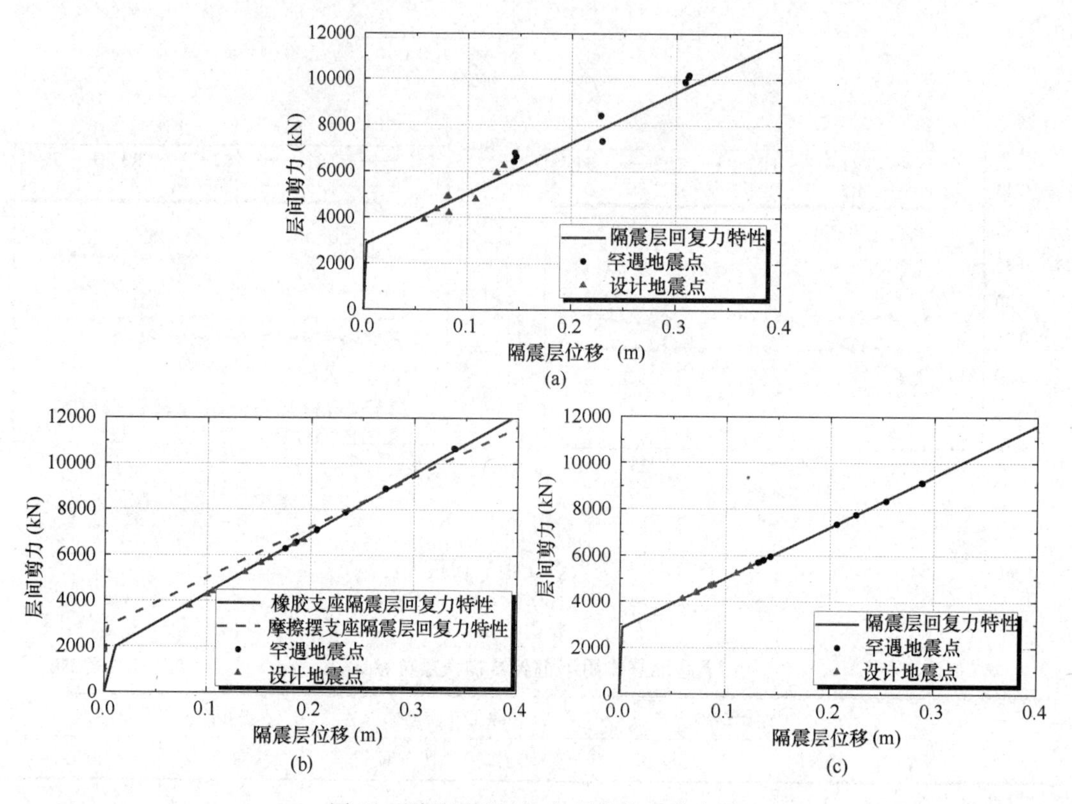

图 18 隔震层层间剪力-平均位移散点图

（a）方案 A；（b）方案 B；（c）方案 C

为进一步了解摩擦摆隔震支座水平回复力模型的适用性，本文又采用了 56 条天然地震动和 35 条人造地震动，对该框架结构（方案 A）进行设计地震和罕遇地震作用下的时程响应分析，并将计算结果与回复力曲线进行了对比，如图 19 所示。为分析竖向地震动真实性对层间剪力-位移计算结果的影响，图 19（a）的分析过程采用了真实的竖向地震动，而图 19（b）则采用水平地震动乘以 0.65 作为竖向地震输入。

由图可知当采用天然地震动（真实竖向地震动）进行时程分析时，摩擦摆支座隔震层的层间剪力-位移点同样围绕水平回复力曲线分布，且绝大多数位于 80%～120%回复力的包络线内。当竖向地震输入采用缩比的水平向地震动时，约 15%的隔震层层间剪力-位移点位于包络线外，水平回复力模型计算误差较大。这是由于水平向地震动较竖向地震动具有更长的卓越周期，且水平向地震动和竖向地震作用处于实时的同相位，放大了竖向地震对结构地震反应的影响。对于采用人造地震动进行的时程响应分析，设计地震条件下的层间剪力-位移点均处于包络线内，而罕遇地震条件下的层间剪力-位移点均位于包络线外，且高于120%回复力。这是由于人造地震动的在各个频段的能量均较为饱满，且水平向地震动和竖向地震动处于实时同相位状态，比天然地震动的影响更加显著。

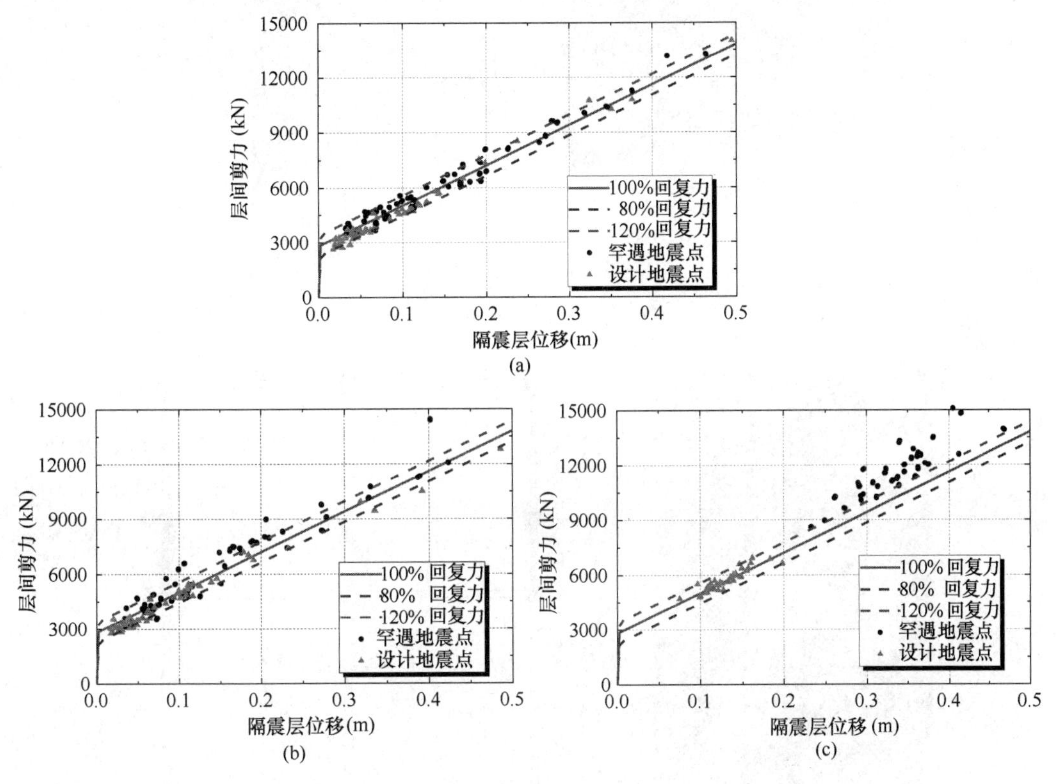

图 19 隔震层剪力-位移响应

(a) 天然地震动 1；(b) 天然地震动 2；(c) 人造地震动

综上可见，式（3）的隔震层水平回复力模型对于真实的天然地震动的计算具有很高的可靠性，结果保证率为 93.75%，具有很高的计算精度。当采用同相位的水平地震动和竖向地震动进行计算时，发现隔震层的反应明显增大，且随着竖向地震动加速度幅值增大其效果越显著，此部分与地震动输入相关内容有待于进一步深入研究和讨论。

4. 摩擦摆支座振动台试验

4.1 试验设计

原型结构为高层框架-剪力墙结构形式，抗震设防烈度为 9 度，设计地震分组第一组，Ⅱ类场地，场地特征周期 0.35s。地上 12 层，结构高宽比为 2.7。为便于振动台试验，将原结构 X 向的 5 跨简化为 2 跨，将 X 向的剪力墙去除，保留 Y 向的两片连肢剪力墙，仅研究结构在 Y 向和竖向（Z 向）地震作用下的响应，具体见图 20。通过 1∶9 的几何比例对结构进行缩尺，具体试验设计及尺寸参数见文献［15］，简化模型隔震层布置有 13 个支座。振动台模型见图 21，隔震层布置见图 22。

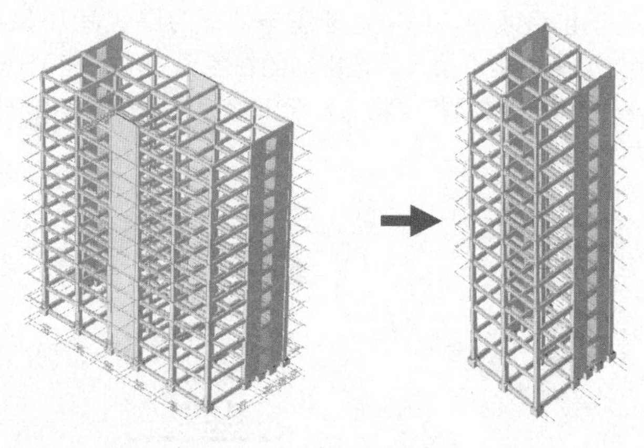

图 20　结构简化图

图 21　振动台模型

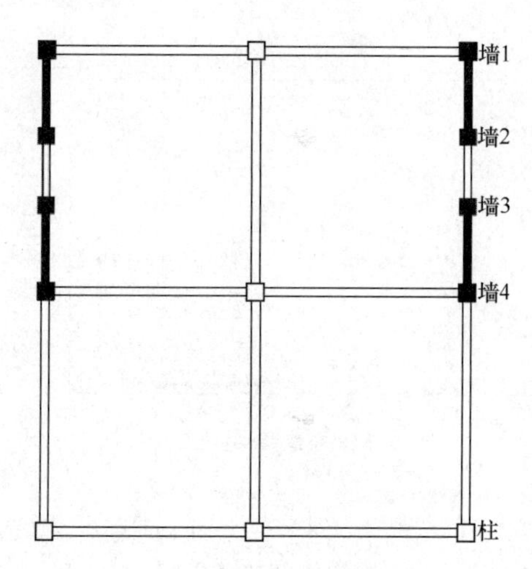

图 22　振动台模型隔震层布置

4.2　隔震层反应

将检测得的台面波作为地震动输入，计算试验模型有限元隔震反应。将测得的隔震层位移、剪力反应与试验结果进行对比。取 9 度设防地震作用下的地震动，对有限元分析结果和试验结果进行对比，位移时程曲线和隔震层剪力时程曲线见图 23。可见，摩擦摆隔震结构隔震层反应时程分析与有限元计算结果具有较好的拟合精度。其中位移时程记录采用的是拉线位移计，隔震层位移响应较有限元分析结果较小，但峰值误差基本在 20%左右；隔震层剪力二者具有极高的拟合精度，峰值点误差小于 5%。采用有限元分析方法从隔震层的角度可以很好地还原振动台试验。

同时为了验证隔震层反应与重力荷载代表值下的隔震层本构之间的关系，将所有试验工况结果与隔震层本构曲线进行对比，具体见图 24。可见，隔震层反应基本围绕在隔震层本构骨架曲线周边，分布在±30%误差范围内。

4.3　隔震支座的提离现象

摩擦摆隔震支座由于其支座结构是由三块相互分离的组件组成，因此当支座提离时会

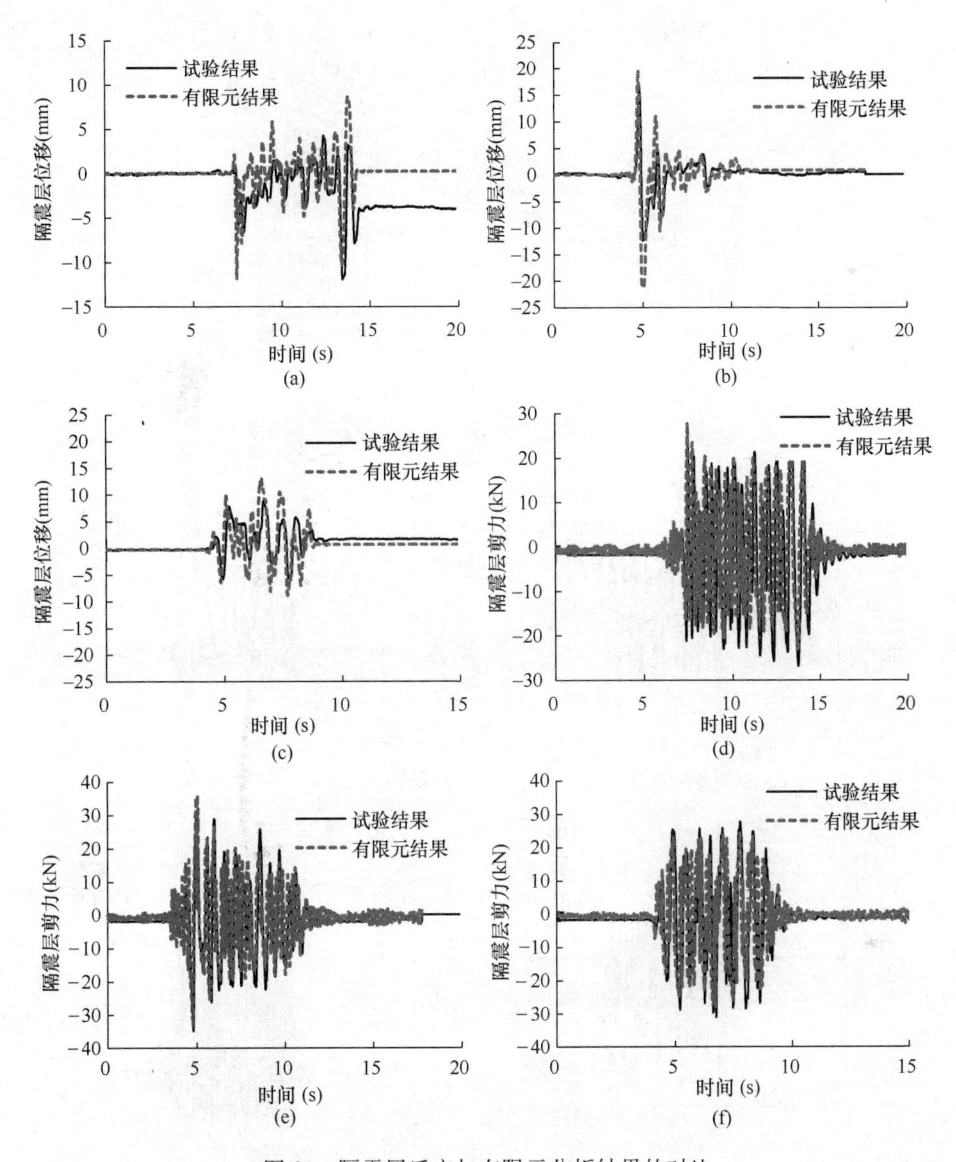

图 23　隔震层反应与有限元分析结果的对比

（a）IEW 波位移时程；（b）IPT 波位移时程；（c）人工波位移时程；（d）IEW 波剪力时程
（e）IPT 波剪力时程；（f）人工波剪力时程

出现个别支座完全不能承受水平剪力，而受压一侧的支座竖向力会有较为明显的提高，从而影响整个结构的反应。为简单起见，我们仅对比图 22 中墙 1 和柱这两个点支座 9 度罕遇地震下 IEW 波的竖向位移和竖向力时程。考虑具体结果见图 25，其中图（c）、图（d）是支座实际竖向力时程，考虑到支座由于安装误差导致初始竖向力与有限元模拟结果不同，去除初始竖向力仅考虑竖向力变化时程情况见图（e）、图（f）。

由图可见，柱下的竖向位移以及力变化时程有限元分析结果与试验结果拟合精度较高，峰值误差在 10%左右；但是墙下的支座竖向位移以及竖向力二者差异较大，试验结果远比有限元分析结果要大，这是由于联肢剪力墙中的转换梁以及连梁的损伤导致刚度下降，联肢剪力墙内力重分布，导致支座提离高度变大、支座竖向力变化幅值增大。因此在

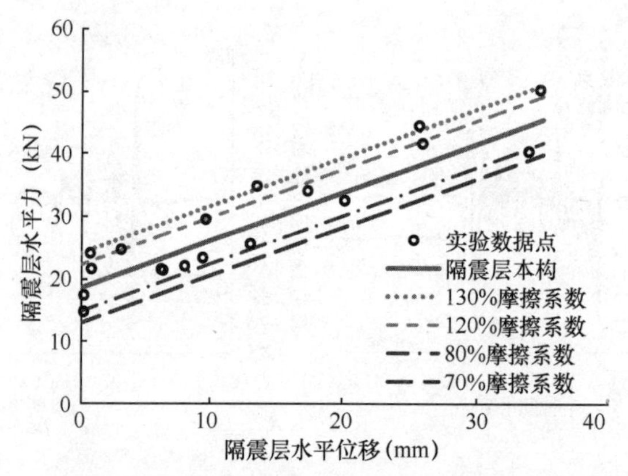

图 24　隔震层反应与重力荷载代表值下隔震层本构的对比

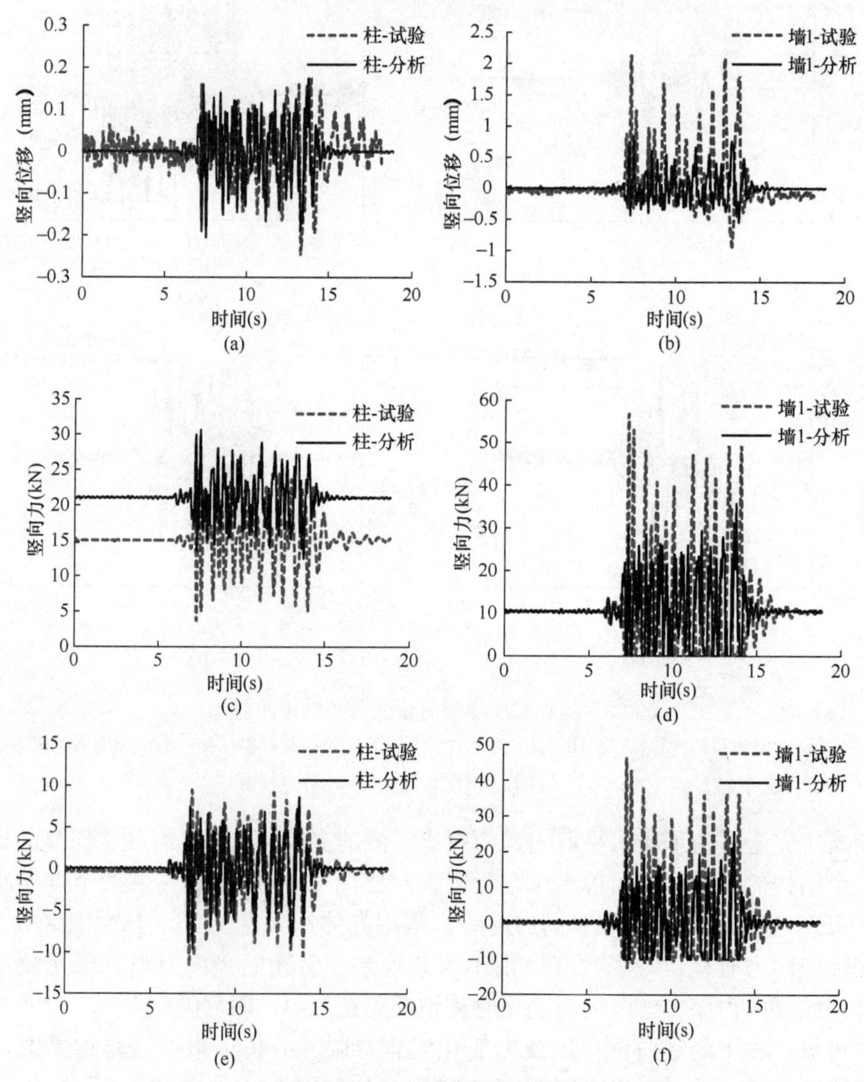

图 25　罕遇地震下支座的竖向位移和竖向力时程

(a) 柱支座竖向位移时程；(b) 墙 1 支座竖向位移时程；(c) 柱支座竖向力时程；(d) 墙 1 支座竖向力时程；
(e) 柱支座竖向力变化时程；(f) 墙 1 支座竖向力变化时程

设计联肢剪力墙隔震结构时，因充分考虑由于连梁、转换梁损伤导致的支座竖向力变化带来的影响，应适当增加剪力墙周边支座的竖向承载力，避免大震作用下受损。

5. 隔震支座的连接构造以及支座受弯问题

隔震支座的上下连接板分别与上下结构通过螺栓连接，该螺栓连接应该采用可拆换性的外插入螺栓连接方案，如图 26 所示。所有连接螺栓或锚固钢筋，均按罕遇地震作用下产生的水平剪力、弯矩进行强度验算。其中对于双摆型支座，其上下支墩受弯矩一致，见式（7）；当支座为单摆型时，支墩弯矩见式（8）。

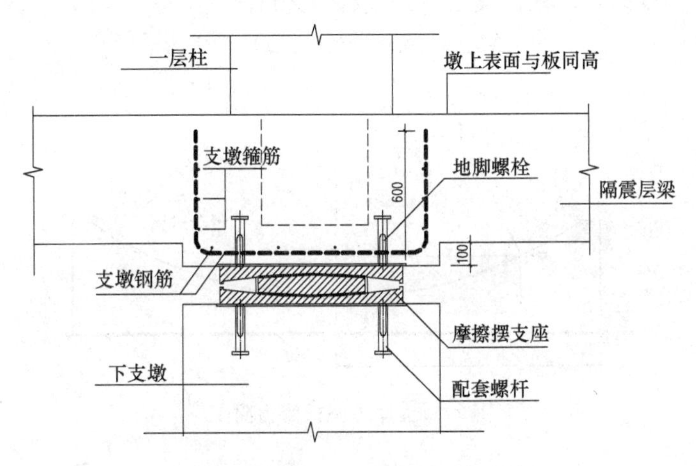

图 26　隔震支座上下连接示意图

$$M=\frac{P\Delta+Vh}{2} \tag{7}$$

$$\begin{cases} M_{有位移}=P\Delta+\dfrac{Vh}{2} \\ M_{无位移}=\dfrac{Vh}{2} \end{cases} \tag{8}$$

其中，M 为支墩弯矩，P 为支座所受竖向力，Δ 为支座水平位移，V 为支座水平剪力，h 为支座高度。

当采用橡胶隔震支座方案时橡胶隔震支座的连接螺栓需要计算在偏心受压弯矩下的拉力[16]。此时计算得 LRB600 橡胶隔震支座的螺栓锚固长度应为 450mm。而摩擦摆隔震支座具有和橡胶隔震支座不同的受力机制，支座构造及下座板受力分析图见图 27，其中近似认为竖向力和水平力均经过球冠板正中。当考虑罕遇地震作用下，支座下座板受摩擦力和竖向力的共同作用，其受力平衡见式（8）。

$$\begin{cases} f\cos\theta+N\sin\theta=F \\ -f\sin\theta+N\cos\theta=P \end{cases} \tag{9}$$

其中

$$\begin{cases} f = \mu N \\ \mu = 0.02 \sim 0.05 \\ \sin\theta = \dfrac{d}{R_1} \end{cases}$$

可得水平回复力与竖向压力关系：

$$F = \frac{\mu\cos\theta + \sin\theta}{-\mu\sin\theta + \cos\theta}P \tag{10}$$

考虑绕O点逆时针方向转动的弯矩见式（11），由于$d < 0.85r$，因此式（11）的不等号在建筑支座中恒成立，即支座不会产生逆时针方向的转动，连接螺栓不受拉。螺栓预埋件考虑最小构造长度即可，根据文献［1］取250mm。

$$M_O = Fh - P(r-d) < 0 \tag{11}$$

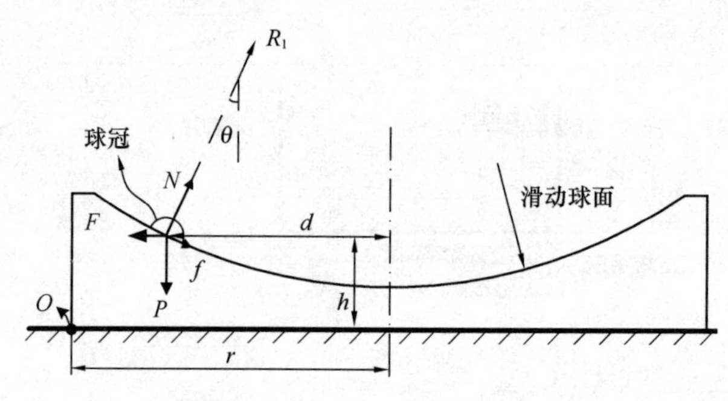

图27　支座下座板受力简图

6. 结论

本文通过对一个钢筋混凝土框架结构工程采用摩擦摆支座隔震方案进行了设计与分析，并与普通橡胶隔震支座方案进行了对比，得到以下结论：

（1）摩擦摆隔震支座的摩擦系数呈现与压应力和速度的相关性，可以通过参数化的方式进行表达，隔震支座的试验结果与数值模拟参数符合度较高。

（2）摩擦摆支座的竖向性能与摩擦材料的弹性变形有直接关系，在长期竖向荷载作用下，具有很小的竖向变形和长期的蠕变稳定性。

（3）摩擦摆隔震结构在天然地震动下的隔震层反应，与其重力荷载代表值下的稳定本构具有很好的符合性。但在人造地震动大震工况下，吻合性较差，这与地震动的输入特性有关，将进行更深入的研究。

（4）摩擦摆支座的隔震层水平刚度中心与上部结构的竖向作用力呈正相关关系，在地震作用下隔震层处的质心与刚心重合度较高，能够有效抑制结构的整体扭转，与叠层橡胶隔震支座相比具有显著优势。

（5）摩擦摆支座隔震结构振动台试验结果表明，摩擦摆支座隔震结构的实际地震反应能够采用有限元分析很好的拟合，其中对于联肢剪力墙下的隔震支座竖向力变化以及竖向提离问题需要充分考虑转换梁和连梁的损伤。

（6）根据简化力学分析，摩擦摆隔震支座在受到罕遇地震作用下，其连接螺栓不会出现受拉作用，在计算中仅需考虑水平受剪验算，螺栓预埋件长度仅满足构造长度即可，大大降低了支座安装难度。

参考文献

[1] 建筑抗震设计规范 GB 50011—2010[S]. 北京：中国建筑工业出版社，2016.

[2] 叠层橡胶隔震支座隔震技术规程 CECS 126：2001[S]. 2001.

[3] Zayas V，Low S and Mahin S. The FPS earthquakeresisting system[R]. Technical Report UCB/EERC-87 /01，University of California at Berkeley，1987.

[4] Bridges S O，Staff S，Bridges T O S O. Guide specifications for seismic isolation design[M]. AASHTO，2010.

[5] Minimum design loads for buildings and other structures：ASCE/SEI 7-16[S]. New York：American Society of Civil Engineers，2017.

[6] Anti-seismic devices：EN 15129：2009[S]. European Committee for Standardization，2009.

[7] Structural bearings：part 2：sliding elements：EN 1337-2：2004[S]. European Committee for Standardization，2004.

[8] 庄军生. 桥梁支座[M]. 北京：中国铁道出版社，2008.

[9] 庄军生. 桥梁减震、隔震支座和装置[M]. 北京：中国铁道出版社，2012.

[10] 公路桥梁摩擦摆式减隔震支座 JT/T 852—2013[S]. 2013.

[11] 桥梁双曲面球型减隔震支座 JT/T 927—2014 [S]. 2014.

[12] 建筑摩擦摆隔震支座 GB/T 37358—2019[S]. 北京：中国建筑工业出版社，2019.

[13] Sarkisian M，Lee P，Long E，et al. Experiences with Friction Pendulum™ seismic isolation in California[J]. Earthquake Resistant Engineering Structures IX，2013，132：357-368.

[14] CSI 分析参考手册[M]. Computers & Structures，Inc，2007.

[15] 周博威. 摩擦摆隔震高层框剪结构抗震性能试验研究 [D]. 中国地震局工程力学研究所，2019.

[16] 橡胶支座　第 3 部分：建筑隔震橡胶支座 GB 20688. 3—2006[S]. 2006.

10 减隔震组合技术在高烈度地震区中的应用研究*

丁洁民[1,2]

（1. 同济大学建筑设计研究院（集团）有限公司，上海；2. 同济大学土木工程学院，上海）

摘　要：为提高高烈度地震区建筑的安全性及抗震性能，国家和地方颁布了一系列政策，鼓励并推广使用减隔震技术。以云南省某规划展览馆为工程背景，提出了一种新型减隔震组合技术——基础隔震＋上部结构减震，并将其首次应用于国内工程设计中。分析了该新型减隔震组合技术的减震效率及结构在罕遇地震作用下的抗震性能。分析结果表明：建议的减隔震组合技术具有良好的减震效果；在罕遇地震作用下，上部结构基本保持弹性，结构表现出良好的抗震性能。同时，针对新型减隔震组合结构，提出了叠加分析法。通过算例分析，认为叠加分析法计算结果用于结构设计偏于安全。提出了适用于新型减隔震组合结构的设计流程，以期为类似的工程设计提供参考。

关键词：高烈度地震区；减隔震组合技术；叠加分析法；设计流程

Application research of seismic isolation technology combined with energy-dissipation technology in high seismic intensity region

Ding Jiemin[1,2]

（1. Tongji Architectural Design (Group) Co.，Ltd，Shanghai；

2. College of Civil Engineering，Tongji University，Shanghai）

Abstract：In order to improve the seismic performance and safety of buildings in high intensity area，national and local governments have promulgated a series of policies，encouraging utilize seismic isolation and energy dissipation technology. This paper takes an exhibition hall locating in Yunnan province as an example，proposes a new kind of seismic isolation technology combined with energy-dissipation technology，and firstly adopt this new technology for engineering design in China. This paper analyzes both the seismic efficiency of the system and the structural seismic performance under rare earthquake. Analysis results indicate that the new system has great seismic efficiency and superstructure of the system remains elastic under rare earthquake，showing great seismic performance. Besides，this paper proposes superposition analysis method for designing the system. According to the calculation results，adopting the results of superposition method to design is safe. Finally，this paper proposes the design process of superposition analysis method for the new system，which provides reference for similar engineering cases.

---

* 丁洁民，涂雨，吴宏磊，王世玉．减隔震组合技术在高烈度抗震设防区的应用研究［J］．建筑结构学报，2019，40（2）：77－87.

Keywords: high seismic intensity region; seismic isolation combined with energy-dissipation technology; superposition analysis method; design process

1. 引言

我国地处环太平洋地震带与欧亚地震带之间，地震断裂带充分发育，是世界上地震发生最频繁的国家之一。近年来，我国发生了多次强震，典型的包括汶川地震（8.0级），玉树地震（7.1级）以及雅安地震（7.0级），导致了大量的人员伤亡及经济损失。从我国地震烈度分布图[1]可以看出，我国高烈度地震区（7度（0.15g）及以上地区）分布广泛，主要集中于西南、西北及中部地区。

为提高高烈度地震区建筑的安全性及抗震性能，住房和城乡建设部印发了《关于房屋建筑工程推广应用减隔震技术的若干意见（暂行）》[2]的通知，对于重点设防类、特殊设防类和位于8度（含8度）以上地震高烈度区的建筑建议采用减隔震技术。此外新疆、陕西、四川、甘肃、云南等省亦陆续颁布相关政策，积极推广并使用减隔震技术。在国家与地方政策的推动下，我国减隔震技术日趋成熟，尤其是隔震技术。同时，随着社会经济的发展与生产技术的提高，隔震支座产品的质量大幅提升，我国隔震建筑得到快速发展。

然而，我国隔震技术的应用水平与日本相比仍有差距。在日本，隔震技术的应用包括：1）隔震技术的单独应用，主要包括铅芯橡胶支座、天然橡胶支座、弹性滑板支座等的单独或组合应用；2）在隔震层内设置减震装置与隔震支座混合应用，如日本清水总部大楼[3]；3）在隔震层外设置减震装置与隔震支座混合应用，如中之岛音乐厅[4]、东京日本桥大楼[5]。其中，主要以后两种方式，即隔震支座与减震装置的混合应用为主[6]。

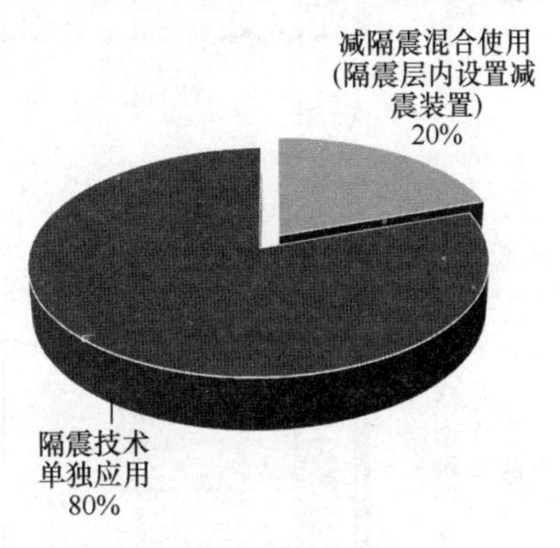

图1　中国隔震建筑减隔震装置使用情况[7]

相较而言，在中国，隔震技术的应用以前两种方式为主，且大多数为隔震技术的单独应用[7]（图1），缺乏对减隔震技术的创新应用。

为此，文中以位于高烈度地震区的某规划展览馆为工程背景，采用在隔震层上部结构布置黏滞阻尼支撑与基础隔震混合使用的新型减隔震组合技术，并将其首次应用于国内工程设计中。通过对新型减隔震组合体系选型的研究，分析结构在罕遇地震作用下的抗震性能，证明新型减隔震组合技术的有效性，并提出适用于在隔震层上部结构设置减震装置的新型减隔震组合结构设计流程，以期供类似工程案例参考。

2. 工程概况

某规划展览馆位于云南省玉溪市澄江县，抗震设防烈度为8度（0.3g），场地类别为Ⅲ类，设计地震分组为第三组，基本风压为0.3kN/m^2，建筑高度为17.7m（不含隔震层），地下1层，地上3层。建筑平面尺寸109m×49.6m，建筑平面及剖面如图2所示。

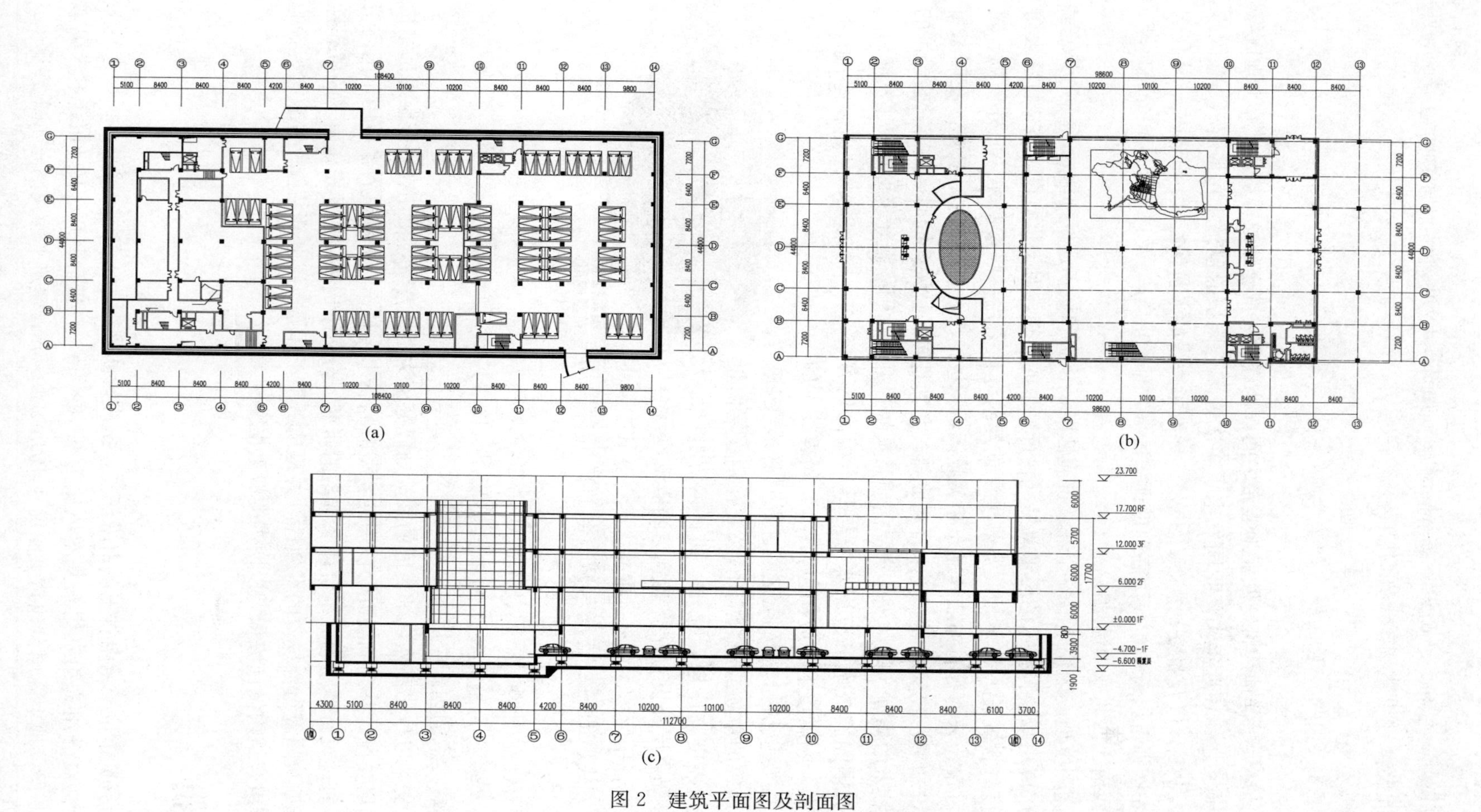

图 2　建筑平面图及剖面图

（a）地下 1 层建筑平面；（b）地上典型楼层建筑平面；（c）建筑剖面

该项目有以下特点：1）属于大型展览馆，抗震设防类别为重点设防类（乙类）；2）地勘报告显示，拟建场地 3km 内存在地震断裂带，根据《建筑抗震设计规范》GB 50011—2010（以下简称《抗规》）[8]，地震作用宜考虑 1.5 倍放大系数，即多遇地震$\alpha_{max}=0.36$，地震作用大于 9 度设防（$\alpha_{max}=0.32$）。因此，对结构应采取有效抗震措施，改善其抗震性能，保证其在地震作用下的安全性。

3. 减隔震组合体系选型

3.1 减隔震组合体系

根据项目特点，结构初步设计拟采用隔震方案，即隔震层上部为钢筋混凝土框架结构，隔震层位于基础与上部结构之间，层高 2m，结构体系组成如图 3 所示。

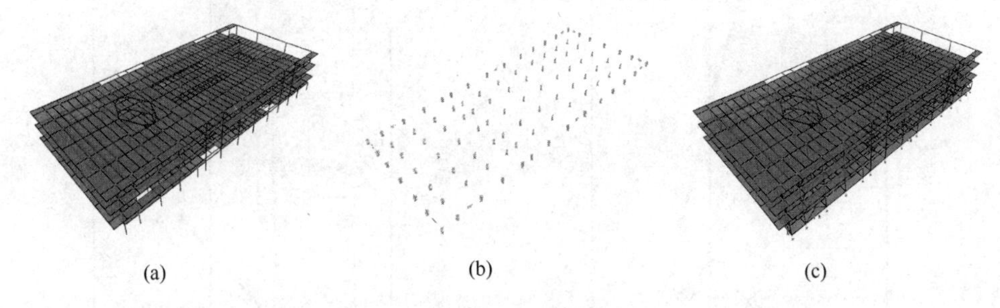

图 3 结构体系组成

（a）上部钢筋混凝土框架；（b）隔震层；（c）整体结构

分析时，从Ⅲ类场地土地震波库中选取 7 组地震波（包括 2 组人工波，5 组天然波）对隔震方案进行动力时程分析。经验算，所选时程波满足《抗规》[8]“弹性时程分析时，每条时程曲线计算所得结构底部剪力不应小于振型分解反应谱法计算结果的 65%，多条时程曲线计算所得结构底部剪力的平均值不应小于振型分解反应谱法计算结果的 80%”的要求。地震波信息见表 1。

地震波信息 **表 1**

编号	地震波	类型	持续时间（s）
RG1	RG1X/RG1Y	人工波	30.01
RG2	RG2X/RG2Y	人工波	30.01
TR1	BJHEW/BJHSN	天然波	49.23
TR2	RSN1116-SHI000/ RSN1116-SHI090	天然波	40.96
TR3	PRS90L/PRS00L	天然波	40
TR4	VEA50/VEA140	天然波	87.79
TR5	VDM290/VDM200	天然波	112.59

分析结果表明，设防地震作用下，隔震层上部结构最大减震系数为0.22（7条波平均值）。根据《云南省隔震减震建筑工程促进规定实施细则》（以下简称《细则》）[9]，上部结构设计时，减震系数取0.27，即上部结构按设防烈度降一度半设计。

图4给出了上部结构多遇地震弹性设计时的最大层间位移角。可以看出，采用隔震方案时，上部结构最大层间位移角不满足《抗规》[8]限值要求。为此，提出两种解决方案：1）增加上部结构刚度，在上部结构布置刚性支撑，形成隔震与刚性支撑组合方案；2）增加上部结构附加阻尼，在上部结构布置消能减震装置，即布置黏滞阻尼支撑，形成减隔震组合方案。表2给出了两种组合方案的主要分析结果，可以看出，两种方案上部结构均能满足结构刚度需求，但由于刚性支撑具备较大的刚度，在罕遇地震作用下，刚性支撑对应位置处隔震支座拉应力较大，刚性支撑方案部分隔震支座无法满足《抗规》[8]限值要求。

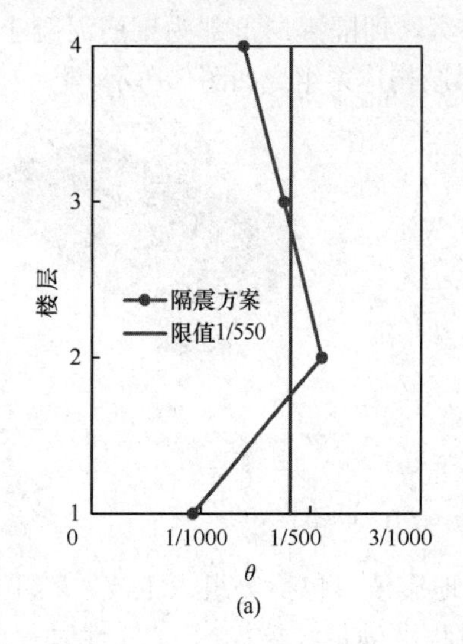

(a)

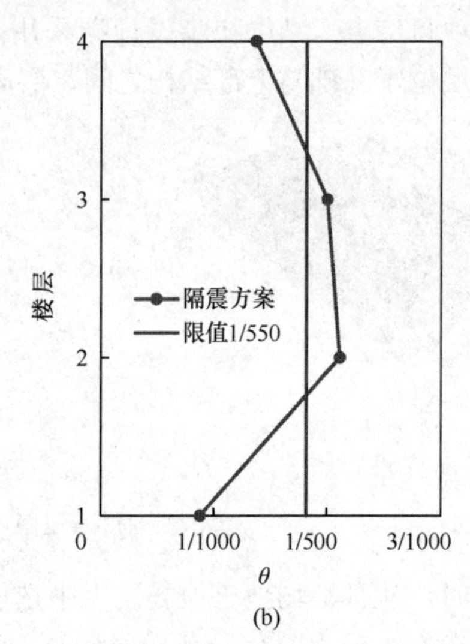

(b)

图4　上部结构层间位移角（多遇地震）

(a) X向；(b) Y向

两种组合方案分析结果　　表2

方案	θ_{max}	$\sigma_{t,max}$
隔震+刚性支撑	1/566（X向）； 1/573（Y向）	1.95 MPa
减隔震组合方案	1/579（X向）； 1/615（Y向）	0.98MPa

注：θ_{max}为多遇地震作用下结构最大层间位移角；$\sigma_{t,max}$为罕遇地震作用下隔震支座最大拉应力。

因此，采用新型减隔震组合体系，即在基础与上部结构之间设置隔震层，在上部结构钢筋混凝土框架中布置黏滞阻尼支撑。结构体系组成如图5所示。

3.2　减隔震装置布置及参数

为满足建筑结构设计需求，创新地采用基础隔震与上部结构黏滞阻尼支撑组合的新型减隔震技术。该技术的应用包括隔震层布置与上部结构黏滞阻尼支撑布置2个关键部分。

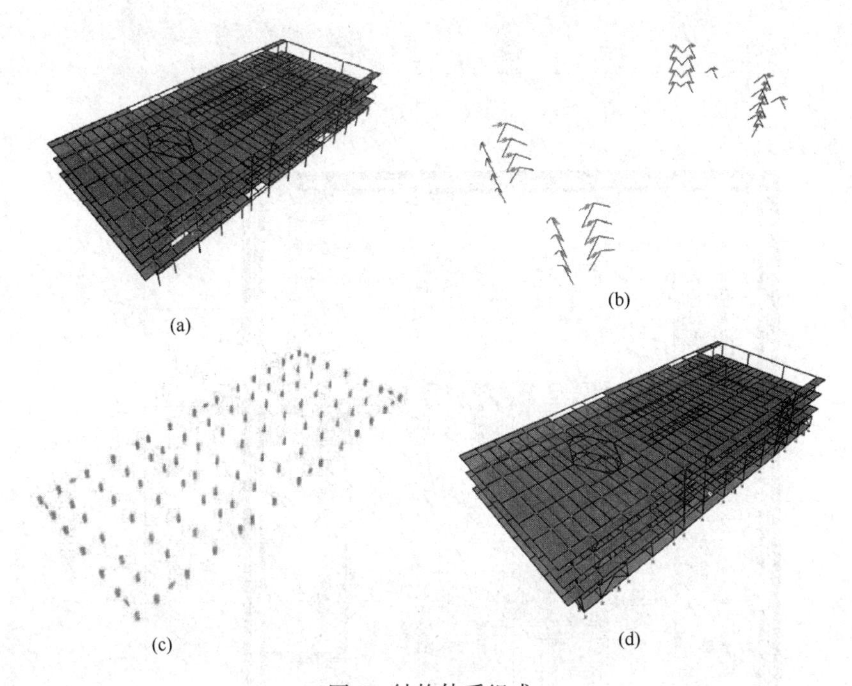

图5　结构体系组成

(a) 钢筋混凝土框架；(b) 黏滞阻尼支撑；(c) 隔震层；(d) 整体结构

3.2.1　隔震层布置及参数

隔震层的布置主要考虑以下因素：1）刚度较大的铅芯橡胶支座沿结构周边布置，刚度较小的天然橡胶支座沿内部布置，以获得更好的扭转刚度；2）重力荷载代表值作用下隔震支座竖向平均应力不应超过《抗规》[7]限值（12MPa，乙类建筑）；3）罕遇地震作用下，隔震支座不宜出现拉应力，当少数隔震支座出现拉应力时，其拉应力不应大于1.0MPa；4）罕遇地震作用下，隔震支座的极限水平变位应小于其有效直径的0.55倍和各橡胶层总厚度3倍二者的较小值。

综合以上因素，隔震层共布置隔震支座100个，黏滞阻尼器10个，如图6所示。表3～表5给出了本项目隔震支座与黏滞阻尼器的设计参数。

铅芯橡胶支座设计参数　　表3

隔震支座型号	LRB800	LRB1000	LRB1300
外径（mm）	800	1000	1300
橡胶总厚（mm）	156.6	197.2	259
第一形状系数	37.0	36.8	46.4
第二形状系数	5.1	5.1	5.0
产品总高（mm）	364.6	435.2	571.0
竖直刚度（kN/mm）	3973	4903	7325
基准面压（N/mm²）	15	15	15
一次刚度（kN/mm）	16.459	20.507	26.279
二次刚度（kN/mm）	1.266	1.577	2.021
屈服荷载（kN）	160	303	423
等效刚度（kN/mm）	2.290	3.114	3.656
阻尼比（%）	26.5	28.9	26.5
数量	13	24	4

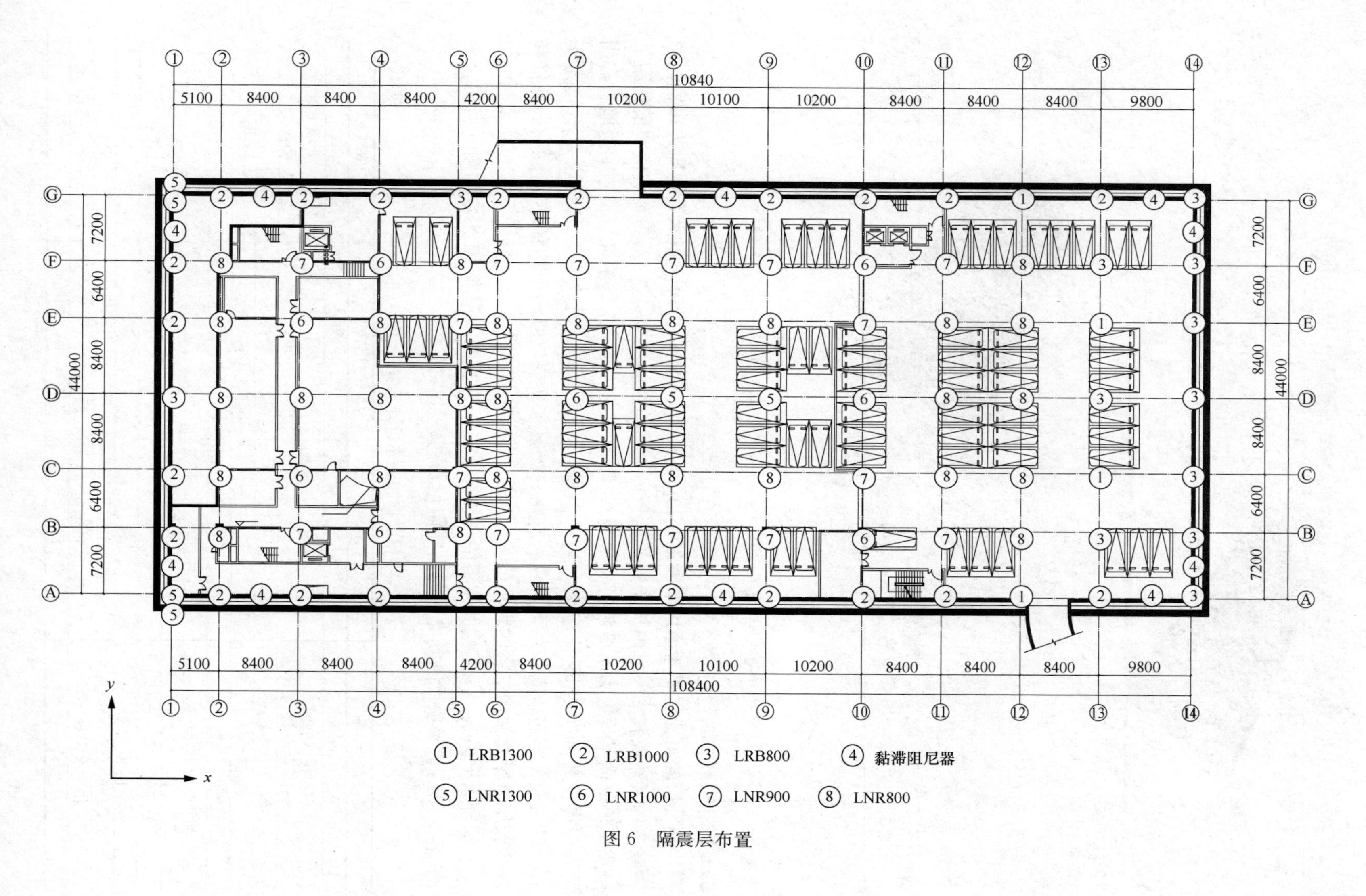
10840
5100
8400
4200
10200
10100
9800
7200
6400
44000
108400
y
x
① LRB1300
② LRB1000
③ LRB800
④ 黏滞阻尼器
⑤ LNR1300
⑥ LNR1000
⑦ LNR900
⑧ LNR800

图 6 隔震层布置

天然橡胶支座设计参数 **表 4**

支座型号	LNR800	LNR900	LNR1000	LNR1300
外径（mm）	800	900	1000	1300
橡胶总厚（mm）	156.6	176.9	197.2	259
S1	35.2	35.0	35.0	44.1
S2	5.1	5.1	5.1	5.0
产品总高（mm）	364.6	398.9	435.2	571
竖直刚度（kN/mm）	3517	3926	4335	6687
基准面压（N/mm^2）	15	15	15	15
等效刚度（kN/mm）	1.238	1.387	1.536	1.977
数量	29	16	8	6

隔震层黏滞阻尼器设计参数 **表 5**

阻尼系数（kN·（m/s）－0.3）	阻尼指数	行程（mm）	数量
1500	0.3	600	10

图 7 和图 8 给出了该项目隔震层的水平恢复力曲线与隔震支座长期面压分布情况。可以看出，隔震层具备足够的初始刚度，能保证结构在 100 年一遇风荷载或其他较小水平荷载作用下的稳定性，且隔震层屈服后水平刚度较小，能保证结构在地震作用下提供良好的隔震效果。同时，重力荷载代表值作用下，隔震支座最大长期面压值为 11.27MPa，满足《抗规》[8] 限值要求。表 6 给出了结构隔震层偏心率计算结果。由表可知，隔震层两个方向的偏心率分别为 0.60%（X 向）、0.59%（Y 向），均小于 3%，说明隔震层布置合理，上部结构重心与隔震层刚心基本重合，隔震支座具备足够的安全性与稳定性。表 7 给出了纯框架结构与单独隔震结构自振周期对比。可见，增设隔震层后，结构自振周期延长约 3 倍。结构周期从 0.97s 延长至 2.97s，远离场地特征周期，有利于减小上部结构地震作用。

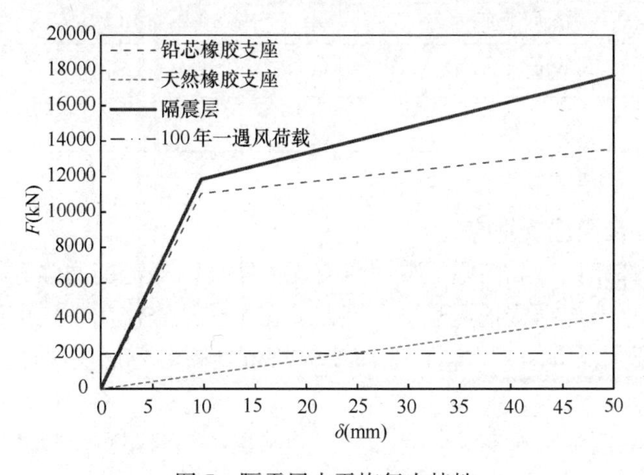

图 7　隔震层水平恢复力特性

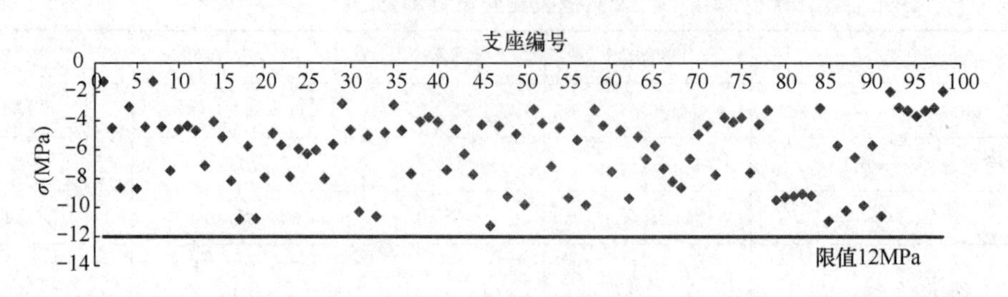

注：图中负值表示支座受压。

图 8　隔震支座长期面压分布

隔震层偏心率计算结果　　表 6

重心位置	X 向，56.345m
	Y 向，25.034m
刚心位置	X 向，56.106m
	Y 向，24.800m
偏心距	X 向，0.234m
	Y 向，0.239m
扭转刚度	3.2E+08 kN·m
回转半径	R_x，39.847m
	R_y，39.847m
偏心率	e_x，0.60%
	e_y，0.59%

纯框架结构与单独隔震结构自振周期对比　　表 7

振型	T/s		η
	纯框架结构	单独隔震结构	
1	0.97（X 向平动）	2.97（X 向平动）	3.06
2	0.95（Y 向平动）	2.96（Y 向平动）	3.12
3	0.88（扭转）	2.71（扭转）	3.08

注：η 为单独隔震结构周期与纯框架结构周期之比。

3.2.2　上部结构黏滞阻尼支撑布置及参数

与传统刚性支撑相比，黏滞阻尼支撑具有无静刚度、耗能能力强等特点，可为结构提供附加阻尼比，减小上部结构地震响应。因此，在罕遇地震作用下，可使隔震支座拉应力控制在《抗规》[8]要求范围内。本项目上部结构共布置黏滞阻尼支撑 34 榀，其中 X 向 16 榀，Y 向 18 榀。

图 9 给出了黏滞阻尼支撑布置示意图，表 8 给出了黏滞阻尼支撑所用黏滞阻尼器设计参数。

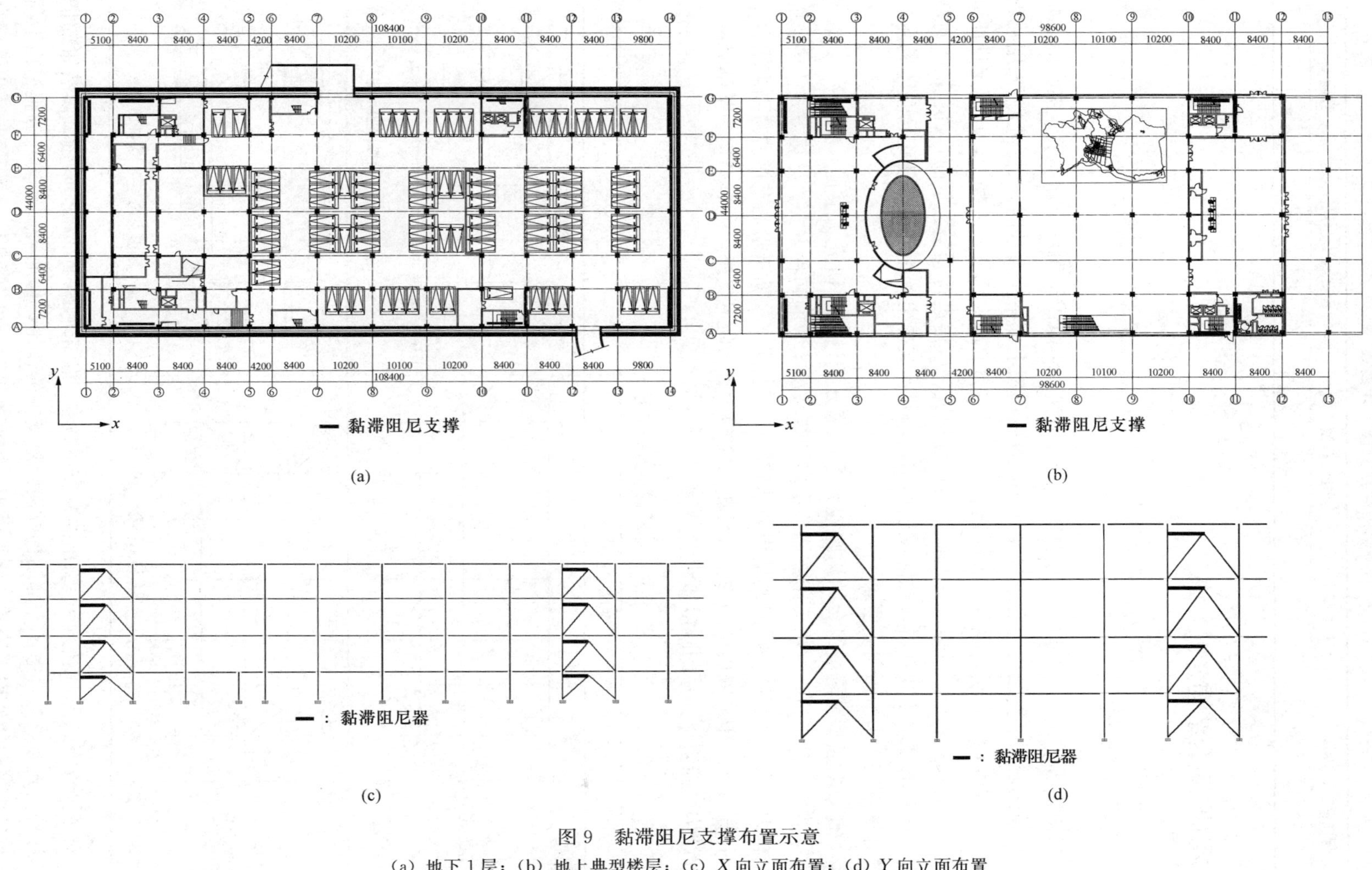

图 9 黏滞阻尼支撑布置示意

（a）地下 1 层；（b）地上典型楼层；（c）X 向立面布置；（d）Y 向立面布置

黏滞阻尼支撑设计参数　　表 8

阻尼系数（kN·（m/s）$^{-0.3}$）	阻尼指数	行程（mm）	数量
2000	0.3	45	34

4. 减隔震组合结构分析方法

4.1　分析方法概述

针对隔震建筑，《抗规》[8]采用分离式设计方法，将隔震结构分为隔震层下部结构，隔震层以及隔震层上部结构分别进行设计。进行隔震层上部结构设计时，《抗规》[8]采用减震系数（按弹性计算所得的隔震与非隔震各层层间剪力的最大比值）的概念，从而定量评价隔震结构的减震效率，并以此为基础，对设计地震反应谱进行调整。目前，《抗规》[8]对减震系数的计算要求主要针对仅含隔震层的结构，对本文提出的新型减隔震组合技术的减震系数计算方法并未给出明确要求。

为此，对新型减隔震组合结构减震系数计算方式进行了研究。根据《抗规》[8]，表 9 给出了新型减隔震组合结构采用整体分析法计算减震系数的两种方式。计算方式 1 中非隔震结构为上部纯框架结构，隔震结构为基础隔震＋上部结构黏滞阻尼支撑的减隔震组合结构；计算方式 2 中非隔震结构为上部纯框架结构＋黏滞阻尼支撑，隔震结构为基础隔震＋上部结构黏滞阻尼支撑的减隔震组合结构。

减震系数计算方式　　表 9

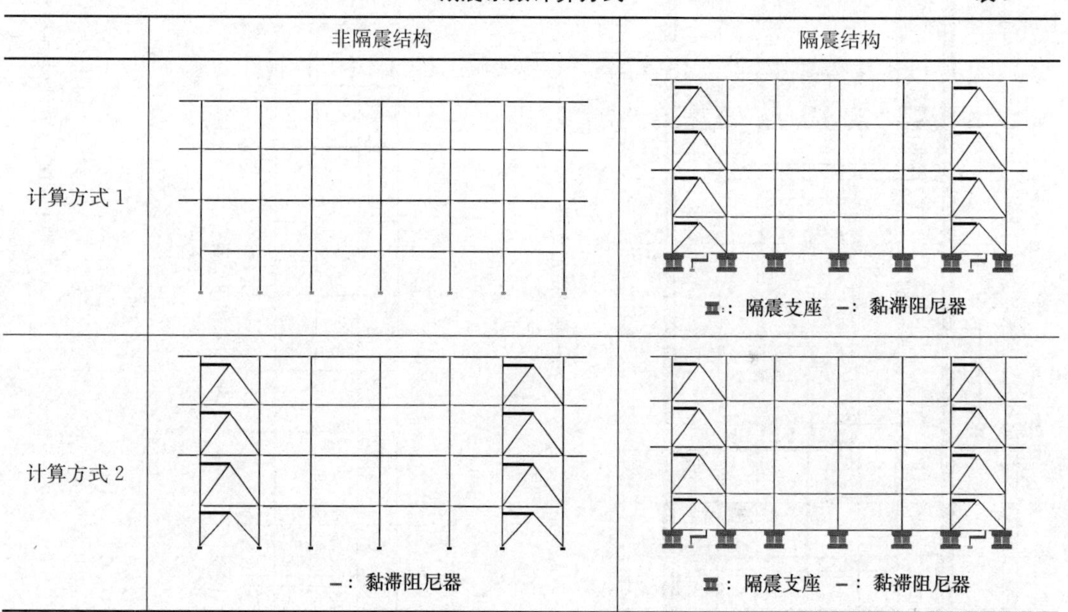

	非隔震结构	隔震结构
计算方式 1		Ⅲ：隔震支座　—：黏滞阻尼器
计算方式 2	—：黏滞阻尼器	Ⅲ：隔震支座　—：黏滞阻尼器

对于计算方式 1，《抗规》[8]给出的减震系数参考值未考虑上部结构附带减震装置的情况；计算方式 2 虽然考虑了上部结构黏滞阻尼支撑的作用，但非隔震结构与隔震结构中黏滞阻尼支撑的耗能作用不等效，所得减震系数亦不准确。可以看出，对于该类减隔震组合

技术，同时考虑隔震层与上部结构黏滞阻尼支撑，对整体结构进行分析，计算减震系数具有一定的局限性，但采用整体分析法可准确地反映结构的减震效率。

因此，文中提出了叠加分析法，用以计算减震系数，设计上部结构，并采用整体分析法对结构整体减震效率进行复核、验算，保证结构的安全性。

4.2 叠加分析法

叠加分析法是根据叠加原理，将减隔震组合体系分为隔震体系与减震体系两部分，分别对其减震效率进行分析。具体分析流程如下：

1）对纯框架结构与单独隔震结构进行隔震效率分析，按《抗规》[8]方法计算得到减震系数；

2）在第一步的基础上，根据所得的减震系数，调整输入到上部减震结构的地震波峰值加速度，对上部减震结构黏滞阻尼支撑的减震效率进行分析。

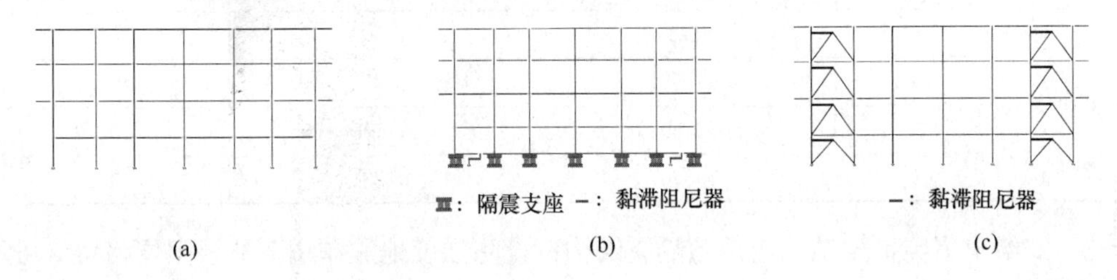

图 10 叠加分析法分析模型

（a）纯框架结构；（b）单独隔震结构；（c）上部减震结构

4.2.1 隔震分析

采用已选好的 7 组地震波（包括 2 组人工波和 5 组天然波），对纯框架结构与单独隔震结构进行设防地震作用（考虑 1.5 倍放大系数）下的时程分析。分析中采用 ETABS 有限元分析软件。梁柱采用 frame 单元模拟，橡胶隔震支座采用 isolator1 单元模拟，黏滞阻尼器采用基于 Maxwell 模型的 damper 单元模拟，采用 Ritz 法进行迭代分析，分析结果取 7 条波的平均值。

图 11 给出了纯框架结构与单独隔震结构在设防烈度地震作用下的层剪力分布情况。

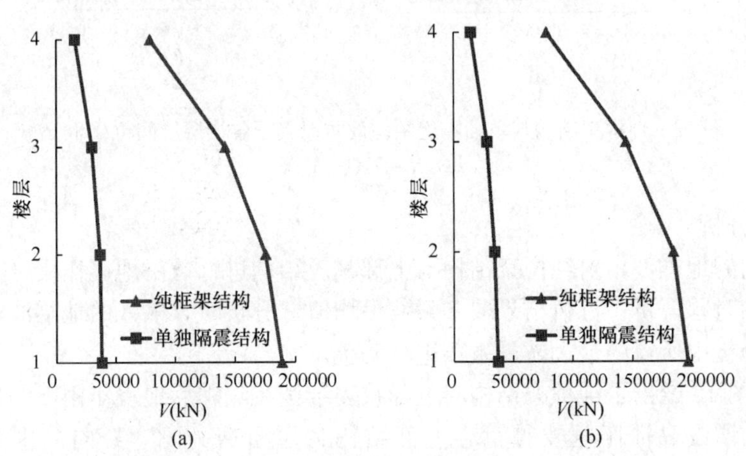

图 11 纯框架结构与单独隔震结构设防地震下各楼层层剪力分布

（a）X 向；（b）Y 向

可以看出，增设隔震层后，上部结构各层层剪力明显减小，隔震体系具有良好的隔震效率。表 10 给出了减震系数计算结果。可见，设防烈度地震作用下，单独隔震结构 X 向最大减震系数为 0.22，Y 向最大减震系数为 0.20。根据《细则》[9]要求，上部结构设计时，减震系数取 0.27，即上部结构按设防烈度降低一度半进行设计。

纯框架结构与单独隔震结构在设防烈度地震作用下各楼层层剪力　　表 10

输入方向	楼层	层剪力 V（kN）		减震系数
		纯框架结构	单独隔震结构	
X 向	4	79362	15534	0.20
	3	142087	29989	0.21
	2	175834	36762	0.22
	1	189264	38147	0.21
Y 向	4	78114	14566	0.19
	3	144571	28278	0.20
	2	184501	34506	0.19
	1	196461	37286	0.20

图 12 给出了纯框架结构与单独隔震结构在设防烈度地震作用下的层间位移角。可见，增设隔震层后，上部结构最大层间位移角 X、Y 向分别为 1/291、1/327，各层层间变形明显减小，结构抗震性能明显改善。

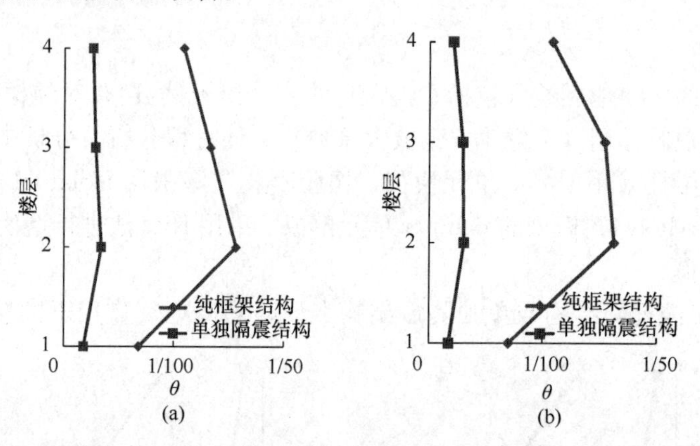

图 12　纯框架结构与单独隔震结构设防地震下各楼层层间位移角分布

(a) X 向；(b) Y 向

4.2.2　减震分析

采用相同的地震波，对纯框架结构与上部减震结构进行设防地震作用（考虑 1.5 倍放大系数）下的时程分析，分析结果取 7 条波平均值。分析时，输入的地震波峰值加速度根据隔震效率分析中所得的最大减震系数进行调整。

表 11 与图 13 给出了纯框架结构与上部减震结构在隔震后地震作用下的层剪力分布情况。上部结构增设黏滞阻尼支撑后，上部结构各层层剪力 X、Y 向分别减小约 22%、21%，黏滞阻尼支撑发挥了良好的耗能作用。

图 14 给出了纯框架结构与上部减震结构在隔震后地震作用下的层间位移角。可以看

出，上部结构增设黏滞阻尼支撑后，各层层间变形 X、Y 向分别减小约 14%、17%，结构具有良好的抗震性能。

注：β_1 为上部减震结构各层层剪力与纯框架结构各层层剪力之比。

表 12 给出了上部结构由黏滞阻尼支撑提供的附加阻尼计算结果。其中，附加阻尼比的计算采用 JGJ 297—2013《建筑消能减震技术规程》[10] 中提供的方法。可以看出，在隔震后地震波作用下，上部结构黏滞阻尼支撑 X、Y 向可分别提供附加阻尼比约 4.99%、4.80%，黏滞阻尼支撑发挥了良好的耗能能力。

纯框架结构与上部减震结构在隔震后地震作用下各楼层层剪力　　表 11

输入方向	楼层	层剪力 V（kN）		β_1
		纯框架结构	上部减震结构	
X 向	4	21386	16133	75%
	3	38519	30232	78%
	2	47157	36385	77%
	1	50847	38494	76%
Y 向	4	19871	15345	77%
	3	36970	29121	79%
	2	47026	35435	75%
	1	49675	38551	78%

注：β_1 为上部减震结构各层层剪力与纯框架结构各层层剪力之比。

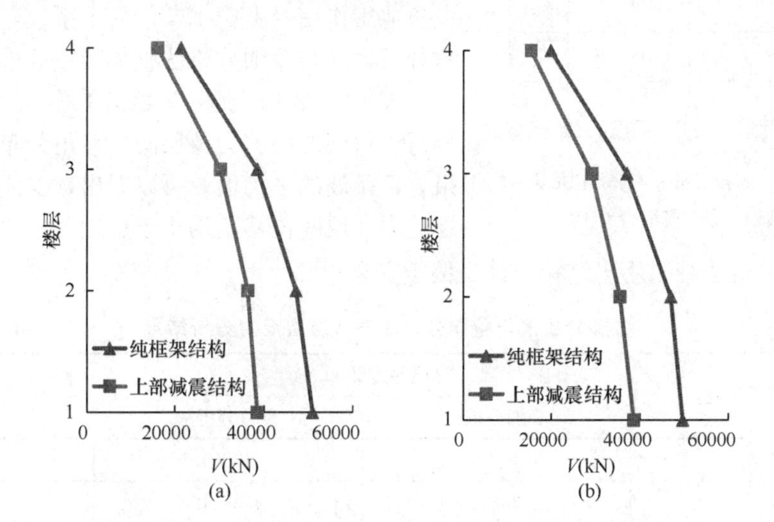

图 13　纯框架结构与上部减震结构隔震后地震作用下各楼层层剪力分布

(a) X 向；(b) Y 向

黏滞阻尼支撑附加阻尼比计算结果　　表 12

输入方向	W_{cj}（kN・m）	W_s（kN・m）	ζ
X 向	5.25×105	0.94×106	4.99%
Y 向	4.44×105	0.81×106	4.80%

注：W_{cj} 为黏滞阻尼支撑耗能；W_s 为结构总应变能；ζ 为结构附加阻尼比。

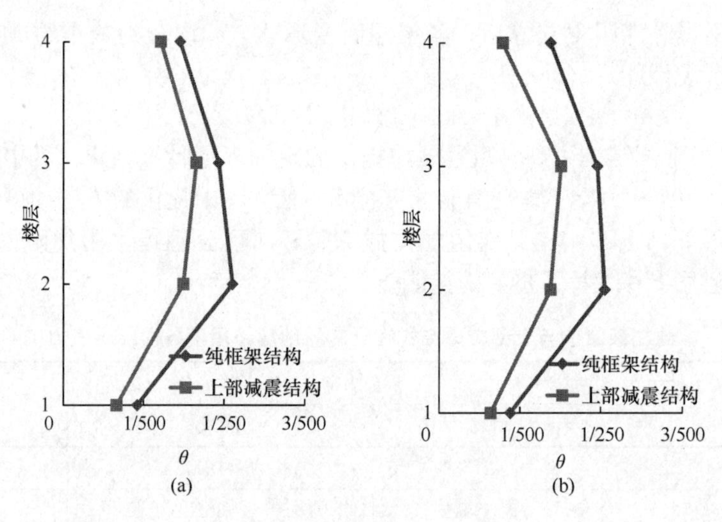

图 14　纯框架结构与上部减震结构隔震后地震作用下各楼层层间位移角分布

(a) X 向；(b) Y 向

4.3　整体分析法

叠加分析法虽然计算过程思路明确，但忽略了上部减震体系与隔震体系间的相互作用，为保证结构的安全性，对同时设有隔震层与上部黏滞阻尼支撑的减隔震组合结构（图 15）进行设防地震作用（考虑 1.5 倍放大系数）下的动力时程分析，并与叠加分析法整体减震效果进行对比。

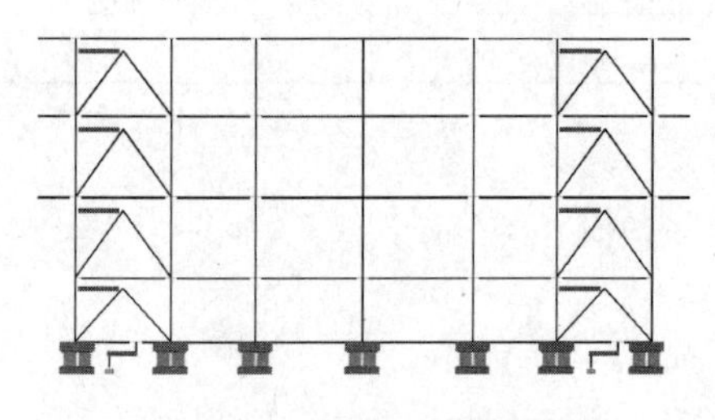

图 15　减隔震组合结构

表 13、图 16 与图 17 给出了叠加分析法与整体分析法得到的层剪力及层间位移角分布情况，其结果为 7 条波的平均值。可以看出，整体分析法各层层剪力、层间位移角均小于叠加分析法的结果。因此，将叠加分析法结果用于结构设计是偏于安全的。

叠加分析法与整体分析法各楼层层剪力分析结果　　**表 13**

输入方向	楼层	层剪力 V（kN）		β_2
		叠加分析法	整体分析法	
X 向	4	16133	11366	142%
	3	30232	23141	131%
	2	36385	30102	121%
	1	38494	35163	109%
Y 向	4	15345	11265	136%
	3	29121	23123	126%
	2	35435	30259	117%
	1	38551	35901	107%

注：β_2 为叠加分析法的各层层剪力与整体分析法的各层层剪力之比。

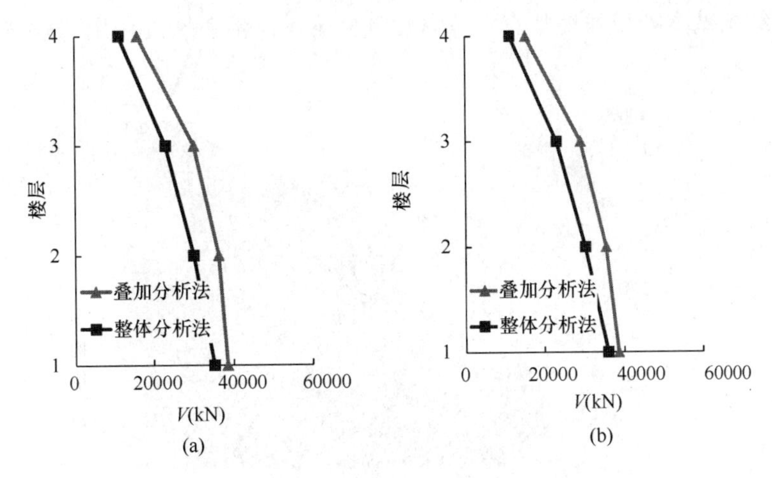

图 16　叠加分析法与整体分析法各楼层层剪力分布

(a) X 向；(b) Y 向

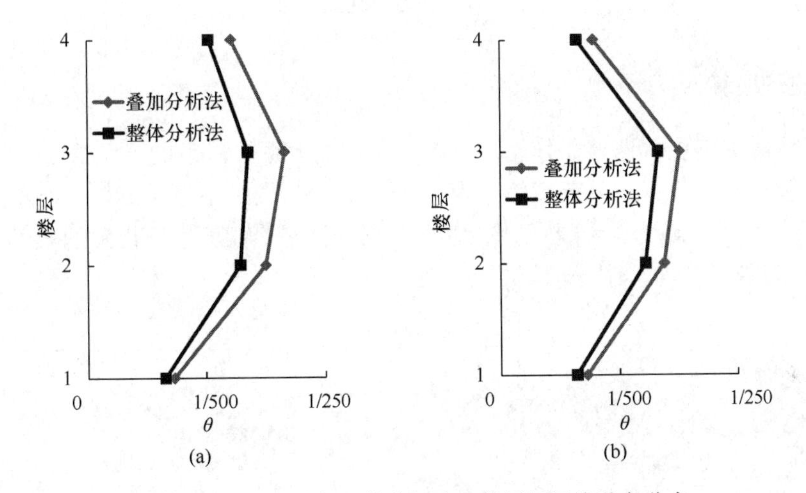

图 17　叠加分析法与整体分析法各楼层层间位移角分布

(a) X 向；(b) Y 向

5. 罕遇地震下结构响应分析

对减隔震组合结构进行罕遇地震作用（考虑 1.5 倍放大系数）下的结构响应分析。分析中采用基于纤维模型理论的弹塑性分析软件 PERFORM-3D V5，框架梁塑性铰采用 M 铰模拟，框架柱塑性铰采用 P-M 铰模拟，橡胶隔震支座采用 isolator 单元模拟，阻尼器采用基于 Maxwell 模型的 damper 单元模拟。

5.1　结构能量时程

图 18 给出了减隔震组合结构在罕遇地震作用（考虑 1.5 倍放大系数）下的能量耗散情况。可以看出，输入到结构的地震能量大部分由隔震层与黏滞阻尼器耗散，其中隔震支座耗能占比约为 23%，黏滞阻尼器耗能占比约为 34%，结构构件塑性耗能占比仅为

0.2%，上部结构基本处于弹性状态，减隔震组合结构在罕遇地震作用下呈现良好的抗震性能。

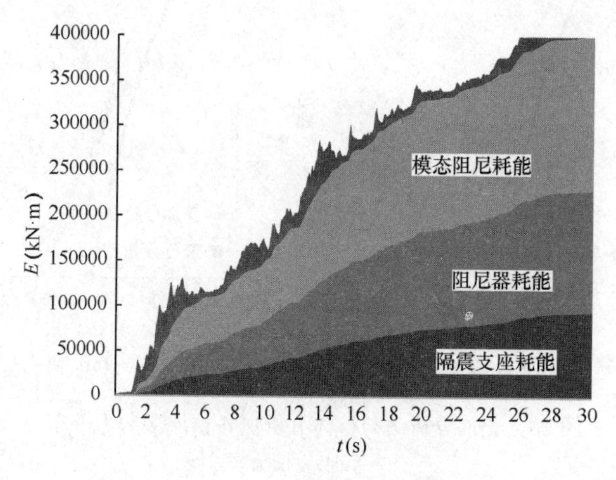

图 18 结构能量耗散分布

5.2 隔震支座验算

为验算隔震支座在罕遇地震作用（考虑 1.5 倍放大系数）下的工作情况，对隔震支座进行了短期极大面压验算，其荷载组合为 1.0×恒载+0.5×活载±1.0×罕遇地震水平地震力±0.5×竖向地震力，结果见图 19。可见，橡胶隔震支座最大的极大值面压为 17.48MPa，满足《抗规》[8]限值要求。

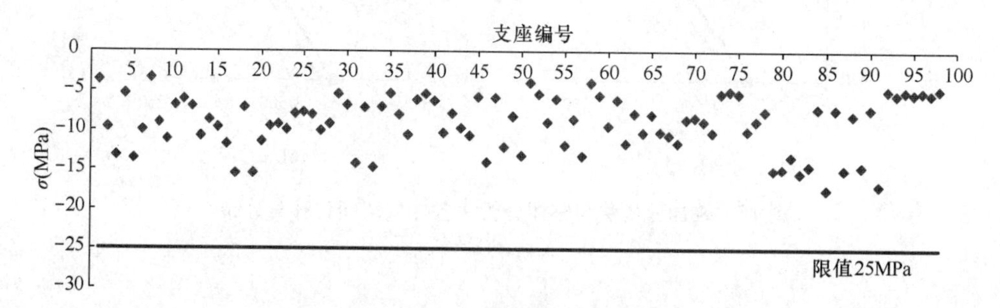

图 19 隔震支座短期极大面压分布

注：图中负值表示支座受压。

此外，对隔震支座在罕遇地震作用（考虑 1.5 倍放大系数）下支座受拉情况，即短期极小面压进行了验算，其荷载组合为 1.0×恒载±1.0×罕遇地震水平地震力±0.5×竖向地震力，结果见图 20。可见，支座最小的极小值面压为 0.98MPa，绝大部分支座未出现受拉的现象，满足《抗规》[8]限值要求。

图 21 给出了罕遇地震作用（考虑 1.5 倍放大系数）下隔震支座 X 向和 Y 向变形。隔震支座在罕遇地震作用下最大水平位移为 431mm，对于最小规格的 LNR800 支座，该水平位移相当于 276%的剪应变，小于 $0.55D$（$0.55D$=440mm）和 300%的剪应变（300% γ=469.8mm），满足《抗规》[8]要求。

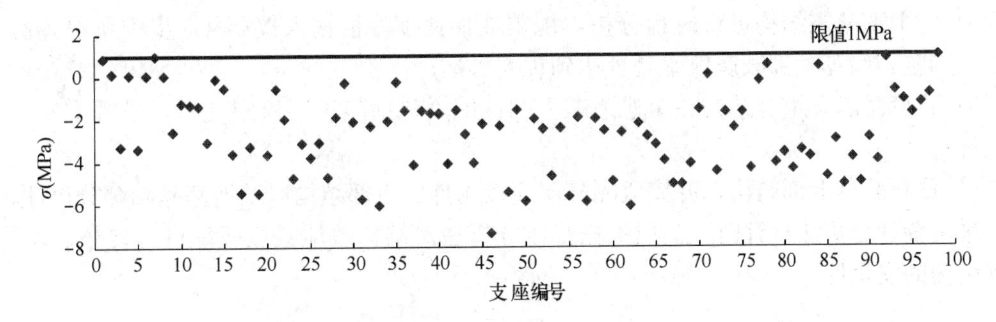

图 20　隔震支座短期极小面压分布

注：图中负值表示支座受压，正值表示支座受拉。

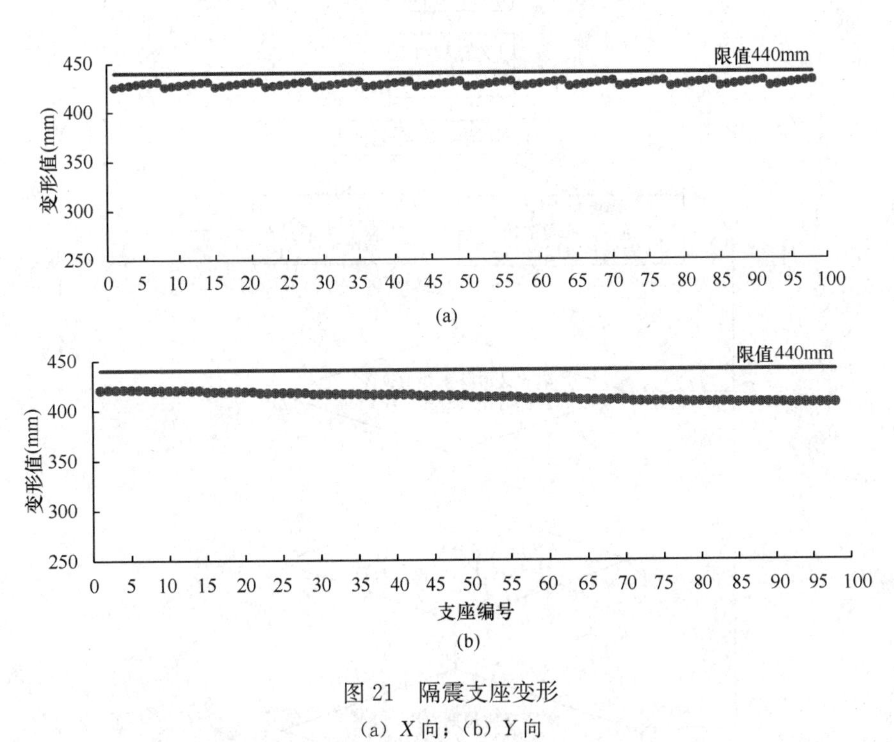

图 21　隔震支座变形

(a) *X* 向；(b) *Y* 向

6. 减隔震组合结构设计流程

根据以上分析结果，总结并提出了采用叠加分析法设计此类新型减隔震组合结构的设计流程，如图 22 所示。设计时，建议按下列步骤进行：

1）根据实际需求，制定合理的设计目标，包括隔震目标与减震目标；

2）同时考虑隔震与减震作用，按反应谱初步设计上部结构；

3）根据设计目标，设计减隔震组合体系，主要包括隔震层设计与上部结构减震体系设计；

4）采用叠加分析法，首先对隔震结构进行时程分析，计算减震系数，然后验算隔震效率是否达到预期目标；

5）对上部减震结构进行时程分析，地震波加速度峰值输入按第4）步中所得减震系数进行调整，验算上部减震效率是否达到预期目标；

6）验算减隔震组合结构在罕遇地震下的性能，包括隔震层极限变形，隔震支座短期面压等；

7）重新验算上部结构，并完成隔震层连接构件、下部结构以及地基基础部分设计。

采用叠加分析法设计时，尚应根据整体分析法对结构整体减震效率进行复核、验算，保证结构的安全性。

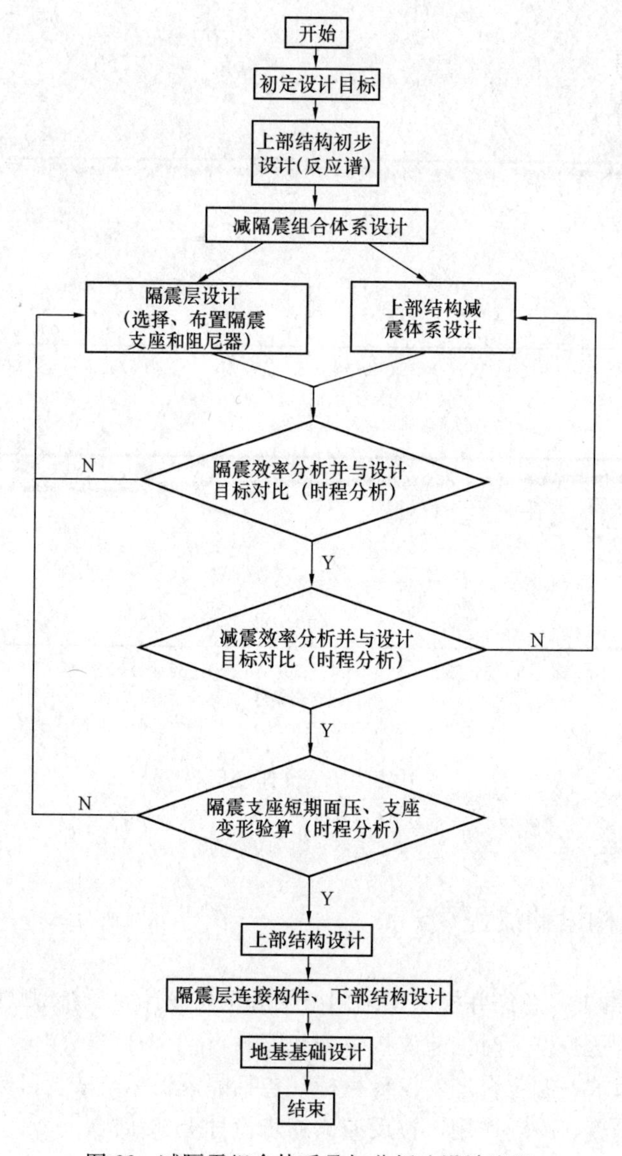

图22 减隔震组合体系叠加分析法设计流程

7. 结论

1）对高烈度地震区建筑，创新地提出了隔震＋上部结构黏滞阻尼支撑的新型减隔震组合技术，并将其首次应用于国内工程设计中。

2）新型减隔震组合技术可显著减小上部结构地震作用。罕遇地震作用下，上部结构基本保持弹性，结构抗震性能明显改善。

3）对于新型减隔震组合结构，提出了叠加分析法。叠加分析法计算过程概念清晰，思路明确，计算结果可用于指导减隔震组合结构的设计。设计时，应按整体分析法对结构最终减震效率进行复核、验算。

4）提出了采用叠加分析法设计新型减隔震组合结构的设计流程，可为类似的工程设计提供参考。

参考文献

[1] 中国地震动峰值加速度区划图 GB 18306—2015[S]. 北京：中国标准出版社，2015.

[2] 中华人民共和国住房和城乡建设部．关于房屋建筑工程推广应用减隔震技术的若干意见(暂行)[Z]. 2014.02.21.

[3] Dai Shimazaki，Kentaro Nakagawa. Seismic Isolation Systems Incorporating with RC Core Walls and Precast Concrete Perimeter Frames[J]. International Journal of High-Rise Buildings，September 2015，Vol 4，No 3，181-189.

[4] Ken Okada，Satoshi Yoshida. Structural Design of Nakanoshima Festival Tower[J]. International Journal of High－Rise Buildings，September 2014，Vol 3，No 3，173-183.

[5] 久次米，曾根朋久．The Structural Design of Tokyo Nihombashi Tower[J]. Structure：journal of Japan Structural Consultants Association，2015：50-51.

[6] 丁洁民，吴宏磊．黏滞阻尼技术工程设计与应用[M]. 北京：中国建筑工业出版社，2017：19.

[7] 刘鹏．高层隔震建筑结构研究与应用[D]. 上海：同济大学，2017.（Liu Peng. Research and Application of High Buildings Using Isolation Technology [D]. Shanghai：Tongji University，2017(in Chinese)）

[8] 建筑抗震设计规范 GB 50011—2010[S]. 北京：中国建筑工业出版社，2016.

[9] 云南省住房和城乡建设厅．云南省隔震减震建筑工程促进规定实施细则[Z]. 2016.12.29.

[10] 建筑消能减震技术规程 JGJ 297—2013[S]，北京：中国建筑工业出版社，2013.

11 屈曲约束支撑研究新进展

周　云，曹邕生，邵鹤天，李　壮
（广州大学土木工程学院，广州）

摘　要：介绍了新型屈曲约束支撑的设计理念，研究团队开发的钢板装配式屈曲约束支撑、开孔钢板装配式屈曲约束支撑、多核心开孔钢板装配式屈曲约束支撑、开孔或开槽式三重钢管屈曲约束支撑、外包钢筋混凝土屈曲约束支撑、屈曲约束耗能腋撑等新型屈曲约束支撑的性能、开孔板式屈曲约束支撑的设计方法、新型屈曲约束支撑框架结构的抗震性能及双向地震作用下屈曲约束支撑框架结构的稳定性能；给出屈曲约束支撑子结构及节点板的设计方法，构建屈曲约束支撑高位转换耗能减震结构体系、屈曲约束支撑耗能减震层结构体系、屈曲约束支撑＋黏滞阻尼器结构体系及屈曲约束支撑框架—核心筒结构体系四种新型结构体系，介绍屈曲约束支撑的标准化进程；指出屈曲约束支撑研究和推广应用中存在的问题，给出今后研究与应用的建议。

关键词：屈曲约束支撑；设计方法；结构体系；标准化

1. 引言

屈曲约束支撑最早在20世纪80年代由日本学者Fujimoto[1]提出。近30多年来，国外学者对屈曲约束支撑及其相关结构体系的性能及设计方法等进行了多方面的研究。我国大陆地区对屈曲约束支撑的研究起步较晚，但发展迅速，特别是2008年汶川大地震之后，减震技术得到了高度重视，屈曲约束支撑凭借其先进的设计理念和优越的耗能性能逐步受到学者及工程界的青睐。国内学者对屈曲约束支撑进行了系统性的研究，开发了不同构造的新型屈曲约束支撑并广泛运用于各类工程之中。本文从新型屈曲约束支撑的设计理念、新型屈曲约束支撑的性能研究、新型屈曲约束支撑的设计方法研究、新型屈曲约束支撑框架结构的抗震性能研究、节点板的设计方法研究、屈曲约束支撑子结构的设计方法研究、屈曲约束支撑工程应用研究及屈曲约束支撑标准化研究多个层面出发，对团队有关屈曲约束支撑领域的研究进行梳理和总结，提出未来需要研究的重点和方向。

2. 新型屈曲约束支撑的设计理念[2]

针对工程中常用的屈曲约束支撑存在屈服位置随机、端部易破坏、普通钢与低屈服点钢焊接质量不可靠等问题，提出“核心单元局部削弱相当于其他部分加强”的新型屈曲约束支撑设计思想，通过局部削弱屈曲约束支撑核心单元，使其有明显的屈服段与非屈服段，屈服位置可以预先设定从而具有定点屈服的功能，改进了屈曲约束支撑的性能。局部削弱的方式有多种，如开槽式、开孔式等，如图1所示。按“核心单元局部削弱”设计思想设计的屈曲约束支撑有以下优点：(1) 核心单元经局部削弱后，核心单元更易屈服，在工作中当受到的拉力和压力达到削弱部位的屈服强度时该部位将率先进入屈服耗能状态耗散地震能量，实现了由同种材料制成的核心单元具有明显的屈服段与非屈服段，避免了使用低屈服点钢，也避免了核心单元使用低屈服点钢与普通钢时的焊接工序及其带来的问题。(2) 由于核心单元的屈服首先发生在局部削弱部位，即通过局部削弱使核心单元实现定点屈服，避免了屈服位置随机，这非常有利于核心单元的设计。(3) 核心单元削弱后，其承载力略微有所下降，根据“强”与“弱”的相对性，这相当于加强了过渡段及连接段，过渡段及连接段不易发生破坏。(4) 当采用多处相同尺寸局部削弱时，可实现多处同时屈服耗能，这样便可充分利用材料，提高核心单元的耗能效率。根据“核心单元局部削弱”的设计理念，团队设计了多种新型屈曲约束支撑。

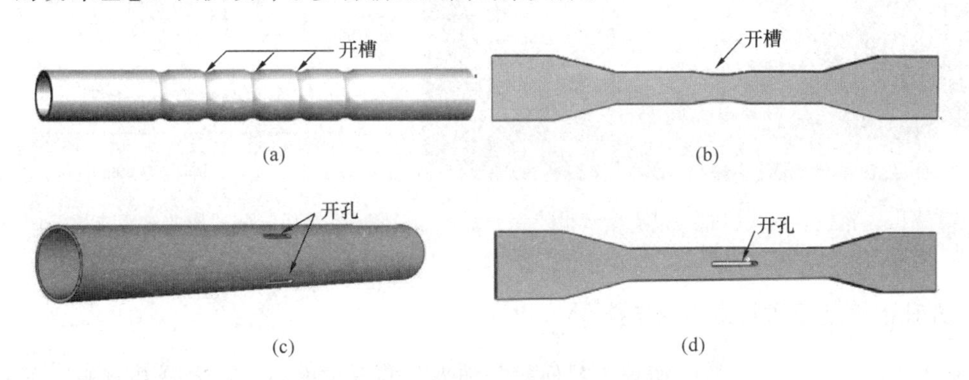

图 1　核心单元局部削弱示意图

(a) 钢管型核心单元开槽；(b) 钢板型核心单元开槽

(c) 钢管型核心单元开孔；(d) 钢板型核心单元开孔

3. 新型屈曲约束支撑开发研究

3.1　钢板装配式屈曲约束支撑[3-6]

传统钢管混凝土屈曲约束支撑质量大，涉及材料类型多，需要灌浆养护，加工工艺复杂，制作周期长。团队利用钢板，提出新型钢板装配式屈曲约束支撑，其构造如图 2 所示。它仅由钢板通过激光切割后，粘贴无粘结材料层，再通过高强螺栓组装而成，仅连接

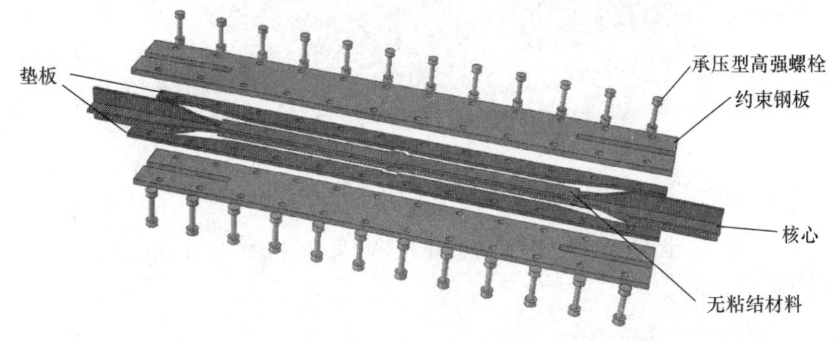

图 2　钢板装配式屈曲约束支撑构造示意图

段加劲肋处需要焊接，具有自重轻，体积小，取材容易，构造简单，加工精度容易控制，可藏于墙体内，现场安装方便等优点，极大地缩短了制作周期。图 3 为钢板装配式屈曲约束支撑的试验装置，图 4 为试验所得到屈曲约束支撑的滞回曲线。研究结果表明，钢板装配式屈曲约束支撑滞回曲线饱满对称，耗能性能优良，承载特性稳定。通过对不同设计参数的钢板装配式屈曲约束支撑进行有限元分析及试验研究，研究了约束比、无粘结材料的有无、加载顺序对其力学性能及耗能性能的影响。

图 3　钢板装配式屈曲约束支撑试验装置

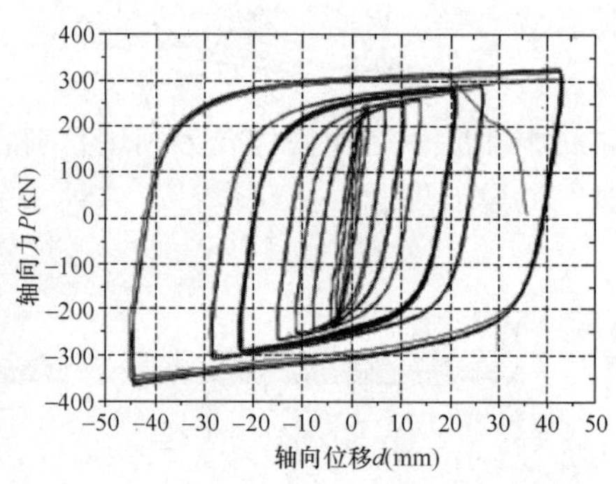

图 4　钢板装配式屈曲约束支撑滞回曲线

3.2　开孔钢板装配式屈曲约束支撑[7,8]

根据“核心单元局部削弱相当于其他部分加强”的设计思想，提出开孔钢板装配式屈曲约束支撑，其构造如图 5 所示。通过对核心单元开孔，使屈服段能够多段同时屈服耗能，充分利用钢材的延性，并采用螺栓代替限位卡进行限位，避免了限位卡与屈服段的截面突变所带来的应力集中现象。图 6 为开孔钢板装配式屈曲约束支撑的滞回曲线。研究结果表明，开孔钢板装配式屈曲约束支撑不仅支撑滞回曲线饱满对称，耗能性能优良，承载特性稳定，而且疲劳性能优越，破坏均发生在屈服段（图 7），达到了定点屈服、定点破坏的设计构想。

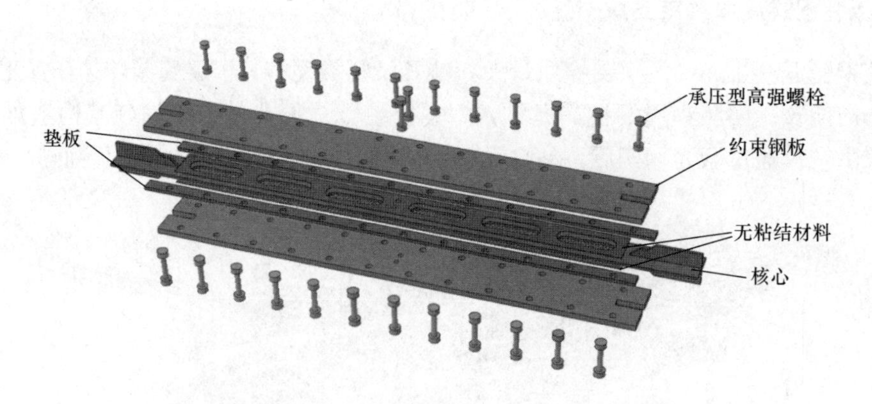

图 5　开孔钢板装配式屈曲约束支撑构造示意图

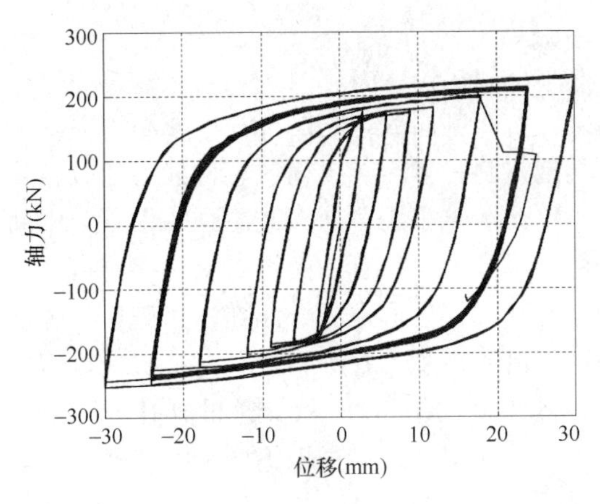

图 6　开孔钢板装配式屈曲约束支撑滞回曲线

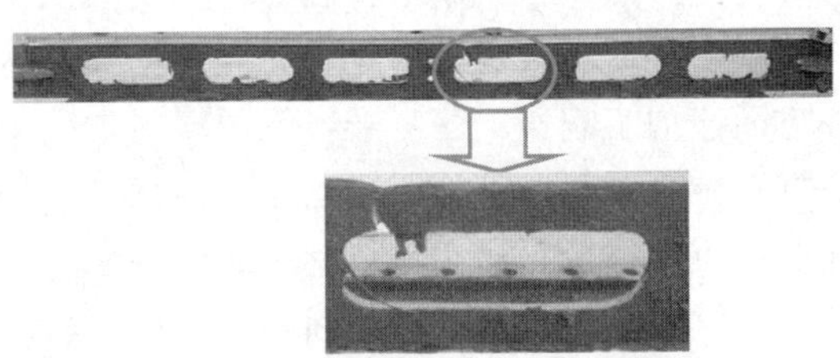

图 7　开孔钢板装配式屈曲约束支撑核心开孔段定点屈服破坏

3.3　多核心开孔钢板装配式屈曲约束支撑[9]

开孔钢板装配式屈曲约束支撑对核心单元开孔，减小了核心板轴向受力截面，降低了支撑的屈服力，不能满足一些承载力需求较大的工程。团队在开孔钢板装配式屈曲约束支撑的基础上提出多核心开孔钢板装配式屈曲约束支撑，其构造如图 8 所示。多核心开孔钢板装配式屈曲约束支撑核心由多块开孔核心板构成，增加了核心单元的受力面积，提高了支撑承载力，满足工程中较大承载力的需求。图 9 为试验到屈曲约束支撑的滞回曲线。研究结果表明，多核心钢板开孔装配式屈曲约束支撑滞回曲线饱满、对称，支撑承载能力高，与核心单元总厚度相同的单核心开孔钢板装配式屈曲约束支撑相比，其疲劳性能更好，端部的抗弯承载力更高。通过对不同设计参数多核心开孔钢板装配式屈曲约束支撑进行有限元分析及试验研究，研究了核心单元厚度、不同的约束形式等参数对其力学性能与耗能性能的影响，提出了该种屈曲约束支撑的设计方法。

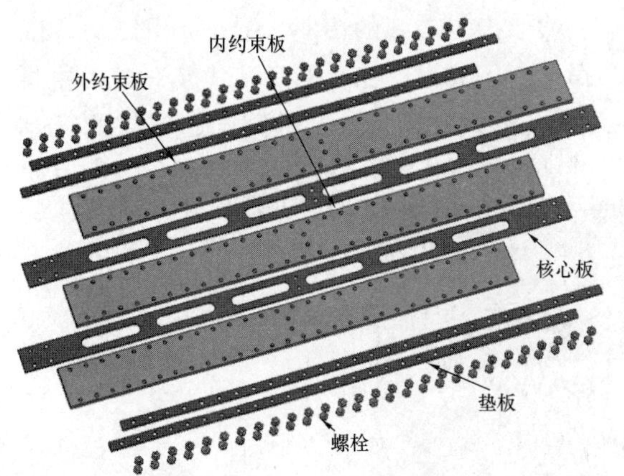

图 8　多核心开孔钢板装配式屈曲约束支撑构造图

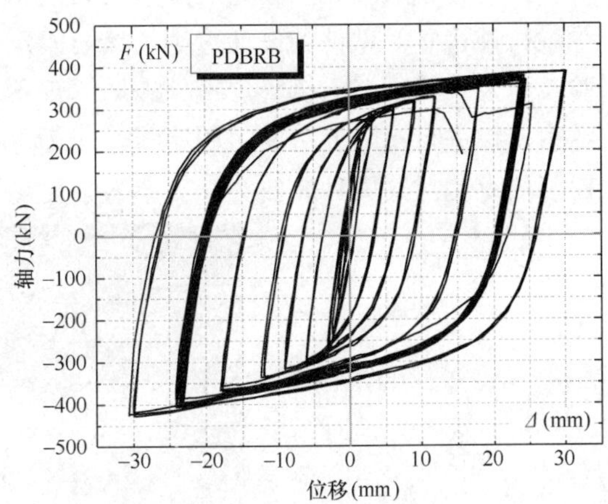

图 9　多核心开孔钢板装配式屈曲约束支撑滞回曲线

3.4 开孔式及开槽式三重钢管屈曲约束支撑[2,11-20]

三重钢管屈曲约束支撑由日本学者 Haginoya[10]研制而成，它的核心单元和约束单元仅由钢管组成，不涉及混凝土、无粘结材料等其他材料，具有加工简便、性能稳定、自重轻等特点。然而，三重钢管屈曲约束支撑的核心单元两端采用普通（或高强）钢管，中间耗能段采用低屈服点钢管，低屈服点钢管与普通（或高强）钢管的两端需要焊接连接，这一方面使加工制作工序复杂，且焊接时的残余应力和焊接变形也易令此处发生破坏。团队依据“核心单元局部削弱相当于其他部分加强”的设计理念，提出了核心开孔式三重钢管屈曲约束支撑及开槽式三重钢管屈曲约束支撑，其构造如图 10 所示。通过开孔或开槽削减核心钢管单元的局部受力面积，降低了核心钢管的屈服力，三支钢管均采用普通型钢制

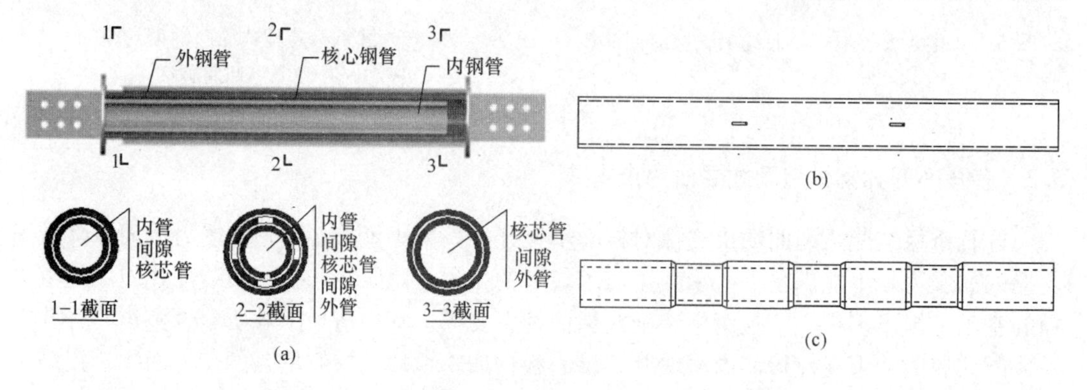

图 10　开孔（开槽）式三重钢管屈曲约束支撑构造图

(a) 开孔（开槽）三重钢管屈曲约束支撑剖面图；(b) 核心钢管开孔示意图；(c) 核心钢管开槽示意图

作也能达到很好的耗能效果。实现了核心段定点屈服，避免使用不同钢材同时也避免了焊接残余变形及残余应力对核心钢管耗能带来的不利影响。图 11 为开孔式及开槽式屈曲约束支撑的试验装置，图 12、图 13 分别为开孔式及开槽式屈曲约束支撑试验得到的滞回曲线。试验结果表明，开孔式及开槽式屈曲约束支撑滞回曲线均饱满对称，耗能性能优良，承载特性稳定。试件的设计达到了核心段定点屈服、定点破坏的预期（图 14、图 15），试验没有发生普通三重钢管焊接处破坏的现象。通过对不同设计参数的开孔式及开槽式屈曲约束支撑进行有限元分析及试验研究，研究了开孔形式、钢管之间的间隙、支撑的径厚比、约束比对这两种屈曲约束支撑力学性能及耗能性能的影响，提出了这两种屈曲约束支

图 11　开孔（开槽）式三重钢管屈曲约束支撑试验装置

撑的设计建议。

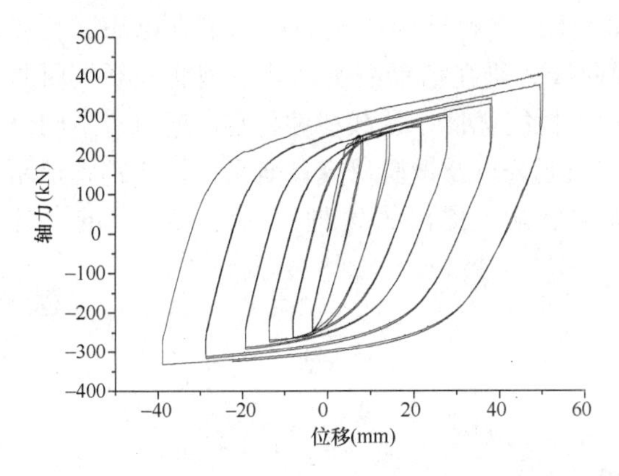

图 12　开孔三重钢管屈曲约束支撑滞回曲线

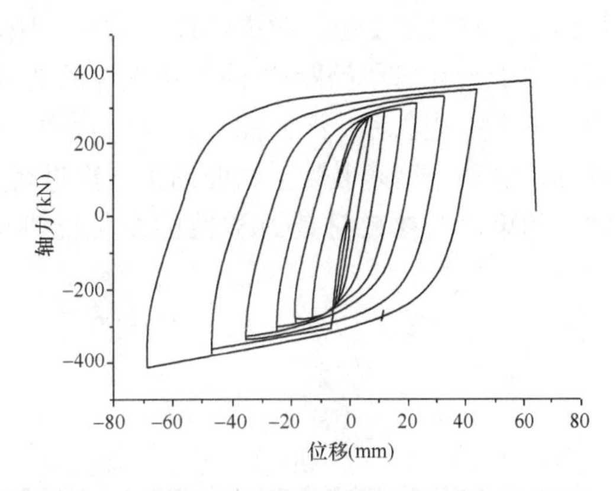

图 13　开槽三重钢管屈曲约束支撑滞回曲线

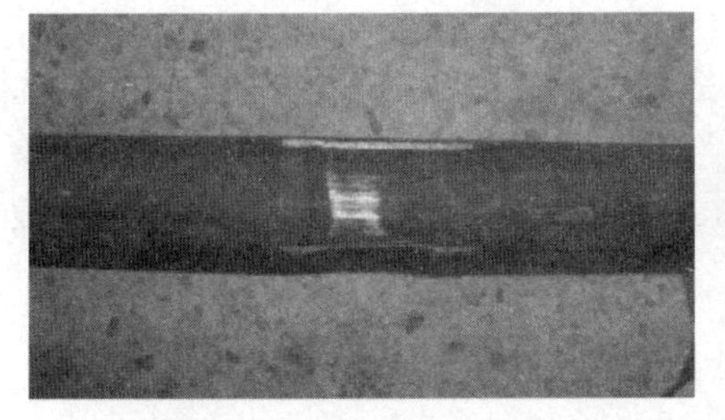

图 14　开孔处定点屈服

图 15　开槽处定点破坏

3.5　外包钢筋混凝土屈曲约束支撑[21]

普通二重钢管屈曲约束支撑由日本学者 Kato[22]提出，其有较为明显的缺点即约束钢管对核心钢管的约束能力不足，易导致支撑发生屈曲破坏，若通过增加截面方法的方式提高约束钢管的约束能力，无疑会大幅增加制造成本。团队利用钢筋混凝土约束能力强的优点对其进行了改进，提出新型外包钢筋混凝土屈曲约束支撑，其构造如图 16 所示。相比传统核心单元为“一”字形和“十”字形的屈曲约束支撑，外包钢筋混凝土管屈曲约束支撑采用钢管作为核心单元，有利于提高支撑的承载能力和抗弯抗扭性能；相比于二重钢管

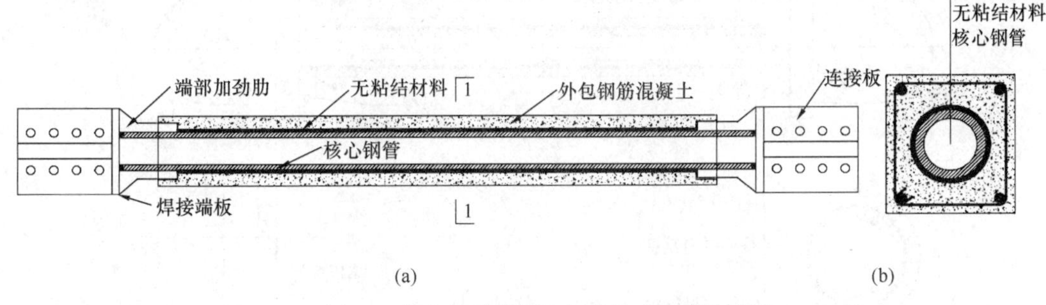

图 16　外包钢筋混凝土屈曲约束支撑构造图

(a) 纵向剖面图；(b) 1-1 剖面图

屈曲约束支撑，外包钢筋混凝土钢管屈曲约束支撑的外约束采用钢筋混凝土，不仅节省了用钢量还解决了支撑的防火性能的要求。图 17 为外包钢筋混凝土屈曲约束支撑的试验装置，图 18 为试验所得到屈曲约束支撑的滞回曲线。研究结果表明，外包钢筋混凝土屈曲约束支撑构造合理，滞回曲线饱满、对称，耗能性能优良，承载特性稳定。通过对不同设计参数的外包钢筋混凝土屈曲约束支撑进行有限元分析及试验研究，研究了核心钢管径厚比、约束比等参数对其力学性能及耗能性能的影响，提出了该种屈曲约束支撑的设计建议。

图 17　外包钢筋混凝土屈曲约束支撑试验安装

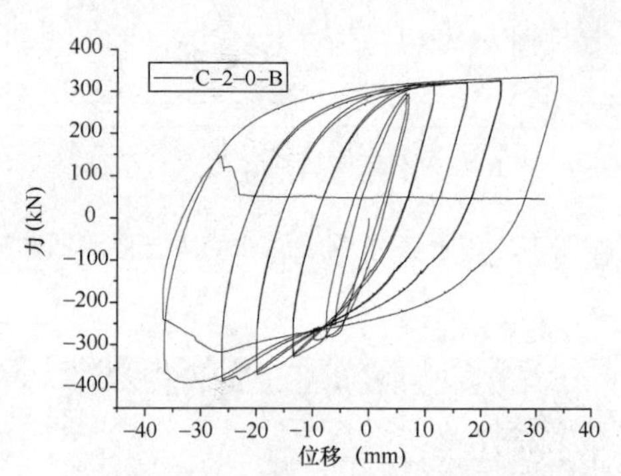

图 18　外包钢筋混凝土屈曲约束支撑滞回曲线

3.6　屈曲约束耗能腋撑[28,29]

屈曲约束支撑通常以节点对角线、“人”字形或“V”字形的方式布置在框架中，会影响子框架的使用空间，对一些框架空间使用要求较高的结构，如商场、停车场等，较难将屈曲约束支撑运用其中。屈曲约束支撑较大的尺寸也不易在结构加固中运用。团队基于“将阻尼器小型化”的思路，提出了一种新型屈曲耗能腋撑，其构造如图 19 所示。其尺寸仅有普通屈曲约束支撑的 1/4，以“八”字形安装在框架上方的两个角。相对于普通屈曲约束支撑框架，其释放了子框架的使用空间，较小的体积也便于在结构加固中运用，同时它改善了框架节点域的受力性能，提高了框架整体的抗震能力。图 20 为带有屈曲约束耗能腋撑的混凝土框架的试验装置，图 21 为试验所得到框架的滞回曲线。试验结果表明，设置屈曲约束耗能腋撑，提高了框架结构的初始刚度、极限强度以及退化后强度和刚度，提高了框架的承载能力及耗能能力，改变了框架结构的受力模式，使塑性铰的发展从梁端

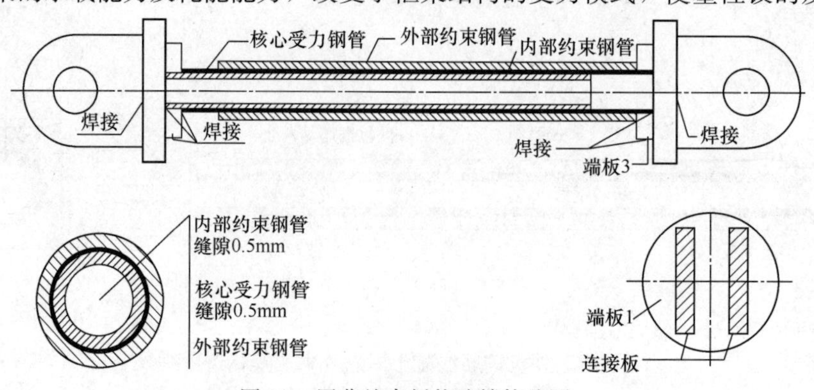

图 19　屈曲约束耗能腋撑构造图

和柱端转到腋撑处，减小了梁柱节点区的受力，有效地保护了节点。

图 20　带屈曲约束耗能腋撑的 RC 框架试验

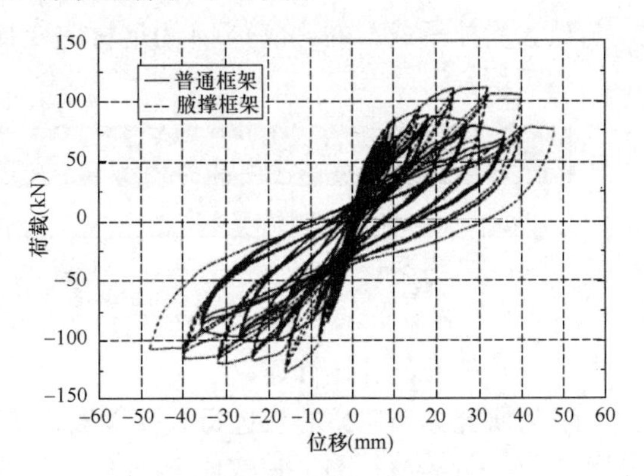

图 21　带耗能腋撑的 RC 框架与普通 RC 框架滞回曲线对比

4. 开孔板式屈曲约束支撑设计方法研究[26,27]

对于板式屈曲约束支撑，对其核心开孔削弱，其核心力的传导方式、受压时的高阶变形及局部稳定性均与普通核心未开孔的屈曲约束支撑不同，意味着传统的屈曲约束支撑设计方法不适用于开孔板式屈曲约束支撑。团队以开孔钢板装配式屈曲约束支撑为例，对开孔板式屈曲约束支撑设计方法进行了研究。

4.1　核心单元设计

（1）核心单元的承载力与刚度

根据核心开孔屈曲约束支撑的特点，在力学性能参数研究的基础上，提出了核心开孔屈曲约束支撑初始刚度、屈服力及极限承载力的计算公式，分别如式（1）～式（3）所示。

$$k_c = \frac{1}{\dfrac{nbR_2}{EBtR_1} + \dfrac{(n-2)L_{c1}}{EA_{c1}} + \dfrac{2L_{c2}}{EA_{c2}} + \dfrac{L_{c3}}{EA_{c3}} + \dfrac{2L_s}{EA_s}} \tag{1}$$

$$F_{yc} = A_y \times \eta \times f_y \tag{2}$$

$$F_{uc} = \beta \times \omega \times F_{yc} \tag{3}$$

式（1）中，k_c为核心单元初始刚度设计值，L_y、L_{c1}、L_{c2}、L_{c3}、L_s分别为开孔段、两孔之间连接区、开孔与连接段之间连接区、限位卡处连接区以及连接段的长度（具体位置如图 22 所示）；A_y、A_{c1}、A_{c2}、A_{c3}和A_s分别为开孔段、两孔之间连接区、开孔与连接段之间连接区、限位卡处连接区以及连接段的横截面面积；n 为开孔段个数；b 为开孔段单边宽度；B 为核心单元宽度；E 为钢材弹性模量；R_1为面积比，$R_1=2b/B$；R_2为开孔段长宽比，$R_2=L_y/b$。

式（2）中，F_{yc}为屈服力设计值；η 为钢材超强系数，即钢材实际屈服强度与名义屈服强度的比值；f_y为钢材名义屈服强度。

式（3）中，F_{uc}为极限承载力理论值；β为拉、压不均匀系数（BRB最大受压承载力与最大受拉承载力的比值），ω为钢材应变强化系数。

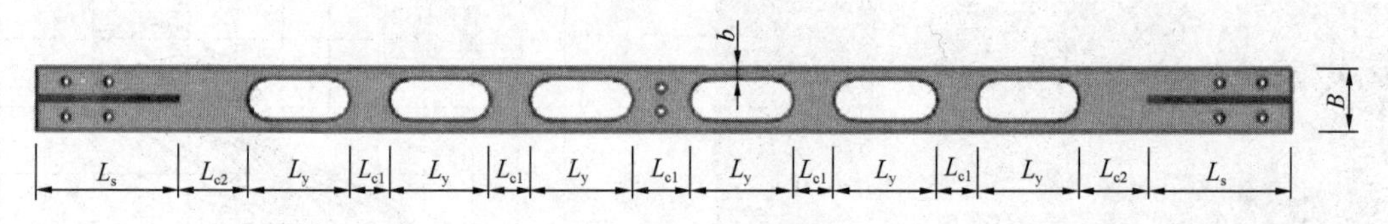

图22　开孔核心参数示意图

（2）核心开孔设计建议

为研究如何对核心板进行开孔以实现支撑具有良好稳定的力学性能及耗能性能，以面积比R_1（$R_1=2b/B$）、长宽比R_2（$R_2=L_y/b$）、间距比R_3（$R_3=L_{c1}/c$）、开孔段个数n及开孔排数为主要控制变量，建立了不同参数多组有限元模型并进行分析。综合各个模型的分析结果，提出了核心单元开孔的设计建议，见表1。

开孔核心单元设计参数　　**表1**

R_1	R_2	R_3	n	开孔排数
0.3～0.6	0～10	1～1.5	开孔段总长和R_1确定的情况下，按较大R_2取值确定开孔个数	单排开孔

4.2　约束单元设计

屈曲约束支撑的约束单元必须达到一定的抗弯刚度才能防止发生整体失稳破坏，同时必须有足够的局部约束能力以防止发生局部失稳破坏。

（1）整体稳定性设计

结合强度-刚度法[28]，提出核心开孔屈曲约束支撑的整体稳定性验算方法，如公式（4）所示：

$$\left(1-\frac{F_{uc}}{P_{eb}}\right)\frac{M_{yb}}{P_{cmax}L}>\frac{\upsilon_0+2C}{L_b} \tag{4}$$

式中，P_{eb}为外约束单元按两端铰接计算得到的弹性屈曲临界力；M_{yb}为外约束单元屈服弯矩；υ_0为外约束单元跨中初始挠度；C为核心单元于外约束单元之间间隙；L_b为外约束单元长度；P_{cmax}为BRB的轴向最大承重力。

（2）局部稳定性设计

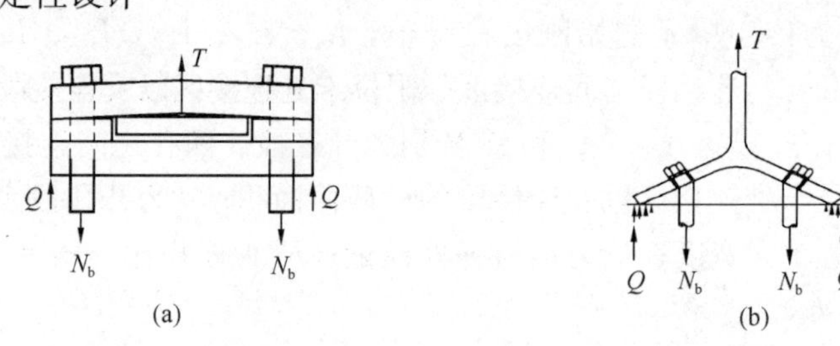

图23　钢板装配式屈曲约束支撑约束单元受力简图

（a）PBRB受压横截面受力；（b）丁形连接

根据核心开孔屈曲约束支撑约束单元的受力特性，提出如图 23（a）所示约束单元的受力简图，及如图 23（ b）所示计算简图，根据受力平衡原则，推导得到开孔钢板装配式屈曲约束支撑的局部稳定性验算方法，如公式（5）、公式（6）所示。

（1）当 $L_b \leqslant L_w$时，需满足：

$$M_c = 0.25L_w \times T \leqslant M_u = \gamma \times \eta \times f_{yb} \times W_r \tag{5}$$

（2）当 $L_b > L_w$时，需满足：

$$M_c = 0.25L_w \times s \times T \leqslant M_u = \gamma \times \eta \times f_{yb} \times W_r \tag{6}$$

式中，L_b为约束单元螺栓间距；L_w为核心单元屈曲半波长；M_c为螺栓间距内外约束单元在核心单元挤压力作用下产生的跨中弯矩；M_u为外约束单元截面抗弯承载力；γ为塑性发展系数；W_r为外约束单元净截面模量。

4.3 开孔钢板装配式屈曲约束支撑设计流程

开孔钢板装配式屈曲约束支撑构件的设计流程如图 24 所示，包括：明确设计目标、确定设计参数、性能设计（力学性能计算和稳定性能验算）、端部构造确定等 4 部分。(1) 确定屈曲约束支撑的空间尺寸、屈服力设计值 F_{yc}、初始刚度设计值 k_c、屈服位移设计值 δ_{yc}。(2) 根据设计目标确定支撑的设计参数，具体为：确定制作构件材料，根据空间尺寸限制和核心单元截面宽厚比要求选择支撑的核心单元横截面尺寸和外约束单元尺寸，结

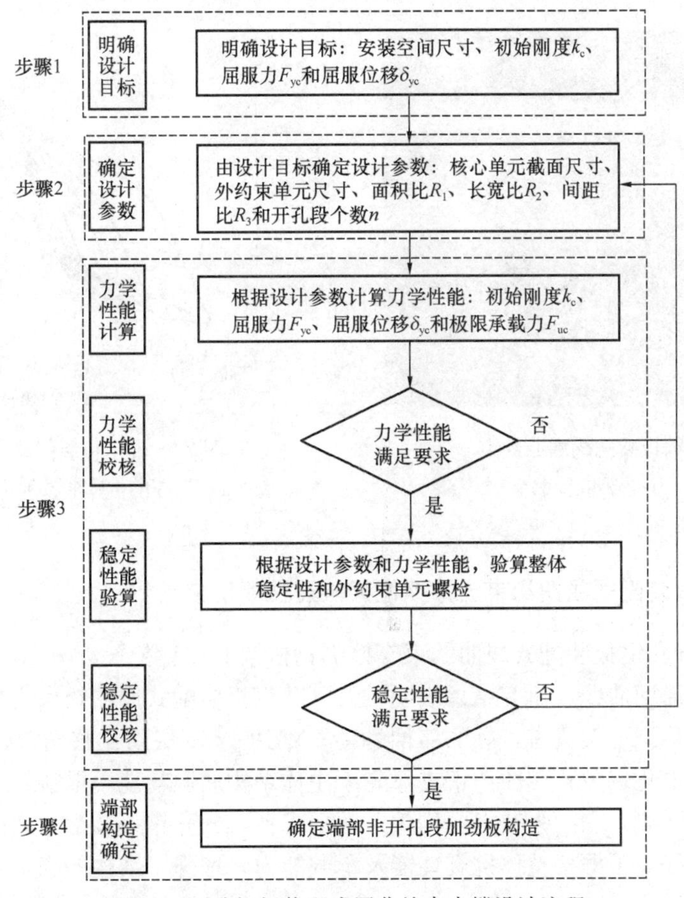

图 24 开孔钢板装配式屈曲约束支撑设计流程

合初始刚度、屈服力和屈服位移需求和表 1 确定核心单元的开孔参数面积比 R_1、长宽比 R_2、间距比 R_3 和开孔段个数 n。(3) 根据确定的设计参数计算复核支撑的力学性能，验算整体稳定性及局部稳定性。(4) 根据现场连接要求确定支撑端部非开孔段加劲板的构造。

5 带新型屈曲约束支撑的框架抗震性能研究

5.1 带钢板装配式屈曲约束支撑混凝土框架抗震性能研究[30]

图 25 为带新型钢板装配式屈曲约束支撑混凝土框架的试验装置，图 26 为试验得到屈曲约束支撑框架的滞回曲线。研究结果表明，带钢板装配式屈曲约束支撑混凝土框架滞回曲线饱满对称，框架性能具有良好的抗震性能，试验中表现出良好的刚度和延性；框架中的屈曲约束支撑均能发挥良好的耗能性能；屈曲约束支撑早于主体结构屈服并耗能，梁钢筋早于柱钢筋屈服，节点板没有发生屈服，符合“强柱弱梁”和“强节点弱构件”的要求；将钢板装配式屈曲约束支撑用于钢筋混凝土框架结构中，提高了框架整体抗震性能及承载能力，增强了结构的变形能力，延缓了梁柱钢筋的屈服及混凝土裂缝的形成。

图 25 带钢板装配式屈曲约束支撑的 RC 框架试验装置

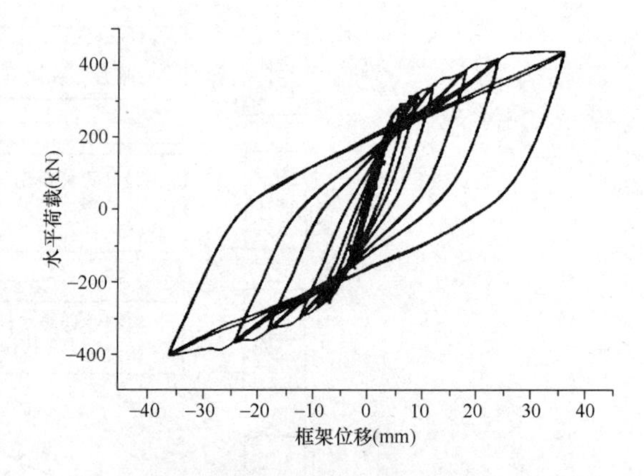

图 26 带钢板装配式屈曲约束支撑的 RC 框架滞回曲线

5.2 带开孔钢板装配式屈曲约束支撑钢框架抗震性能研究[31,32]

图 27 为带开孔钢板装配式屈曲约束支撑的钢框架的试验装置，图 28 为试验得到屈曲约束支撑框架的滞回曲线。研究结果表明，带开孔钢板装配式屈曲约束支撑钢框架滞回曲线饱满且基本对称，框架具有良好的滞回性能；到达 1/30 层间位移角时，框架承载力和刚度均没有明显退化现象；屈曲约束支撑先于主体结构屈服耗能，梁端塑性铰早于柱端形成，符合“强柱弱梁”和“强节点弱构件”的要求；将开孔钢板装配式屈曲约束支撑用于钢框架结构中，提高了框架整体抗震性能及承载能力，增强了结构的变形能力，延缓了梁柱节点塑性铰的发展。

图 27　带核心开孔钢板装配式 BRB 的钢框架试验装置

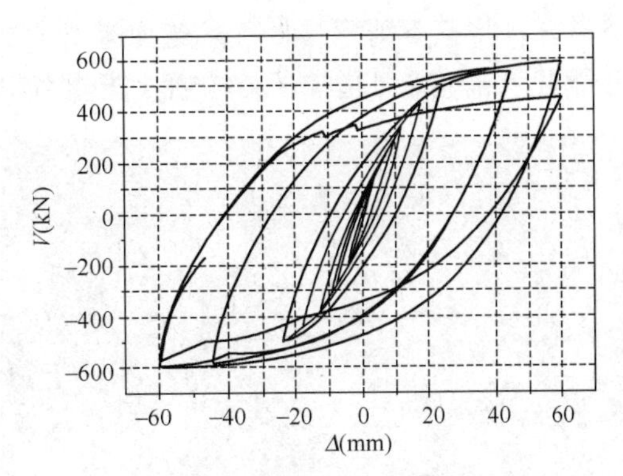

图 28　带核心开孔钢板装配式 BRB 的钢框架滞回曲线

5.3　带三重钢管屈曲约束支撑钢管混凝土框架抗震性能研究[33]

图 29 为带三重钢管屈曲约束支撑的钢管混凝土框架的试验装置，图 30 为试验得到屈曲约束支撑框架的滞回曲线。研究结果表明，带三重钢管屈曲约束支撑的钢管混凝土框架滞回曲线饱满，框架具有良好的滞回性能。三重钢管屈曲约束支撑在结构中率先屈服耗能，而后钢管混凝土框架才进入屈服状态；将三重钢管屈曲约束支撑运用于钢管混凝土框架结构中，提高了框架的承载能力、增强了结构的变形能力，延缓了梁柱节点塑性铰的发展，改善了钢管混凝土框架的抗震性能。

图 29　带三重钢管屈曲约束支撑钢管混凝土框架试验装置

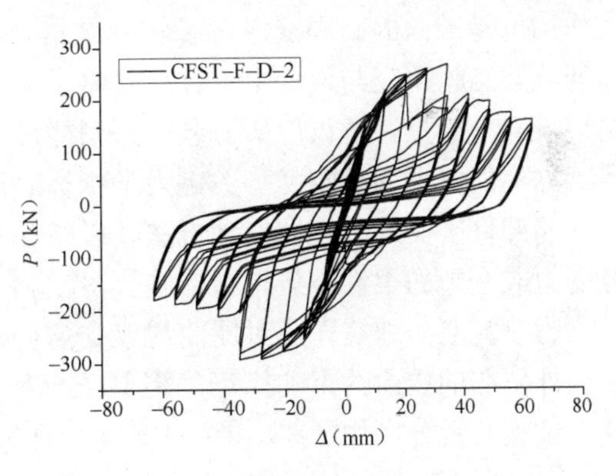

图 30　带三重钢管屈曲约束支撑钢管混凝土框架滞回曲线

5.4　带三重钢管屈曲约束支撑框架-剪力墙结构抗震性能研究[34]

将三重钢管屈曲约束支撑设置于钢管混凝土框架-剪力墙结构中，形成钢管混凝土框架-剪力墙减震结构。图 31 为带三重钢管屈曲约束支撑钢管混凝土框架-混凝土剪力墙结构的试验装置。图 32 为试验得到框架的滞回曲线。试验结果表明，三重钢管屈曲约束支撑设置于钢管混凝土框架-剪力墙结构滞回曲线饱满对称，结构具有良好的滞回性能和耗能性能；三重钢管屈曲约束支撑在结构中率先屈服耗能，而后剪力墙底部出现塑性铰并最终破坏；将三重钢管屈曲约束支撑运用于框架-剪力墙结构中，提高了结构的整体的抗震性能及承载性能，增加了结构的耗能能力，减缓了剪力墙底部塑性铰的形成。

图 31　带三重钢管屈曲约束支撑钢管混凝土-剪力墙结构试验装置

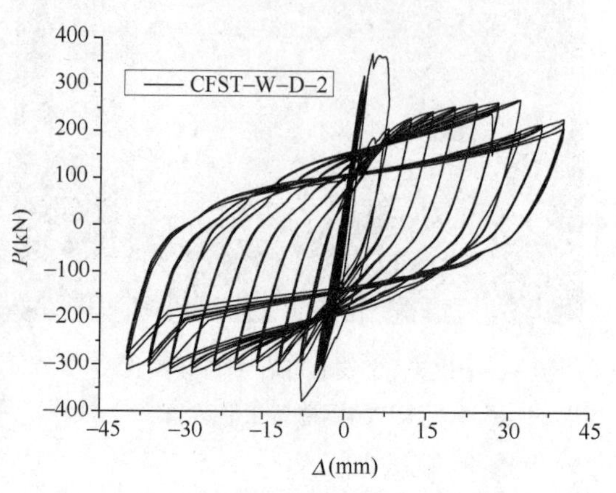

图 32　带三重钢管屈曲约束支撑钢管混凝土-剪力墙结构滞回曲线

6. 双向地震作用下屈曲约束支撑框架结构稳定性能研究[32,35,36]

目前，多数学者对屈曲约束支撑及屈曲约束支撑框架结构的研究均是基于单方向平面内受力的，我国规范对屈曲约束支撑的设计建议及要求也是基于框架及支撑单方向平面内受力提出的，但实际中结构水平方向往往遭受到双向地震作用，屈曲约束支撑所遭受的地震作用也是双向的。周云探讨了实际结构及阻尼器考虑双向地震作用的必要性[35]，认为消能减震结构研究与设计中，往往忽略了消能部件平面外的力学特性，可能为消能减震结构埋下安全隐患，建议应研究各类型消能部件在双向地震作用下的力学特性及破坏模式，找到如消能器平面外刚度等关键力学指标对消能部件平面外稳定性的影响规律；建立各类型消能部件平面外稳定性设计方法，以指导减震结构分析和设计；研究能释放消能器平面外受力或变形的柔性连接构造措施。

为研究双向地震作用对屈曲约束支撑框架受力性能及稳定性的影响，分别对带有不同平面外位移的屈曲约束支撑钢筋混凝土框架及屈曲约束支撑钢框架进行了拟静力试验研究，试验装置分别如图 33、图 34 所示。研究表明，当作用在框架的平面外地震作用较大

图 33　屈曲约束支撑钢框架双向地震作用加载试

图 34　屈曲约束支撑 RC 框架双向地震作用加载试验

时，屈曲约束支撑发生了平面外的屈曲，如图 35 所示；对于屈曲约束支撑钢框架结构，钢框架梁柱的抗扭刚度和平面外抗弯刚度不足，受到支撑平面外的水平分力和弯矩作用框架柱容易产生扭转和平面外弯曲，框架梁发生平面外弯曲，框架平面外变形容易导致屈曲约束支撑系统发生失稳破坏，并且平面外作用越大，失稳破坏发生越早；对于屈曲约束支撑钢筋混凝土结构，框架平面外滞回曲线捏缩严重，耗能能力较差；节点板、支撑外伸段和支撑约束段之间的平面外刚度突变使得它们之间连接位置产生转角突变，影响屈曲约束支撑框架的平面外稳定性。

为进一步研究如何避免屈曲约束支撑发生平面外屈曲，基于变形协调原则，建立了考虑平面外地震作用的屈曲约束支撑简化力学模型，如图 36 所示，推导了特定条件下屈曲约束支撑发生平面外屈曲的条件，通过有限元分析验证所推导得到的公式的准确性。以屈曲约束支撑核心单元外伸段底部截面边缘纤维是否屈服作为抗力设计准则，结合小挠度理论，提出双向地震作用下结构中屈曲约束支撑平面外局部稳定判别方法。对两组屈曲约束支撑钢框架有限元分析模型进行了平面内＋平面外地震作用的加载分析，以判别所提出结构中屈曲约束支撑平面外局部稳定判别方法是否可行。结果表明，该判别方法可以较为准确且保守的预测结构中屈曲约束支撑在极限承载力下是否发生平面外局部失稳破坏。

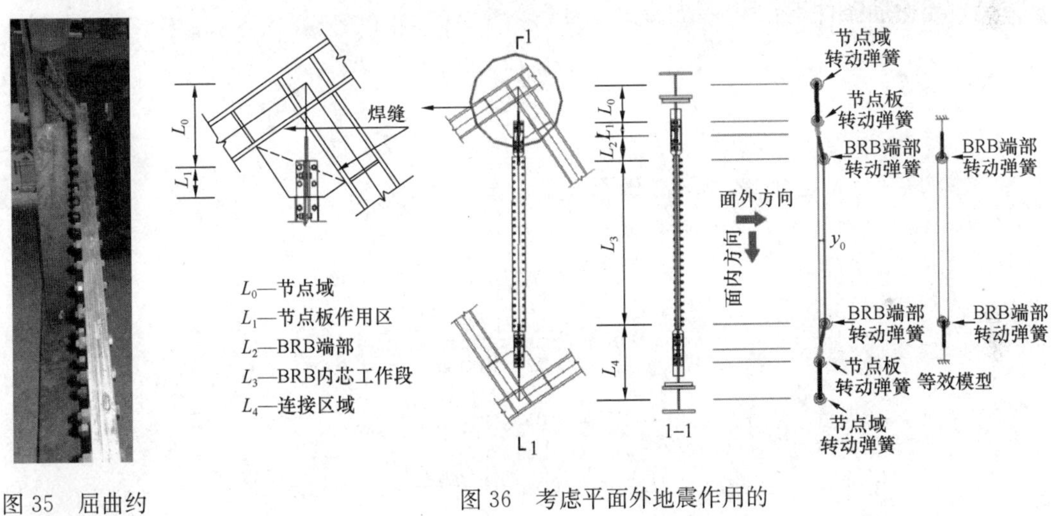

图 35　屈曲约束支撑发生平面外屈曲

图 36　考虑平面外地震作用的屈曲约束支撑简化等效力学模型

为研究不同设计参数对屈曲约束支撑平面外稳定性能的影响，以节点板作用区抗弯线刚度 K_1、支撑端部抗弯线刚度 K_2 及核心单元工作段抗弯线刚度 K_3 为研究变量，建立上述 3 种参数各不相同的 1000 组有限元模型并进行分析，研究上述 3 种参数的变化对屈曲约束支撑稳定性能的影响。研究结果表明：（1）单一增大节点板作用区抗弯线刚度 K_1、支撑端部抗弯线刚度 K_2 或核心单元工作段抗弯线刚度 K_3 对支撑的稳定性都是不利的；（2）当前屈曲约束支撑构件与屈曲约束支撑结构设计中，将节点板作用区、屈曲约束支撑端部和核心单元工作段中某一部分设计的刚度很大是不妥的；（3）对屈曲约束支撑的稳定性设计时应当考虑各部分刚度之间的相互作用影响，当屈曲约束支撑加节点板发生的弯曲变形比较均匀、协调时，支撑才能具有较好稳定性，保证其发挥正常工作性能；（4）当节点板作用区抗弯线刚度 K_1、屈曲约束支撑端部抗弯线刚度 K_2 和核心单元工作段抗弯线刚度 K_3 的刚度较小且比例接近时，屈曲约束支撑加节点板产生的内力较小且变形最为均匀的，但是，各部分的抗弯线刚度最低不可小于“约束比”或“刚度比”等指标的要求。

7. 节点板设计方法研究

7.1 节点板屈服力计算方法研究[37]

屈曲约束支撑通过节点板与结构相连接，节点板的性能决定了屈曲约束支撑能否在工作时稳定发挥作用。《建筑消能减震技术规程》JGJ 297—2013[38] 对节点板的设计要求是在屈曲约束支撑极限力作用下既不能屈服也不能屈曲，始终保持弹性，故节点板屈服力计算的准确性十分重要。然而《建筑消能减震技术规程》中没有明确给出节点板屈服力的计算公式，仅提出了节点板强度验算要求。对此，团队对屈曲约束支撑结构中焊接或螺栓连接的节点板的屈服力计算方法进行了研究，推导得到节点板的屈服力可根据公式（7）～（15）计算：

（1）无边肋条件下：

$$P_y = \alpha_1 f(\eta_1 L_2 t + \eta_2 L_3 t) \tag{7}$$

$$\alpha_1 = 1 + p \times \frac{L_1 - L_1^*}{L_{1\max} - L_1^*} \tag{8}$$

$$L_1^* = \left(\frac{C_j}{f \times t} - \eta_1 \frac{L_r \cos\alpha}{2} - \eta_2 \frac{L_r \sin\alpha}{2}\right) / (\eta_1 \sin\alpha + \eta_2 \cos\alpha) \tag{9}$$

$$\eta_1 = \frac{1}{\sqrt{1 + 2\cos(90^\circ - \alpha)}} \tag{10}$$

$$\eta_2 = \frac{1}{\sqrt{1 + 2\cos\alpha}} \tag{11}$$

$$L_2 = L_1 \sin\alpha + \frac{L_r \cos\alpha}{2} \tag{12}$$

$$L_3 = L_1 \cos\alpha + \frac{L_r \sin\alpha}{2} \tag{13}$$

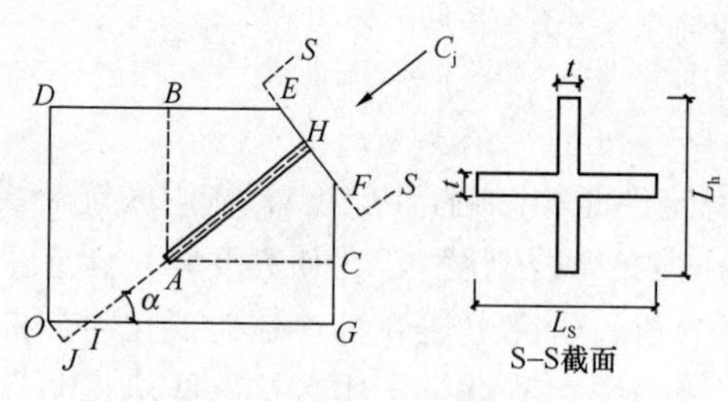

图 37　焊接及螺栓连接中的节点板长度取值示意图

（2）有边肋条件下：

$$P_{y,S} = \alpha_2 f(\eta_1 L_2 t + \eta_2 L_3 t) \tag{14}$$

$$a_2 = q a_1 \tag{15}$$

式中，P_y 为无边肋条件下的节点板屈服力；$P_{y,S}$ 为有边肋条件下的节点板屈服力；α_1 为无边肋条件下中心加劲肋长度影响的修正系数；C_j 为作用于节点板的最大轴力，即屈曲约束支撑的设计承载力；L_1^* 为中心加劲肋长度最小值；L_r 为屈曲约束支撑与节点板相连接的宽度，如图 37 为 $\overline{EF}$ 的长度；t 为节点板厚度；f 为节点板的强度设计值，按照现行钢结构设计规范确定；α 为支撑轴线与梁轴线的夹角；η_1 为 $\overline{AB}$ 的抗拉折减系数，η_2 为 $\overline{AC}$ 的抗拉折减系数；L_r为图 37 中节点板与支撑连接截面$\overline{EF}$的长度；L_{1max}为图 37 中$\overline{lH}$的长度；L_2为图 37 中弯折线$\overline{AB}$的长度；L_3为图 37 中弯折线$\overline{AC}$的长度；p 为长度放大系数，p 取 0.1；α_2 为边肋作用的修正系数；q——屈服力放大系数，q 取 1.08。

7.2 节点板轴向刚度计算方法研究[37]

屈曲约束支撑及节点板（以下简称“屈曲约束支撑系统”）为结构提供的附加刚度 K，应将节点板和屈曲约束支撑的轴向刚度进行“串联”计算，如公式（16）所示。

$$K=\frac{K_J K_{BRB}}{K_J+2K_{BRB}} \tag{16}$$

式中，K 为屈曲约束支撑及节点板为结构提供的附加刚度；K_J为节点板平面内轴向初始刚度；K_{BRB}为屈曲约束支撑平面内轴向刚度。

由于缺乏节点板轴向刚度的计算方法，计算屈曲约束支撑系统为结构提供的附加刚度时，通常将节点板轴向刚度视为无限大，导致计算得到屈曲约束支撑系统为结构提供的附加刚度值偏大，使设计偏于危险。为解决上述问题，团队对屈曲约束支撑结构中焊接或螺栓连接的节点板的轴向刚度计算方法进行了研究，推导得到节点板的轴向刚度可根据公式（17）、（18）计算：

（1）无边肋条件下：

$$K_J=\frac{2Et}{\sqrt{3}}\frac{1}{\ln\left|\frac{\sqrt{3}I_r+\sqrt{3}(L_h-t)+2L_1}{\sqrt{3}L_r+\sqrt{3}(L_h-t)}\times\frac{\sqrt{3}L_r+2L_1+L_j+2L_s}{\sqrt{3}L_r+2L_1}\right|} \tag{17}$$

（2）有边肋条件下：

$$K_{J,S}=\beta\times K_J \tag{18}$$

式中，E 为节点钢材材料的弹性模量；β 为考虑初始缺陷和边肋作用的修正系数，β 取 1.1；L_s为图 37 中$\overline{AI}$的长度；L_j为图 37 中$\overline{IJ}$的长度；L_h为平面外方向连接部分的宽度。

7.3 梁柱间开合效应对节点板性能影响研究[39,40]

结构在地震作用下发生侧向变形时，梁柱间会发生相对转动，导致节点板的受力状态发生变化。为研究节点板与钢框架间的相互作用效应，以不同节点板连接形式、节点板尺寸，以及是否设置自由边加劲肋为参数，建立了含有框架－节点板－屈曲约束支撑的子结构有限元模型并进行分析。研究结果表明：（1）在安装屈曲约束支撑与节点板后，梁屈服位置转移到节点板端部，随节点板尺寸的变化而移动；（2）边加劲肋对框架刚度影响很小，对梁柱等效应力峰值有小幅改善，但对节点板连接边应力状态没有改善作用；（3）梁柱间的开合效应易使节点板发生平面外失稳。

8. 屈曲约束支撑子结构设计方法研究[41,42]

目前消能减震结构设计方法认为只要消能减震结构中的子结构梁柱或者墙满足了罕遇地震作用下极限承载力的验算，就认定消能部件能实现罕遇地震下的耗能和保护主体结构安全。但是，子结构中受力最复杂的不是梁、柱或墙体，而是消能子结构中的节点。《建筑消能减震技术规程》JGJ 297—2013[38]指出"与消能部件相连接的主体结构构件与节点应考虑消能器在最大输出阻尼力作用，从而保证消能器在罕遇地震作用下不丧失功能"，如何保证消能子结构节点核心区在罕遇地震作用下不先于阻尼器到达极限承载力发生剪切破坏，是消能减震结构设计的关键，也是消能器能否充分发挥保护主体结构安全作用的根本。团队提出消能子结构弹性设计法，即利用弹性分析下结构各构件内力完成消能子结构节点设计与验算的弹性系数法，用弹性分析下结构各构件内力 V_e 乘以一个与抗震等级和分项系数有关的系数 ϕ 得到的值作为子结构节点核心区的剪力设计值 V_d，完成子结构节点核心区的设计；用弹性分析下结构各构件内力 V_e 乘以一个与地震影响有关的系数 ρ 得到的值作为罕遇地震下子结构梁柱在节点引起的剪力值 V_j^*，验算该剪力值是否满足子结构节点核心区控制方程。消能子结构节点弹性系数法的设计流程如图 38 所示。

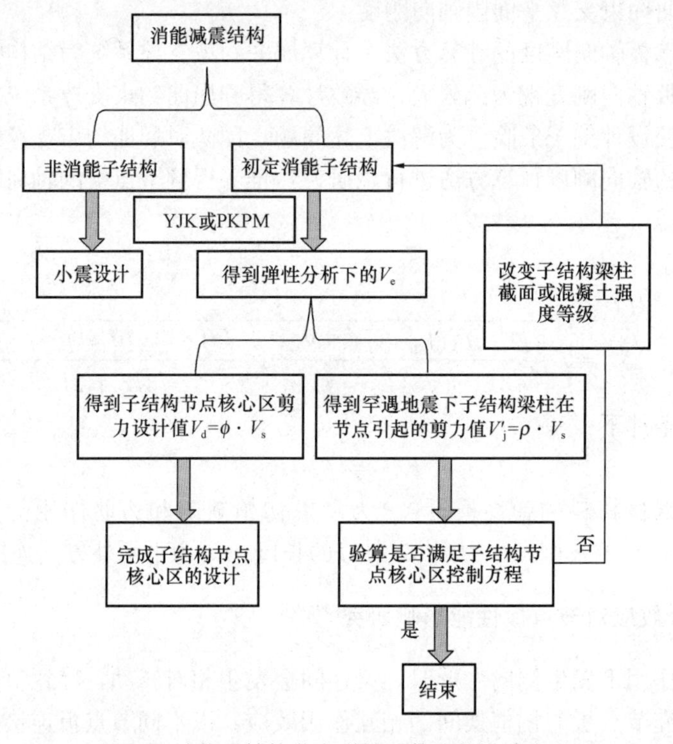

图 38　消能子结构节点弹性系数法的设计流程

以带屈曲约束支撑的钢筋混凝土框架为研究对象，建立了子结构节点核心区控制方程如下公式（19）：

$$V_{BRB}^* \leqslant V_u^* + V_B - V_j^* \tag{19}$$

$$V_j^* = \rho V_j \tag{20}$$

式中，V_{BRB}^* 为屈曲约束支撑的极限承载力在节点引起的剪力；V_u^* 为子结构梁柱节点的抗震受剪极限承载力；V_B 为各设计准则下屈曲约束支撑的内力在节点引起的剪力；V_j 为各设计准则下子结构梁柱在节点引起的剪力（包括屈曲约束支撑的影响），直接由软件数据求得；V_j^* 为罕遇地震作用下子结构梁柱在节点引起的剪力（包括屈曲约束支撑的影响）；ρ 为各设计准则到罕遇地震的放大系数。

以抗震等级作为反映结构或构件延性能力的指标，综合考虑构件延性和构件承载力的作用以实现子结构作为重要构件的设计目标，探讨了满足《建筑消能减震技术规程》JGJ 297—2013 的组合工况，研究了各个工况组合学下重要参数 ρ、V_j^*、V_b等的取值。通过实际工程案例，验证了所提出的消能子结构节点弹性系数法的可行性与合理性，荷载组合工况的合理性，以及子结构节点核心区控制方程中的参数取值方法的可行性。

9. 屈曲约束支撑新型结构体系研究

屈曲约束支撑构造简单，耗能原理明确，耗能性能优良，值得在工程中推广与运用。团队基于工程经验及需求，将屈曲约束支撑巧妙地应用于建筑结构中，提出了多种屈曲约束支撑新型结构体系。

9.1 高位转换耗能减震结构体系[44-46]

对于高位转换高层建筑，由于转换层上、下楼层 竖向构件不连续，结构竖向刚度发生明显变化，转换层上、下楼层的构件内力容易发生突变，对抗震极为不利[43]。针对高位转换结构存在的问题，团队提出“从改变结构特性的角度出发解决高位转换结构存在的抗震问题”的思路，将屈曲约束支撑作为耗能支撑引入高位转换结构中，从而形成支撑型高位转换耗能减震结构体系，其结构示意图如图 39 所示。在高位转换结构转换层以下加入屈曲约束支撑，一方面可提供附加刚度，使结构竖向刚度变得均匀，另一方面，在地震作用下，屈曲约束支撑屈服耗散地震能量，减小结构的地震反应。屈曲约束支撑高位转换耗能减震结构分析结果表明，屈曲约束支撑高位转换耗能减震结构体系设计合理，结构竖向刚度均匀，各楼层的构件内力突变小，满足规范要求，将屈曲约束支撑应用于高位转换结构中并形成高位转换耗能减震结构新体系是合理，可行的。

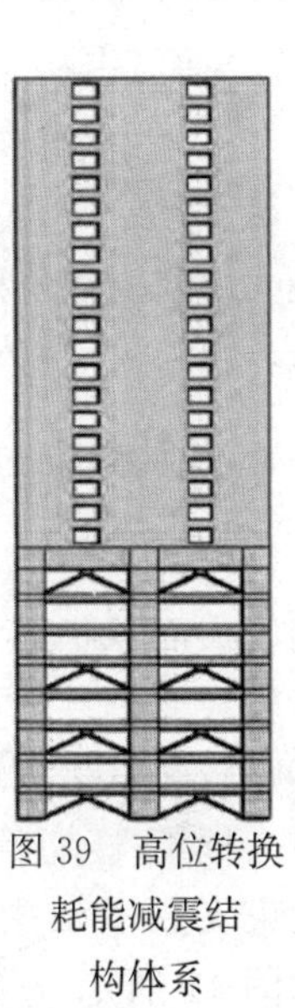

图 39 高位转换耗能减震结构体系

9.2 耗能减震层高层结构体系[45,47,48]

随着高层建筑功能的需要，高层结构体型越来越复杂，采用现有结构体系进行地震与风振控制的难度越来越高。团队提出耗能减震层的概念，将屈曲约束支撑引入到高层结构加强层中，将加强层伸臂桁架和环带桁架中的普通支撑用屈曲约束支撑代替，使其成为耗能减震层，形成耗能减震层高层结构体系，其结构示意图如图 40 所示。屈曲约束支撑耗能减震层结构分析结果表明，屈曲约束支撑耗能减震层结构能够有效地控制结构地震反应，减小结构地震力突变，将屈曲约束支撑用于耗能减震层高层结构体系的设计理念是合理、可行的。

9.3 屈曲约束支撑＋黏滞阻尼器结构体系[49]

屈曲约束支撑框架结构体系中的屈曲约束支撑，在多遇地震作用下处于弹性阶段，在

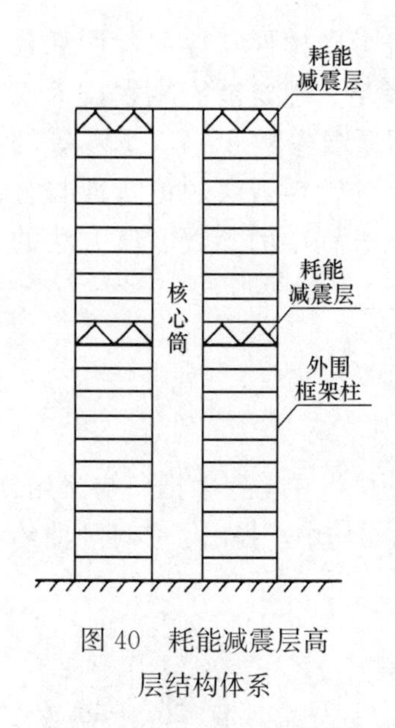

图 40　耗能减震层高层结构体系

罕遇地震下屈服耗能，即小震作用下不提供阻尼。这种体系虽然提高了结构刚度，但也提高了输入结构的地震作用，当屈曲约束支撑布置较多时，经济性较差。结构中布置黏滞阻尼器通过滞回耗能吸收地震能量，可提高结构阻尼比，然而当主体结构太弱时，需要为结构布置较多的阻尼器才能满足层间位移角要求，经济效益也不明显。团队结合两种阻尼器各自的优势，提出屈曲约束支撑和黏滞阻尼器混合使用（BRB+VD）的新型耗能结构体系，实现了刚柔结合，采用适量屈曲约束支撑增大主体结构的侧向刚度，再采用黏滞阻尼器整体提高结构的减震效果。团队对 BRB+VD 体系的减震原理、体系的周期合理限值、体系的层间位移角限值进行了探讨，提出了 BRB+VD 体系的设计方法。根据实际工程，对比了 BRB+VD 体系、纯屈曲约束支撑结构体系、纯黏滞阻尼器结构体系的减震效果。研究结果表明，将屈曲约束支撑和黏滞阻尼器布置在框架结构中后，能明显减小结构最大层间位移角和层间剪力；相比单纯的屈曲约束支撑框架结构体系或黏滞阻尼器结构体系，BRB+VD 体系减震效果更明显，更适用于高烈度地区复杂框架的减震。

9.4　屈曲约束支撑框架-核心筒结构体系[50]

传统框架-核心筒结构体系的核心筒与外框架抗震性能不够协调，易导致核心筒因刚度大而承担绝大部分地震力进而首先发生较大变形甚至严重破坏，外框架因刚度不足随之发生失稳。团队将屈曲约束支撑引入框架-核心筒结构结果中，形成屈曲约束支撑框架-核心筒结构体系，其结构如图 41 所示。通过在外框架布置一定数量的屈曲约束支撑，提高了外框架的刚度，改善了核心筒与外框架间的地震力分配，同时形成二道防线，有效保障结构的安全。屈曲约束支撑框架-核心筒结构分析结果表明，屈曲约束支撑框架-核心筒结构的核心筒与外框架间地震力分配均匀、合理，外框架的受力性能得到改善，结构整体的抗侧能力提高；屈曲约束支撑能先于主体结构屈服并耗散地震能量，提高了结构的阻尼，降低了整体结构的地震反应，保护主体结构的同时，实现了框架-核心筒结构体系的多道抗震防线。

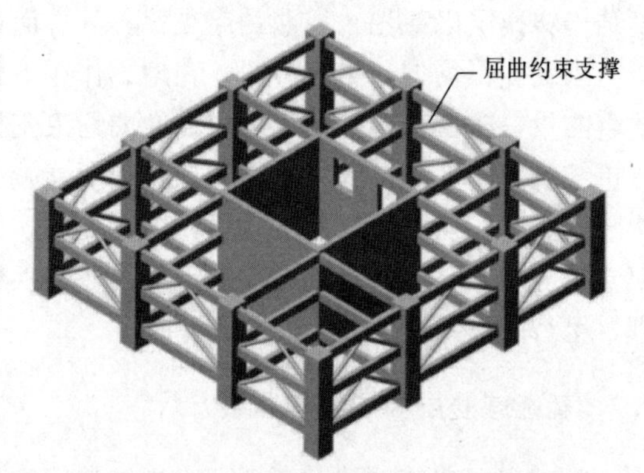

图 41　屈曲约束支撑框架-核心筒结构体系

10. 屈曲约束支撑标准化研究

随着工程界对屈曲约束支撑运用需求的不断增加，屈曲约束支撑的标准化也在逐步推进中。2013年，广州大学会同有关单位，编制了《建筑消能减震技术规程》JGJ 297—2013，其中单独给出了屈曲约束支撑的技术性能指标、构造设计及结构设计方法。同年，广州大学参编的广东省标准《高层建筑混凝土结构技术规程》DBJ 15—92—2013[51]颁布实施，为屈曲约束支撑在高层结构中的运用给出指导。2020年，广州大学会同有关高等院校、研究机构、设计单位及施工单位，编制了全国性的行业标准《屈曲约束支撑应用技术规程》[52]，该规程对屈曲约束支撑的设计与制作、性能指标、检测标准提出要求，给出了屈曲约束支撑在结构中的设计方法与建议，对屈曲约束支撑及节点板的提出稳定性要求，为屈曲约束支撑的施工、安装、验收、检查与维护等工作提供指导。《屈曲约束支撑应用技术规程》的出版进一步保证屈曲约束支撑及其附属子构件在结构中的性能，确保屈曲约束支撑结构的安全性，规范屈曲约束支撑在结构中的使用，促进屈曲约束支撑在工程中推广应用，具有重要的工程意义。

11. 展望

本文总结了近年来团队有关屈曲约束支撑领域的研究成果，屈曲约束支撑具有减震机理明确、构造简单、工作性能稳定、耗能能力和耗能效果好的特点，既可用于新建结构的消能减震设计，又可用于既有建筑的抗震加固，具有广阔的应用前景。为了使屈曲约束支撑能够在我国得到推广应用以提高我国建筑的抗震性能，应在以下几方面加强研究：

(1) 目前消能减震技术研究和消能减震结构设计大多仅基于消能器平面内方向受力与变形，忽略了平面外方向的力学特性与破坏模式。现行研究已表明，平面外的地震作用对框架及屈曲约束支撑的力学性能、耗能性能及稳定性均有非常大的不利影响。未来应着重研究屈曲约束支撑双向受力下的力学特性及破坏模式，采取必要措施减小平面外的影响，并建立屈曲约支撑平面外稳定性设计方法；

(2) 进一步研究屈曲约束支撑减震结构在巨震、主余震等地震作用下的抗震、抗倒塌性能以及在地震、台风等多灾害耦合作用下结构安全性能；

(3) 着重研究基于性能的屈曲约束支撑抗震设计方法，综合考虑温度、湿度、火灾等环境因素对屈曲约束支撑性能及可靠性的影响；

(4) 屈曲约束支撑在地震作用后有较大变形，如何观测屈曲约束支撑在地震作用后变形及对其残余性能的评估，如何在震后保证结构主体安全的情况下对其更换，可展开相关方面的研究；

(5) 开发出屈曲约束支撑相关选型设计软件或插件，帮助设计院快速选择合理的支撑类型及支撑布置形式，便于设计人员开展相关设计。

参考文献

[1] Fujimoto M., Wada A., Saeki E, etc. A study on the unbonded brace encased in buckling-restraining concrete and steel tube. Paper Collection of Structural Engineering[J]. 1988, B-34B: 249-258.

[2] 周云，钱洪涛，褚洪民，邹征敏. 新型屈曲约束支撑设计原理与性能研究[J]. 土木工程学报，2009，42(04)：64-71.

[3] 周云，钟根全，陈清祥，龚晨．不同构造钢板装配式屈曲约束支撑性能试验研究[J]．土木工程学报，2017，50(12)：9-17.

[4] 尹绕章，邓雪松，周云．新型钢板装配式屈曲约束支撑有限元分析[J]．地震研究，2014，37(02)：288-292＋324.

[5] 周云，尹绕章，张文鑫，纪宏恩，邓雪松，韩素玲．钢板装配式屈曲约束支撑性能试验研究[J]．建筑结构学报，2014，35(08)：37-43.

[6] 王浩，邓雪松，周云．全钢型板式屈曲约束支撑无粘结层的有限元分析[J]．土木工程学报，2013，46(S1)：13-18.

[7] 周云，龚晨，陈清祥，钟根全．开孔钢板装配式屈曲约束支撑减震性能试验研究[J]．建筑结构学报，2016，37(08)：101-107.

[8] 邓雪松，纪宏恩，张文鑫，尹绕章，周云，韩素玲．开孔板式屈曲约束支撑拟静力滞回性能试验研究[J]．土木工程学报，2015，48(01)：49-55.

[9] 田时雨，邓雪松，周云，龚晨．开孔双核心钢板装配式屈曲约束支撑性能分析研究[J]．建筑科学，2019，35(07)：8-13.

[10] Haginoya M.，Nagao.，Taguchi T.，Takeita K. Studies on buckling-restrained bracing using triple steel tubes：part 1：outline of triple steel tube member and static cyclic loading tests. Summaries of Technical Papers of Annual Meeting Architectural Institute of Japan. C-1，Structures III，Timber Structures Steel Structures Steel Reinforced Concrete Structures，2005：1011-1012.

[11] 周云，陈真，邓雪松，邹征敏，杨叶斌．开孔与开槽式三重钢管屈曲约束支撑设计方法研究[J]．土木工程学报，2012，45(02)：20-25.

[12] 周云，邓雪松，钱洪涛，褚洪民．开孔式三重钢管屈曲约束支撑性能试验研究[J]．土木工程学报，2010，43(09)：77-87.

[13] 邓雪松，陈真，周云．开孔三重钢管屈曲约束支撑性能试验研究[J]．建筑结构学报，2012，33(06)：42-49.

[14] 邓雪松，陈真，周云．开孔三重钢管屈曲约束支撑影响因素分析[J]．振动与冲击，2012，31(02)：101-108.

[15] 邓雪松，陈真，周云．间隙与支座类型对开孔三重钢管屈曲约束支撑性能的影响[J]．地震工程与工程振动，2010，30(03)：64-69.

[16] 陈真，褚洪民，邓雪松，周云．开孔三重钢管屈曲约束支撑有限元分析[J]．中山大学学报(自然科学版)，2010，49(03)：140-145.

[17] 邓雪松，邹征敏，周云，陈真，杨叶斌．开槽式三重钢管屈曲约束支撑试验研究[J]．土木工程学报，2011，44(07)：37-45.

[18] 邓雪松，邹征敏，周云．开槽式三重钢管屈曲约束支撑试验研究与有限元模拟[J]．土木工程学报，2010，43(12)：41-49.

[19] 吴从永，吴从晓，周云．新型开槽式屈曲约束支撑力学模型研究及应用[J]．土木工程学报，2010，43(S1)：397-402.

[20] 邓雪松，褚洪民，钱洪涛，周云，陈麟．三重钢管屈曲约束支撑性能的有限元模拟分析[J]．华中科技大学学报(城市科学版)，2008(03)：99-103

[21] 邓雪松，杨叶斌，周云，陈真，邹征敏．新型外包钢筋混凝土钢管屈曲约束支撑性能[J]．土木建筑与环境工程，2012，34(01)：21-28.

[22] Kato M.，Kasai A.，Ma X.，etc. An analytical study on the cyclic behavior of double-tube type buckling-restrained braces[J]. Paper Collection of Structural Engineering，2004，50A：103-112.

[23] 杨叶斌．新型外包钢筋混凝土钢管防屈曲耗能支撑有限元模拟与性能试验研究[D]．广州：广州

大学 . 2010.
[24] 周云，尹庆利，林绍明，邓雪松 . 带防屈曲耗能腋撑钢筋混凝土框架结构抗震性能研究[J]. 土木工程学报，2012，45(11)：29-38.
[25] 尹庆利 . 耗能腋撑钢筋混凝土框架减震性能研究[D]. 广州大学，2011.
[26] 龚晨，周云，钟根全，田时雨 . 开孔参数对装配式开孔钢板屈曲约束支撑性能影响研究[J]. 建筑结构学报，2018，39(S2)：328-335.
[27] 周云，龚晨，钟根全，田时雨 . 开孔钢板装配式屈曲约束支撑设计方法研究[J]. 土木工程学报，2019，52(12)：57-65.
[28] 赵俊贤，吴斌 . 防屈曲支撑的工作机理及稳定性设计方法[J]. 地震工程与工程振动，2009，29(3) ：131-139.
[29] 钢结构设计标准 GB 50017—2017. 北京：中国建筑业出版社，2018.
[30] 袁宇鹏 . 屈曲约束耗能支撑钢筋混凝土框架抗震性能试验研究 [D]. 广州大学，2015.
[31] 周云，钟根全，龚晨，陈清祥 . 开孔钢板装配式屈曲约束支撑钢框架抗震性能试验研究[J]. 建筑结构学报，2019，40(03)：152-160.
[32] 钟根全 . 屈曲约束支撑框架平面外抗震性能研究[D]. 广州大学，2018.
[33] Fengming Ren，Yun Zhou，Jiebiao Zhang，Shaoming Lin. Experimental study on seismic performance of CFST frame structures with energy dissipation devices，Journal of Construction & Steel Research[J]，2013，90：120-132.
[34] Fengming Ren，Yun Zhou，Guangming Chen，Jianwei Liang. Experimental study on seismic performance of CFST frame-shear wall structure with BRBs.，The structural design of tall and special buildings，2015，24：73-95.
[35] 周云，陈清祥，龚晨，钟根全，田时雨 . 消能部件平面外稳定性问题探讨[J]. 建筑结构，2019，49(07)：111-115+121.
[36] 龚晨 . 开孔钢板装配式 BRB 设计方法及双向地震作用下结构中 BRB 平面外局部稳定性研究 . 广州大学，2018.
[37] 曾政伦 . 消能支撑节点板平面内屈服力和刚度设计方法研究[D]. 广州大学 . 2020.
[38] 建筑消能减震技术规程：JGJ 297—2013. 北京：中国建筑工业出版社，2013.
[39] 孔瑜文 . 考虑钢框架开合效应的防屈曲支撑角部节点板连接受力性能研究[D]. 广州大学，2016.
[40] 孔瑜文，赵俊贤，周云 . 防屈曲支撑角部节点板与钢框架的相互作用效应[J]. 土木工程学报，2016，49(S1)：107-113.
[41] 周云，王贤鹏，陈清祥，高冉 . 消能子结构节点设计方法研究[J]. 工程抗震与加固改造，2019，41(04)：1-13.
[42] 王贤鹏 . 带 BRB 钢筋混凝土框架子结构节点核心区设计方法研究[D]. 广州：广州大学，2018.
[43] 王森，魏琏 . 不同高位转换层对高层建筑动力特性和 地震作用影响的研究[J]. 建筑结构，2002，32(8)：54-58.
[44] 周云，吴从永，邓雪松，等 . 高位转换耗能减震结构新体系[J] . 工程抗震与加固改造，2006，28(5)：72-77.
[45] 周云，邓雪松，吴从晓 . 高层建筑耗能减震新体系概念与实现[J]. 工程抗震与加固改造，2007，29(6)：1-9.
[46] 吴从晓 . 高位转换耗能减震结构体系分析研究[D] . 广州：广州大学，2007.
[47] 丁鲲，周云，邓雪松 . 框架-核心筒结构耗能减震层的 减震效果分析[J] . 工程抗震与加固改造，2007，29 (3) ：35-40，29.
[48] 邓雪松，丁鲲，周云 . 耗能减震层对框架-核心筒结构的减震效果及其影响分析[J] . 工程抗震与

加固改造，2008，30(1)：1-8.
[49] 陈斯聪，周云．BRB＋VD消能减震结构体系分析研究[J]．建筑结构，2016，46(11)：85-90.
[50] 韩家军．防屈曲支撑框架-核心筒结构减震分析研究[D]．广州大学，2009.
[51] 高层建筑混凝土结构技术规程 DBJ 15—92—2013.
[52] 屈曲约束支撑应用技术规程．2020.

12 减隔震技术在航站楼大跨结构中的应用

朱忠义[1] 束伟农[1] 柯长华[1] 卜龙瑰[1] 杨玉臣[1] 秦 凯[1] 黄 嘉[2] 王 毅[1] 张 琳[1]
（1. 北京市建筑设计研究院，北京；2. 重庆市建筑设计研究院，重庆）

摘 要：建筑结构基础隔震设计在国内外得到了极大的发展和应用，技术已经比较成熟，为工程防震减灾领域带来了新的思路和方法。目前采用隔震技术的结构大多为较规则的混凝土框架、框剪、剪力墙结构，而且单体规模较小。对于单体超大，体型复杂的大跨度空间结构采用隔震技术的工程较少，相应的研究也不多。本文结合昆明新机场航站楼和深圳机场 T3 航站楼工程减隔震结构设计的具体情况，介绍减隔震设计在航站楼大跨结构中的应用，得到了一些有意义的结论，可为同类结构的设计研究提供借鉴。

关键词：航站楼；大跨结构；隔震；阻尼器；弹簧

Application of the seismic isolation and mitigation technologies in Large span sturcture of airport terminal buildings

Zhu Zhongyi[1], Shu Weinong[1], Ke Changhua[1], Bu Longgui[1], Yang Yuchen[1], Qin Kai[1], Huang Jia[2], Wang Yi[1], Zhang Lin[1]
（1. Beijing Institute of Architectural Design，Beijing；
2. Chongqing Architectural Design Institute，Chongqing）

Abstract: Seismic base isolation design for building structures has a fast development and wide application in China and abroad. Such technology has already become mature, rendering new ideas and solutions to the earthquake prevention and disaster reduction in structural engineering. At present structures that employ base isolation technique mainly include RC frames, frame—shear wall and shear wall structures with regular configurations, which are usually of small scale for a single building. In comparison, there are fewer researches and applications of the seismic isolation technology in large span structures with giant and complex configuration. This paper introduces the seismic isolation design on the terminal building of Kunming new airport and the Terminal T 3 of Shenzhen Baoan Airport, and thereby reviews the present applications of the seismic isolation and mitigation technologies in large span structures of airport terminal buildings. Some practical conclusions have been reached, providing reference to the design and research of other structures of the same kind.

Keywords: terminal buildings; large span structures; seismic isolation; damper; spring

1. 引言

地震发生时，地面运动引起结构的反应，结构吸收了大量的地震能量，转换为动能或热能的形式。传统的抗震结构体系，容许结构及支承构件（柱、梁、节点等）在地震中出现损坏，这一损坏过程就是能量的耗散过程，而结构及构件的严重损坏或倒塌，就是能量转换或消耗的最终完成。

结构消能减震技术是在结构物的某些部位，设置耗能装置，通过耗能装置产生摩擦，弯曲（或剪切、扭转）弹塑性滞回变形来消耗或吸收地震输入结构中的能量，以减小主体结构的地震反应。在中震、大震及强震作用下，耗能装置率先进入耗能状态，产生较大的阻尼，耗散地震输入结构的大部分能量，并迅速衰减结构的动力反应，而主体结构不出现明显的弹塑性变形，从而确保其在强震或强风作用下的安全性和正常使用性。

隔震技术的基本原理是在建筑物上部结构与基础之间，设置隔震支座和阻尼器，以延长结构周期、增大结构阻尼，阻隔地面运动向上部结构的传递，从而减少上部结构的地震反应。隔震技术已经广泛应用在建筑、桥梁等结构上，目前有大约三十多个国家在开展基础隔震应用的研究。美、日、新西兰、法、意和我国已相继建成了一些隔震建筑，目前全世界已有 2 千余栋橡胶支座隔震建筑。少量隔震建筑已经受了地震的考验，显示出良好的隔震效果，特别是 1994 年美国加州北岭 6.7 级地震中的洛杉矶南加州大学医院，1995 年日本阪神 7.2 级地震中的日本西部邮政大楼都表现出良好的隔震性能，使得隔震技术越来越为广大的工程技术人员和社会所接受。

本文结合昆明新机场航站楼和深圳机场 T3 航站楼结构设计，介绍减隔震技术在大跨结构中的应用。

2. 昆明新机场航站楼大空旷结构隔振设计

2.1 航站楼核心区结构体系

昆明新机场航站楼建筑面积为 54.8 万 m^2，建成后为国内第四大机场，如图 1 所示。

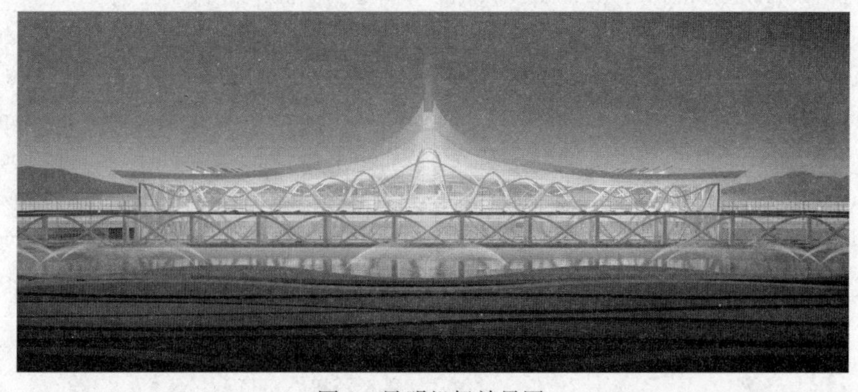

图 1 昆明机场效果图

航站楼距小江断裂带仅 12km，小江断裂带为世界上活动级别最高的断裂带之一，500 年来，平均 150 年发生一次近 8 级地震，至今 170 年没有 7 级以上地震，形势愈来愈严峻。

结构缝将航站楼划分为 7 块（区），其中核心区面积最大，其东西方向尺寸 337m，南北方向尺寸 275m，单体投影面积近 90000m²。核心区屋顶为双曲面，采用变厚度四角锥网架结构，网架上、下表面均为空间曲面，最大跨度 72m，最大悬挑 36m。屋顶主要支承结构为 7 榀钢彩带及锥形钢管柱和变截面箱形摇摆柱，彩带结构复杂，为空间弯扭构件，钢结构下部为 6 层混凝土框架结构，屋顶及其支承结构如图 2 所示。

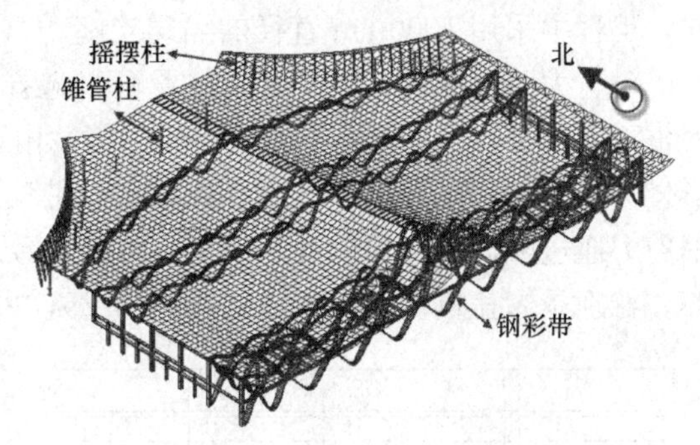

图 2　核心区大空旷结构支撑体系

由于昆明新机场地处高烈度区、地震多发区、距小江断裂带只有 12km，且核心区建筑功能多、结构复杂、造价高、活动人员多，因此针对核心区采用了隔震设计。

2.2　减隔震方案的选择

在房屋结构中设置消能装置，通过其局部变形提供附加阻尼，可以消耗输入上部结构的地震能量，达到预期防震目标。昆明新机场航站楼属于大型复杂结构，如产生较大的变形，对彩带结构的稳定不利，也可能导致幕墙玻璃破碎。另外，目前我国在消能减震工程实践中，多采用阻尼支撑形式，而昆明新机场航站楼采用大跨结构、玻璃幕墙，如采用支撑形式，既影响视觉效果，又妨碍交通组织。同时由于结构体量、自重大，抗震设防烈度高，采用消能减震设计方案时，设计阻尼力大，而我国目前大吨位的黏滞阻尼墙、黏弹性阻尼墙等产品不成熟，尚无工程实践经验，现阶段不便在重大工程中应用。因此，在昆明新机场航站楼设计中不适合采用消能减震技术。

国际上，在地震高烈度区的大跨建筑，有许多采取隔震设计的成功经验。如美国旧金山国际机场，距圣安德烈斯断裂只有 5km，圣安德烈斯断裂，总长超过 965km，是世界著名的活动断层。旧金山国际机场候机主楼长 350m，宽 150m，采用隔震技术，可抵抗该断层最近点可能发生的 8 级地震，同时大幅降低了造价，成为世界典型的隔震建筑。土耳其 Antalya International Terminal Building 1，采用隔震设计方案，共采用 130 个 FPS 隔震支座。日本除在 50 层的高层建筑中采用隔震技术外，还建设了多座 300m×100m 的隔震仓库。

因此，昆明新机场航站楼工程采用隔震技术是可行的。

2.3　隔震层布置

多种隔震支座与消能装置综合使用的组合隔震消能技术是解决大型复杂结构抗震问题的研究趋势。昆明新机场航站楼的隔震层由橡胶隔震支座、铅芯橡胶隔震支座和阻尼器组成。由于本工程支座的荷载较大，应当采用大直径的橡胶隔震支座。而目前我国橡胶隔震支座的生产能力有限，很难生产较大直径橡胶隔震支座。基于我国目前隔震支座的生产现状，工程中采用1000mm直径的隔震支座。

对于大跨度结构，隔震层的位置可以选择在支承屋盖的柱顶、±0.00附近或地下室底板顶面等部位。对于昆明新机场航站楼，由于支承钢结构屋盖的是非对称布置的钢结构彩带，其间还镶嵌着幕墙玻璃。如果在彩带顶设置隔震支座，大屋盖受地震作用产生的水平剪力通过隔震支座传递到彩带上，可能导致彩带失稳，或由于彩带变形引起幕墙玻璃破碎。特别是屋盖重量仅占整体结构的1/25，采用屋盖隔震意义不大，因此隔震层不宜布置在彩带顶部。当隔震层布置在±0.00位置时，由于建筑使用功能的要求，结构首层楼面开洞较多，影响了隔震层以上楼板的整体刚度。对大底盘隔震来说，需要通过提高隔震层以上楼面的整体刚度，来克服可能出现的地基不均匀沉降导致的不利影响，同时确保上部结构作整体运动。另外，由于机场功能复杂，安装自动扶梯、电梯等设备，布置幕墙玻璃、锥形钢管柱等，不适合穿过隔震层。因此，隔震层没有设置在±0.00处。在结构地下室底板顶面布置隔震层时，不影响整体建筑的功能布置，也可充分发挥隔震效果，降低因上部结构采用复杂结构形式带来的风险，从根本上提高整体结构的抗震性能。同时，又由于昆明新机场航站楼工程的基础底板位于填方区，增加隔震层还可以减小土方的回填量。综合以上因素，昆明新机场航站楼工程的隔震层设置在基础底板上部，位于－14.2m以下，在地下三层之下与基础顶面之间，如图3所示。

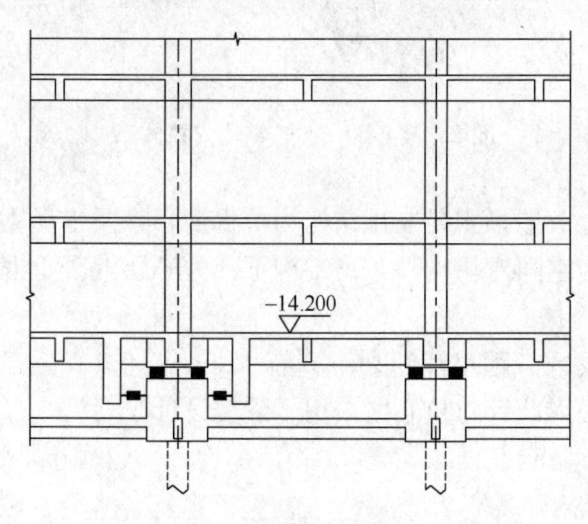

图3　隔振层位置

隔振层布置如图4、图5所示，其中叠层橡胶隔震垫共1156个，布置在中心区；铅芯橡胶隔震垫共654个，布置在外围；黏滞阻尼器共104个，南北和东西各52个，均匀布置。这种隔振层布置可以提高隔震效率，限制隔振层位移过大，增强隔震层抵抗结构偶然偏心的能力，避免隔震结构的扭转。橡胶垫参数如表1，阻尼器的阻尼系数为1500 kN/(m/s) 0.4、阻尼指数为0.4。隔振层刚度如图6所示。

橡胶垫参数　　**表1**

型号	剪切模量（N/mm²）	类型	1次形状系数	2次形状系数	有效面积（cm²）	水平刚度（kN/m）			竖向刚度（kN/mm）
						屈服前刚度	屈服力	屈服后刚度	
RB1000	0.55	无铅芯橡胶垫	38.0	5.4	7834	2540			5779
LRB1000	0.55	铅芯橡胶垫	41.7	5.2	7854	28900	208	2670	6030

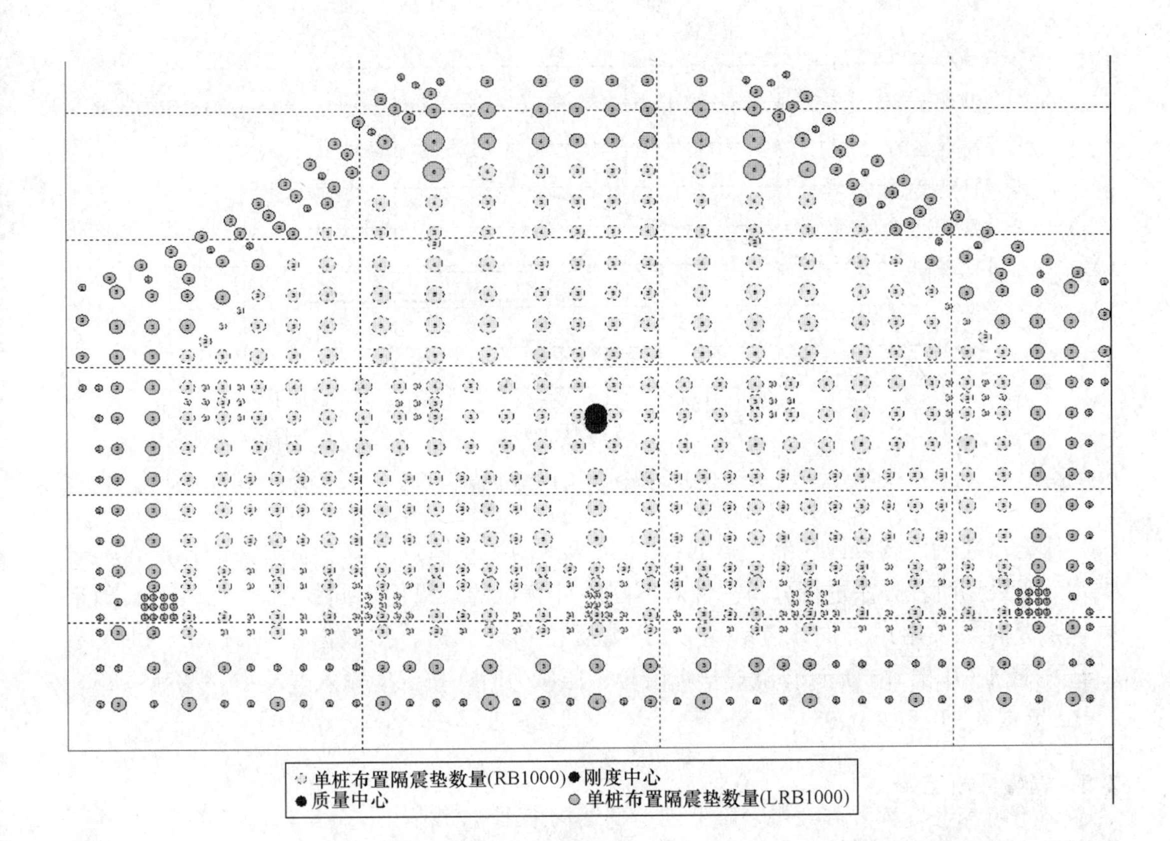

图 4　隔振层橡胶垫布置

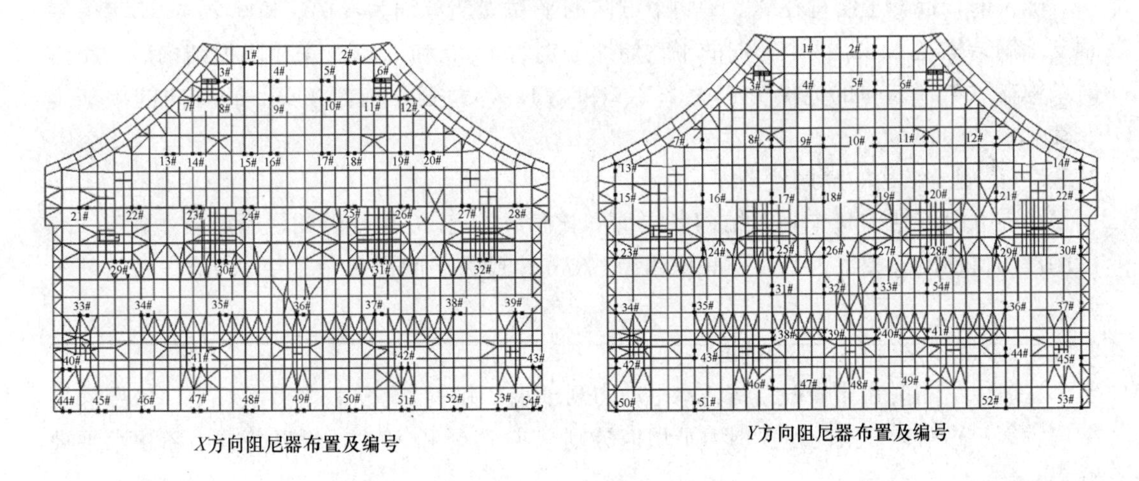

X方向阻尼器布置及编号
Y方向阻尼器布置及编号

图 5　隔振层阻尼器布置

2.4　计算模型和分析方法

隔震分析模型为整体结构计算模型，模型中包括下部混凝土结构、屋顶支承钢结构、屋顶结构以及隔震层。隔震层由普通叠层橡胶垫、铅芯橡胶垫和黏滞阻尼器组成。其中，普通叠层橡胶垫计算模型取为线性模型；铅芯橡胶垫计算模型为非线性模型；黏滞阻尼器采用 Maxwell 模型。

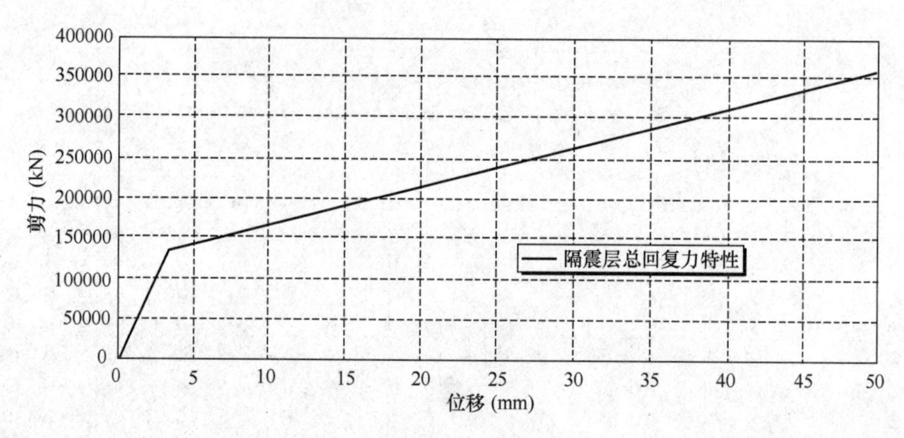

图 6　隔振层阻尼器布置

依据《建筑抗震设计规范》第 12.2.2 条给出的关于隔震计算分析方法的规定，隔震结构计算采用时程方法进行分析。针对不同的计算目的，采用不同的输入方式：1）确定减震系数时，采用单方向水平输入；2）大震位移计算时，按双向水平输入（X∶Y=0.85∶1或 1∶0.85）；3）计算隔震垫极限拉、压应力时，按三向输入（X∶Y∶Z=0.85∶1∶0.65 或 1∶0.85∶0.65）

2.5　隔震分析结果

2.5.1　隔震结构的周期

隔振前，混凝土结构在 X、Y 两个方向的平动周期分别为 0.75s 和 0.78s，彩带等屋顶支承钢结构在 X、Y 两个方向的平动周期分别为 1.03 和 0.66；采用隔振结构后，小震时结构在 X、Y 方向的周期分别为 2.21s 和 2.19s，大震时结构在 X、Y 方向的周期为 2.82s 和 2.80s。

2.5.2　混凝土结构的隔振效果

小震下隔震结构与非隔震结构各层剪力比值均小于 0.46，因此，隔震后能达到 7.5 度的设防目标。大震下，最大层间位移为 17mm，最大层间位移角 1/294，结构基本处于弹性。

2.5.3　彩带等屋顶支承结构的隔振效果

屋顶传给隔振和非隔振结构的水平剪力列于表 2 和表 3。从表中可以看出，采用隔振结构后 X、Y 两个方向的屋顶剪力平均值分别减小 74%和 70%，可显著减小屋顶水平地震力。

屋顶支承结构顶部剪力比（X 方向）　　**表 2**

	人工 61 号波	人工 62 号波	1 号天然波	2 号天然波	3 号天然波	平均值
非隔震地震剪力（kN）(1)	30072	27917	28402	27394	30117	28780
隔震地震剪力（kN）(2)	7090	7511	6937	6509	9008	7411
剪力比（2）/（1）	0.24	0.27	0.24	0.24	0.30	0.26

屋顶支承结构顶部剪力比（*Y* 方向）　　表 3

	人工 61 号波	人工 62 号波	1 号天然波	2 号天然波	3 号天然波	平均值
非隔震地震剪力（kN）(1)	43414	45633	38536	46469	47310	44272
隔震地震剪力（kN）(2)	14763	11147	11833	13926	15733	13481
剪力比（2）/（1）	0.34	0.24	0.31	0.30	0.33	0.30

大震下，彩带的最大层间位移角为 1/120、悬臂柱最大层间位移角为 1/124，远小于钢框架结构 1/50 的要求，主要支承构件在大震下的性能有了很大提升。

2.5.4　隔振层的性能

大震下，隔振层位移平均值为 536mm，阻尼力平均值为 1468kN。

2.6　相关试验研究

针对本项目的隔振结构，业主组织进行了以下试验研究：

（1）隔震模型和非隔震模型的振动台试验，验证了结构具有良好的隔震性能；

（2）隔震支座的力学性能试验，包括竖向力学性能、水平力学性能、耐久性等各种相关性能；

（3）为高精度、高效率、低成本安装隔震支座，进行了隔震支座安装方法试验研究；

（4）对航站楼在施工和运营期间进行监测，包括基础沉降监测 、隔震层监测、钢结构监测；

（5）建立了强震监测系统，可以采集地震输入以及基础、隔震层、上部结构的地震动响应，可以用于隔震评估、为复杂结构的抗震研究提供资料；

（6）在受载状态下，对支座更换进行试验研究，形成了受载支座的成套更换技术。

2.7　结论

（1）虽然本工程为混凝土结构和钢结构组成的复杂结构体系，特别是屋顶支承结构相对混凝土结构刚度较小，但是屋顶支承钢结构仍然可以取得预期的隔震效果，采用整体隔震体系是可行的。

（2）支承屋顶的彩带结构体系复杂，两个方向刚度差别较大，属于大空旷钢结构，采用隔震设计后，可以满足隔震目标，大震下基本处于弹性状态，提高了这类关键结构构件的抗震性能，为类似工程的应用提供了经验。

（3）多点输入对隔震结构隔震层的位移影响较小，多点输入的位移较单点输入的位移稍小；多点输入对隔震结构的内力影响较小，柱子剪力较单点输入的剪力稍小；由于地震动输入存在相位差，多点输入对结构的扭转影响较大，结构扭转效应。

（4）对于大尺度底盘的隔震结构，隔震支座安装时，应考虑底板混凝土的收缩徐变的影响。

（5）根据施工中发现的问题，建议进一步完善隔震垫的检验要求。

3. 深圳机场T3航站楼指廊区屋顶结构减隔振设计

3.1 工程概况及指廊结构体系

T3航站楼外形为飞鱼形式，建筑面积达50万m^2左右，为目前国内最大单体面积的航站楼之一，图7为航站楼的效果图。

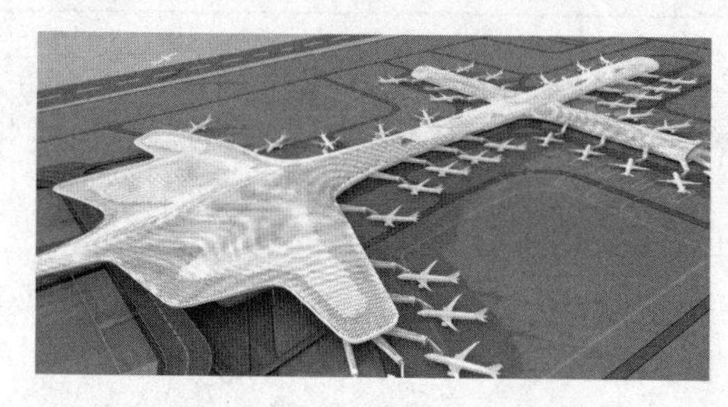

图7 深圳机场T3航站楼效果图

主体混凝土结构南北长1038m、东西宽588m，下部主体结构采用钢筋混凝土框架结构，分为10块。屋顶为自由曲面，长1128m、宽640m，采用网架结构，分为7块，包括主指廊D1、次指廊G和H、交叉指廊C、过渡区B以及大厅A共七块，如图8所示。

主指廊D1区屋顶采用带加强桁架的斜交斜放双层网架。网架曲面延伸到标高4.4m的二层楼面，与下部混凝土支承结构对应，屋顶结构每隔18m设一支座铰接于混凝土异形柱，并且在与支座对应的屋顶部位设置两榀加强桁架作为主要受力体系。沿结构跨度方向，支座间距为44.8m，结构最宽处为61.1m左右。屋顶曲面有凹陷区，沿结构纵向曲面变化较大，网壳沿纵向和横向均变厚度，其中网壳最厚处为8.8m。沿结构纵向，斜交斜放形成的菱形网格对角线长度为9m，另一方向对角线长度在6m左右，局部网格加密。图9～图11是主指廊平面、剖面图。

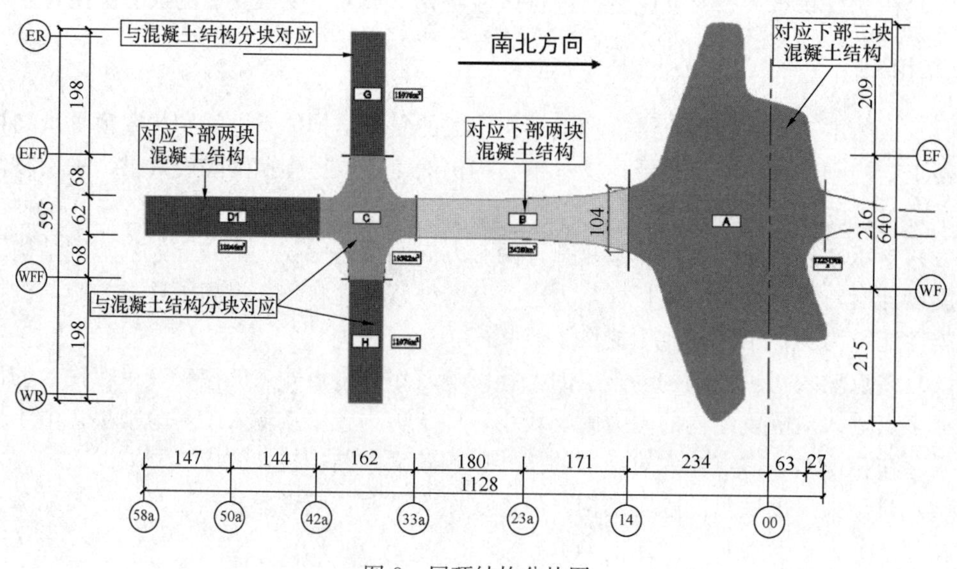

图8 屋顶结构分块图

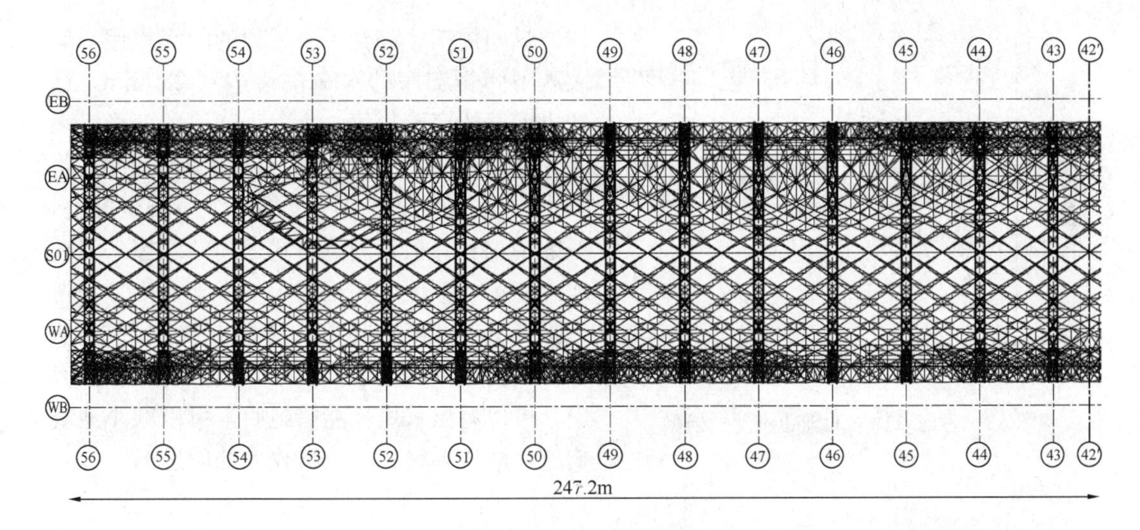

图 9　D1 区屋顶平面图

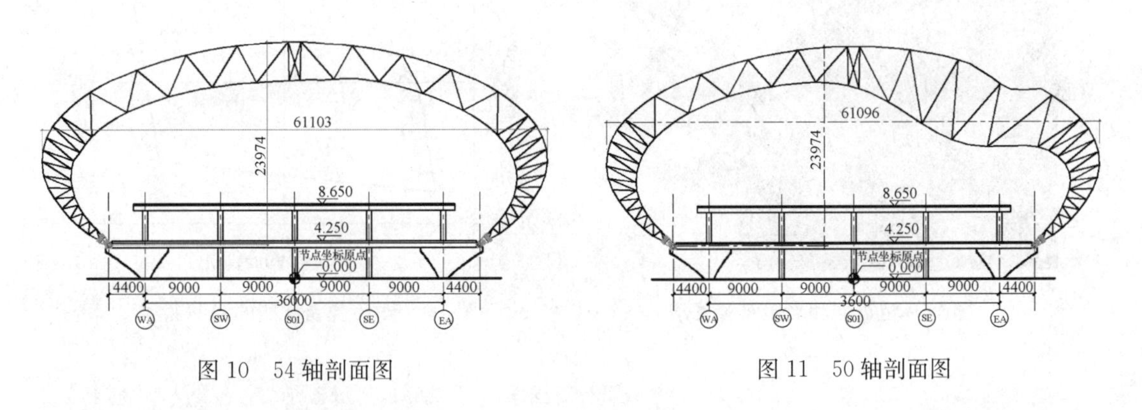

图 10　54 轴剖面图　　　　图 11　50 轴剖面图

3.2　指廊 D1 区减隔震设计

指廊屋顶采用带加强桁架的斜交斜放网架，结构面内刚度较大，屋顶纵向较长，且跨越不同混凝土分块，为减小屋顶结构的温度效应，以及下部不同混凝土分块在地震、温度下由于变形不协调对屋顶结构影响，采用了固定铰支座和弹簧支座结合得形式。D1 区有 14 榀加强桁架，28 个支座。其中全景窗处网架开洞较多，刚度有较大削弱，这些部位的加强桁架支座为固定铰支座，共 10 个；其余部位的支座为纵向弹簧支座。根据结构的受力要求，弹簧支座满足以下要求：1）满足各方向转动要求；2）弹簧刚度 6000～8000kN/m；3）静摩擦系数不大于 10%，动摩擦系数不大于 3%；4）弹簧位移量 60mm。

比较不同边界条件下，升温 30℃支座的反力如下：1）全部支座采用固定铰支座，产生 1788kN 的水平推力；2）采用本工程确定的边界条件，但不考虑支座销轴和耳板的摩擦力，产生 164kN 的水平推力；3）采用本工程确定的边界条件，考虑 10%的摩擦力，克服摩擦力后弹簧提供刚度，按照如图 12 的理想刚塑性模型计算，产生最大 583kN 的水平力。采用弹簧支座可以大幅度减小温度效应，同时在混凝土分缝处，减小由于地震作用带来的屋顶支座反力的突变。另外在计算温度效应时，应该考虑摩擦力的影响。

本工程纵向采用弹簧支座后，减小了结构支承刚度，降低了温度作用，延长了屋顶结

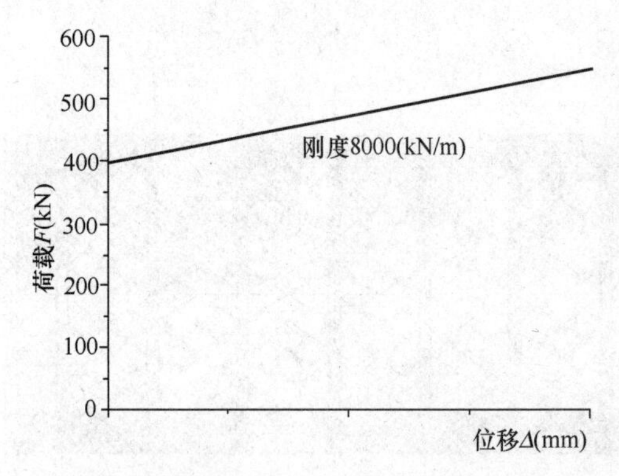

图 12　考虑摩擦力的理想刚塑性模型

构的纵向周期，起到了对屋顶的隔震作用，但大震下弹簧支座的水平位移达到 220mm，远超出弹簧的变形限值。为减小弹簧的变形，采取和考虑了以下措施：1）设置黏滞阻尼器，减小位移，如图 13 所示；2）作为安全储备，考虑销轴和耳板自润滑材料之间的摩擦，消耗地震能量，降低结构相应，减小弹簧变形。考虑弹簧刚度和摩擦耗能的滞回曲线如图 14 所示。结构在大震下主要结果如表 4 所示。从表中，可以看出采用耗能措施后，可以减小地震响应，解决弹簧支座位移较大的问题。

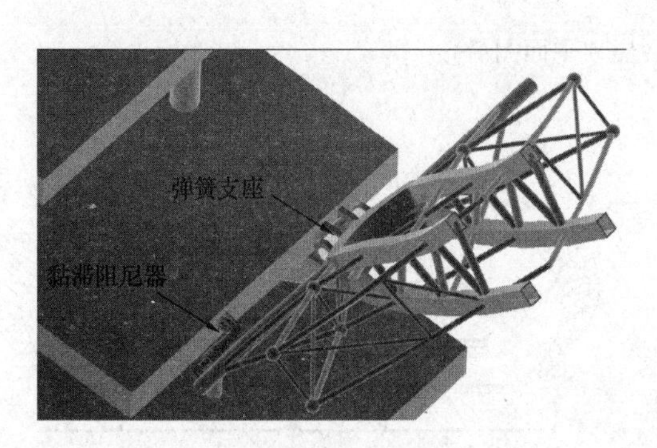

图 13　阻尼器和支座布置关系

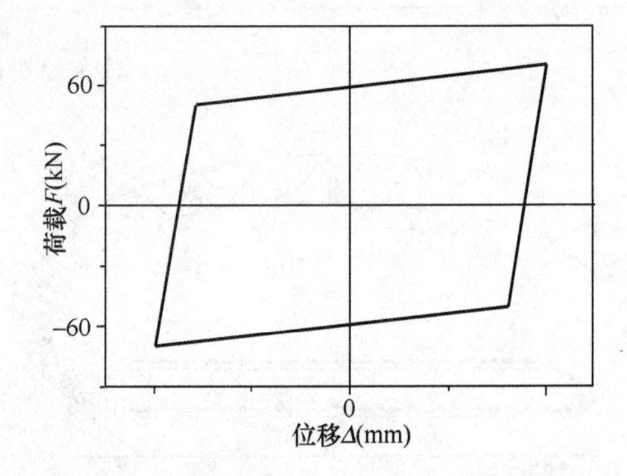

图 14　摩擦-弹簧耗能的滞回曲线

减震计算结果　　**表 4**

模型	支座处剪力（kN）	阻尼器剪力（kN）	总剪力（kN）	最大位移（mm）
弹簧支座	20634	0	20634	220
弹簧支座 ＋ 阻尼器	6406	7834	13415	54
弹簧支座 ＋摩擦耗能	9717	0	9716	75
弹簧支座 ＋ 阻尼器 ＋ 摩擦耗能	6943	6748	13709	46

4. 结语

通过对昆明新机场航站楼基础隔振和深圳机场 T3 航站楼屋顶减隔震分析研究和工程实践，表明采用减隔震技术可以大幅降低大跨结构的地震力，提高结构的抗震性能。随着各种耗能阻尼器产品的成熟，减隔震技术在大跨结构中的应用会越来越多，也会进一步推动减隔震技术在大跨结构中的应用水平。

参考文献

[1] 束伟农，朱忠义，柯长华等．昆明新机场航站楼工程结构设计介绍．建筑结构，2009.5(2) Vol. 39 No. 5.

[2] 朱忠义，束伟农，卜龙瑰等．昆明新机场大空旷结构隔振性能研究．第十三届空间结构学术会议论文集，2010.12，深圳．

[3] 柯长华，朱忠义，秦凯等．深圳宝安国际机场 T3 航站楼钢结构设计．第十三届空间结构学术会议论文集，2010.12，深圳．

[4] 唐家祥，刘再华．建筑结构基础隔震[M]．武汉：华中理工大学出版社，1993.

[5] 周福霖．工程结构减震控制[M]：第 33 页．北京：地震出版社，1997 年 4 月．

[6] 周锡元，阎维明，杨润林．建筑结构的隔震、减震和振动控制[J]．建筑结构学报，2002，23(2)：2-12，26.

13 循环荷载作用下预制混凝土结构中装配式梁柱钢质节点试验研究*

李祚华，齐一鹤，滕　军

（哈尔滨工业大学（深圳）土木与环境工程学院，深圳）

摘　要：针对预制混凝土结构，提出一种新型装配式带阻尼器的梁柱钢质节点。阻尼器由低屈服点钢开孔形成的耗能段和非耗能段构成。通过对装配式钢质节点进行循环荷载加载足尺试验，研究装配式钢质节点承载力、滞回耗能、位移延性、承载力退化及耗能能力等抗震性能指标，研究非耗能段厚度和耗能段长度对装配式钢质节点抗震性能的影响。试验结果表明，相较于现浇节点，装配式钢质节点的滞回曲线更饱满，耗能能力更强，位移延性更好，承载能力退化更缓慢；该装配式钢质节点的极限承载力低于现浇节点，能有效控制混凝土损伤，避免梁端混凝土发生弯曲破坏，实现梁端“塑性可控”；装配式钢质节点承载力随非耗能段厚度增加和耗能段长度减小而增加，耗能能力随非耗能段厚度增加和耗能段长度减小而降低。

关键词：装配式梁柱钢质节点；塑性可控；循环荷载加载试验；抗震性能

1. 引言

装配式混凝土结构具有生产效率高、施工速度快、构件质量好以及节约资源和保护环境等优点[1-3]，是未来建筑结构建造的发展方向。国内外学者已针对装配式结构进行了研究并取得了一定的成果[4]。历史震害调查表明，已有装配式混凝土结构在地震中的严重破坏多因节点破坏造成[5-8]，突出了梁柱节点在装配式混凝土结构中的重要作用。装配式混凝土结构中传统梁柱节点为施工现场预制构件连接处后浇整体式节点，利用对角钢筋[9]、套筒灌浆[10]、组合钢桁架[11]等方式对传统梁柱节点进行连接，确保装配式混凝土结构整体性。研究表明传统后浇整体式节点的滞回性能与现浇节点类似[12]，但仍存在施工现场需额外材料及人工成本、不利于保护环境等问题。

20 世纪 90 年代起，美国和日本开展预应力连接方式的预制混凝土结构体系研究[13]，预制混凝土结构体系由钢绞线提供节点预应力，结构残余位移可忽略不计。李灿军，周臻等[14]提出一种自复位梁柱节点，利用超弹性形状记忆合金能够有效降低节点残余变形，与传统预应力连接方式节点相比，利用摩擦耗能器能有效增加节点耗能能力。Morgen 和 Kurama[15, 16]开发了一种新型无粘结后张法预制梁柱节点，该节点引入摩擦阻尼器以提供显著的能量耗散，并具有良好的自复位能力。此外，也有学者在自复位节点中采用可更换

* Li ZH, Qi YH, Teng J. Experimental investigation of prefabricated beam-to-column composite joints for precast concrete structures under cyclic loading. [J] . Engineering Structures, 2020, 209: 110217.

低碳钢[1]、竹形阻尼器[17]等耗能构件提高节点的耗能能力。Ersoy 和 Tankut[18]通过在预制构件连接处顶部和底部引入钢板，对节点进行焊接连接有利于保证装配式节点完整性，研究表明节点刚度、耗能和强度与现浇节点相似。

同时，装配式混凝土结构中塑性铰位置的确定也引起学者广泛关注[19-21]。Eom 等[22]提出了装配式混凝土结构中塑性铰位置重新定位方法，通过对钢筋加固或截面削弱方法，对塑性铰位置外移梁柱节点位置或处于梁柱节点位置，研究表明梁柱节点的剪切开裂和钢筋粘结滑移均有所减小。Yang 等[23]提出一种带有延性节点的新型混合梁体系，塑性铰位置发生与混凝土梁上而非 H 型钢，试验过程中连接处未出现劈裂拉伸裂缝或剪切裂缝。Choi 等[24]人对节点内灌入纤维混凝土以提高节点抗剪韧性和抗拉强度，控制塑性铰位置处于节点外混凝土梁上。

综上所述，现有的装配式节点设计仍然沿用传统节点的设计方法，通过保证其强度、刚度和延性以达到“等强代替现浇”的基本原则，但节点核心区通常成为抗震薄弱环节，塑性发展不可控。即使以加强节点连接区或削弱混凝土梁端区方式，实现塑性铰位置控制在节点连接范围外的混凝土梁上，但易造成混凝土梁损伤严重，且少有考虑在通过人为加入塑性铰，控制构件塑性发展过程，保护混凝土梁避免损伤。

鉴于此，本文提出了一种新型装配式梁柱钢质节点，该新型节点具有抠孔形式的阻尼器耗能区。地震时节点耗能区先进入塑性，并充分耗能，控制梁端混凝土损伤，提高梁柱节点整体抗震性能。为研究该新型节点的抗震性能，对三个装配式梁柱钢质节点进行了拟静力加载足尺试验，并与现浇节点足尺试验进行对比，分析其滞回性能、耗能能力、位移延性等指标，验证该新型节点是否满足梁端“塑性可控”抗震概念设计。同时研究装配式梁柱钢质节点非耗能段厚度和耗能段长度对节点抗震性能的影响。

2. 装配式梁柱钢质节点的提出

本文提出的装配式梁柱钢质节点（以下简称：装配式节点），如图 1 所示。装配式节点由阻尼器、耳板和高强销轴组成。预制混凝土梁端部焊接钢板并预先开孔，用于与阻尼器螺栓连接。预制混凝土梁内部纵筋和角钢共同焊接，且纵筋和角钢端部均与预制混凝土梁端部钢板焊接连接。预制混凝土柱端部焊接钢板并预先开孔，用于与柱柱节点螺栓连接。预制混凝土柱内焊接 H 型钢。柱柱节点内焊接 H 型钢，该 H 型钢与预制混凝土柱内 H 型钢尺寸、翼缘方向和钢材型号均一致。柱柱节点内焊接防屈曲肋板，抑制柱柱节点在受弯时外侧钢板发生屈曲。柱柱节点外侧由封板等钢板焊接连接，并预先开设螺栓孔与

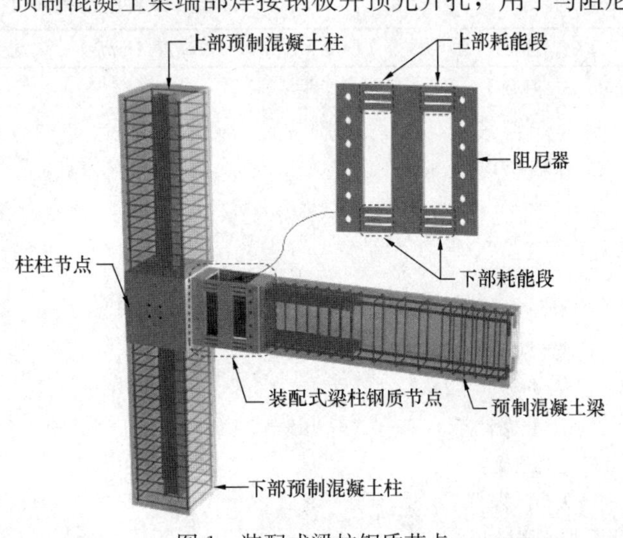

图 1　装配式梁柱钢质节点

上下预制混凝土柱螺栓连接。柱柱节点和预制混凝土梁均焊接耳板，并通过高强销轴进行铰接。对低屈服点钢开矩形孔形成耗能段和非耗能段的阻尼器，通过螺栓与柱柱节点和预制混凝土梁连接。装配式节点受力明确，节点抗弯承载能力由阻尼器提供，节点抗剪能力由高强销轴提供。

装配式节点具有易安装、可拆卸及保护环境等优点，可用于高层建筑以缩短建筑施工工期并降低成本。阻尼器采用低屈服点钢和耗能段设计，在地震作用下耗能段产生较大拉压变形，增加节点在地震作用下耗能能力和延性。此外，装配式节点能够控制梁柱构件塑性铰位置。传统钢筋混凝土梁柱节点的破坏是钢筋屈服和混凝土压碎，地震能量是通过混凝土梁中塑性铰的转动变形进行消耗，从而传统钢筋混凝土结构中混凝土梁破坏严重。为此，装配式节点中阻尼器通过开矩形孔，削弱节点抗弯承载能力，使节点在地震作用下塑性变形集中于阻尼器内，类似在混凝土梁端设置塑性铰，实现梁端“塑性可控”，保护混凝土梁免受损伤。

3. 试验概况

3.1 试件设计

按足尺比例设计了三个装配式节点试件（PSJ），同时设计了一个梁柱配筋和尺寸一致的现浇混凝土节点试件（MJ）以作对比。试件尺寸截断取自结构的反弯点位置，试件柱高柱高 3740mm，梁长 2820mm，柱截面尺寸为 600mm×600mm，梁截面尺寸为 300mm×600mm。现浇节点和装配式节点的配筋见图 2 和图 3，装配式节点试件的型钢和低屈服点钢的尺寸见表 1 和表 2。PSJ 试件的阻尼器尺寸为 600 mm×640 mm×25 mm，耗能段高度为 28mm，PSJ-1、PSJ-2 和 PSJ-3 的耗能段长度分别为 130mm、130mm 和 150mm。PSJ-1、PSJ-2 和 PSJ-3 的非耗能段厚度分别为 25 mm、30 mm 和 30 mm。所有试样的设计均符合《混凝土结构设计规范》GB 50010—2010[25] 的规定。

PSJ 试件的型钢尺寸 **表 1**

构件	厚度（mm）	高度（mm）	宽度（mm）	长度（mm）	直径（mm）
耳板 1	30	—	—	—	—
耳板 2	40	—	—	—	—
高强销轴	—	—	—	100	60
端板 1	10	560	570	—	—
端板 2	15	560	600	—	—
端板 3	20	600	600	—	—
柱内 H 型钢	—	—	—	1500	—
柱柱节点内 H 型钢	—	—	—	560	—
肋板 1	10	560	210	—	—
肋板 2	10	560	270	—	—
垫板	10	200	210	—	—
方钢管	30	560	300	120	—
高强螺栓 10.9	—	—	—	—	32

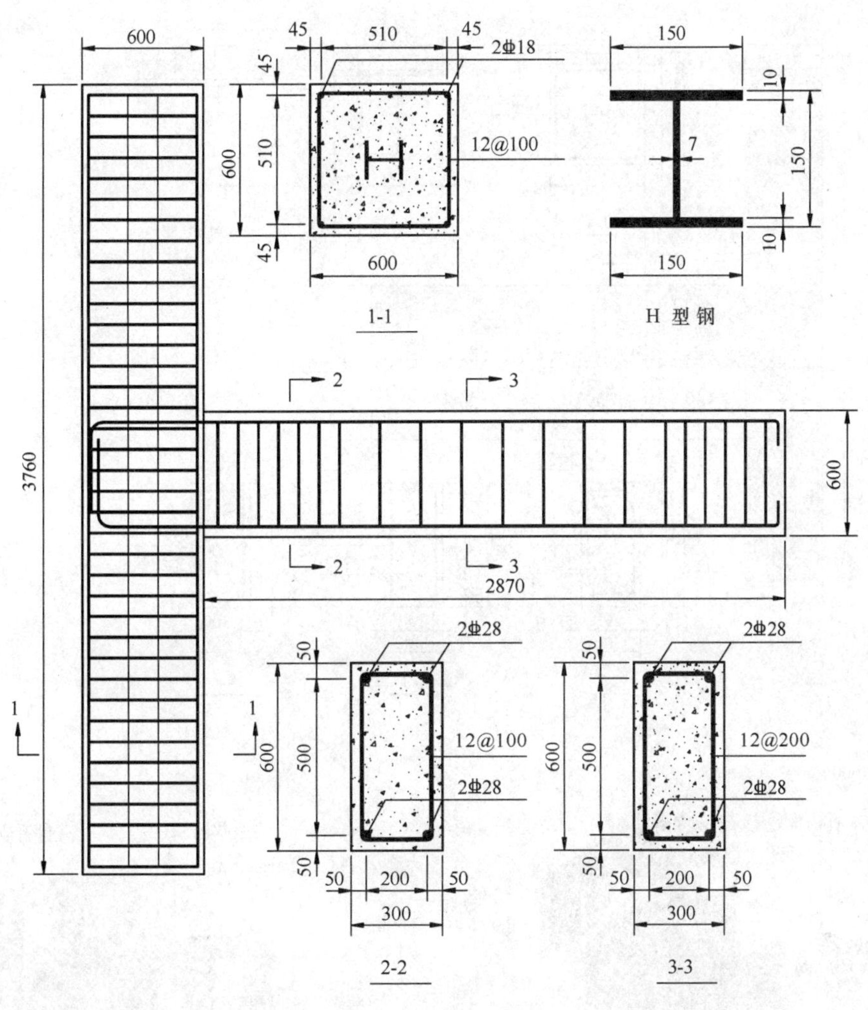

图 2　现浇节点配筋图

PSJ 试件的低屈服点型钢尺寸　　**表 2**

试件	厚度（mm）	高度（mm）	宽度（mm）	耗能段高度（mm）	耗能段长度（mm）	非耗能段厚度（mm）
PSJ-1	25	600	640	28	130	25
PSJ-2	25	600	640	28	130	30
PSJ-3	25	600	640	28	150	30

3.2　材料性能

阻尼器采用 LYP 100 低屈服点钢，试件其他钢材均为 Q345 钢，螺栓为高强螺栓，钢筋为 HRB400。按《金属材料　室温拉伸试验方法》GB/T 228—2010 标准[26]对钢材和钢筋进行了材性试验，钢板拉伸试件尺寸如图 4 所示，采用位移控制，拉伸速率为 1.2mm/min。钢筋和钢板材性试验数据见表 3 所示。按《混凝土物理力学性能试验方法标准》GB/T 50081—2002[27]对 3 个混凝土立方体试块进行抗压强度试验，实测混凝土性能见表 4。

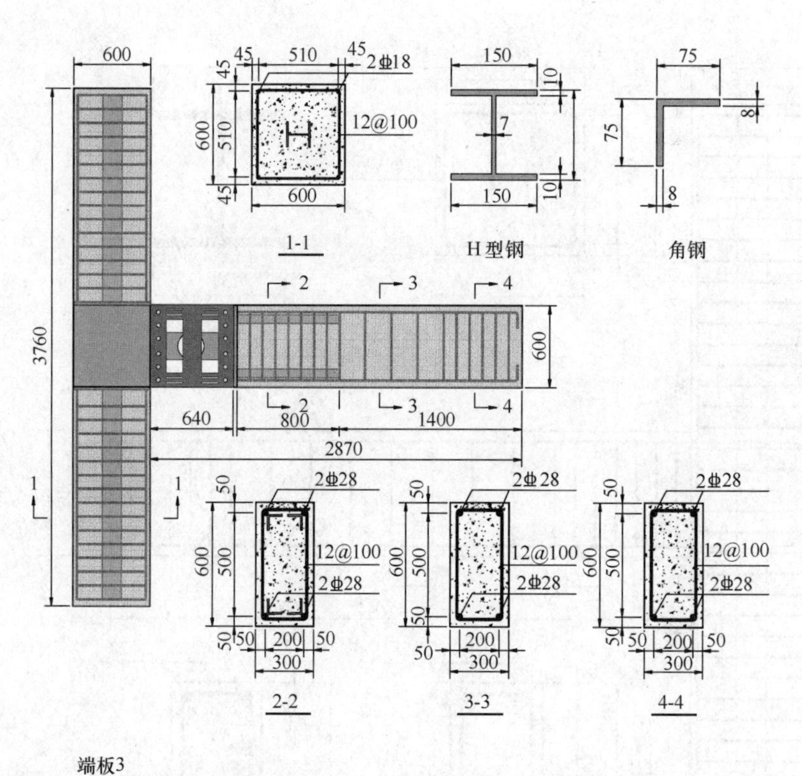
600
45
510
45
2Φ18
12@100
1-1
150
10
7
H 型钢
75
8
角钢
3760
2
3
4
640
800
1400
2870
1
2-2
3-3
4-4
2Φ28
500
50
200
300

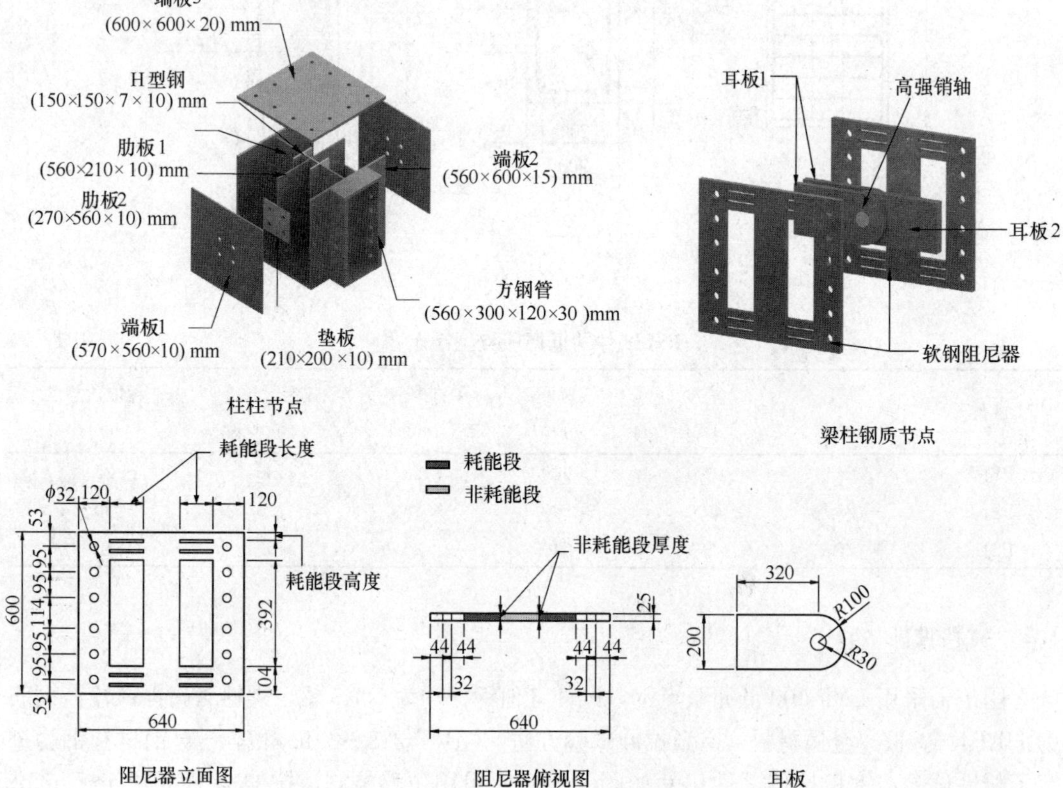
端板3
(600×600×20) mm
H 型钢
(150×150×7×10) mm
肋板1
(560×210×10) mm
肋板2
(270×560×10) mm
端板2
(560×600×15) mm
方钢管
(560×300×120×30)mm
端板1
(570×560×10) mm
垫板
(210×200×10) mm
柱柱节点
耳板1
高强销轴
耳板2
软钢阻尼器
梁柱钢质节点
耗能段长度
耗能段
非耗能段
φ32
120
耗能段高度
392
104
640
非耗能段厚度
25
44
32
320
R100
R30
200
阻尼器立面图
阻尼器俯视图
耳板

图 3 装配式节点配筋图

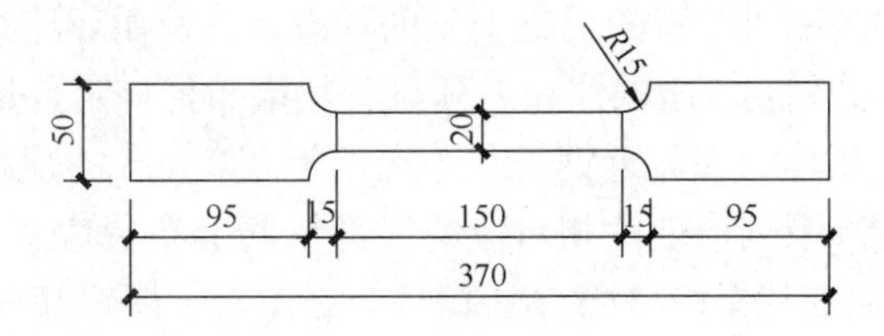

图 4　LYP100 钢材性试验尺寸

钢材材料性能　　表 3

钢材牌号	直径（板厚）(mm)	屈服强度 f_y (MPa)	抗拉强度 f_u (MPa)
Q345	10	361	495
LYP100	25	81	223
HRB400	28	440	554

混凝土材料性能　　表 4

混凝土强度等级	极限荷载 f_c (kN)	立方体抗压强度 f_{cu} (MPa)	弹性模量 E_c (MPa)
C50	1206.87	53.63	34996

3.3　加载装置和加载制度

试验加载装置如图 5 所示。试件上下柱端铰接，采用千斤顶对柱端施加 1375kN 轴向压力，柱顶通过水平支撑杆与反力墙连接，约束柱顶水平位移，梁端采用量程为 500kN 的 MTS 液压伺服作动器施加往复循环荷载。

试验加载方式采用力—位移混合控制，如图 6 所示。节点未达到屈服荷载前采用力控制加载，试件屈服前，荷载每级增加 15kN，每级循环 1 次。以试件的滞回曲线是否出现拐点判断试件是否屈服，试件屈服后采用位移加载，以屈服位移为增量逐级加载，每级循环 3 次。直至试件承载能力下降至极限承载能力的 85%，停止加载。

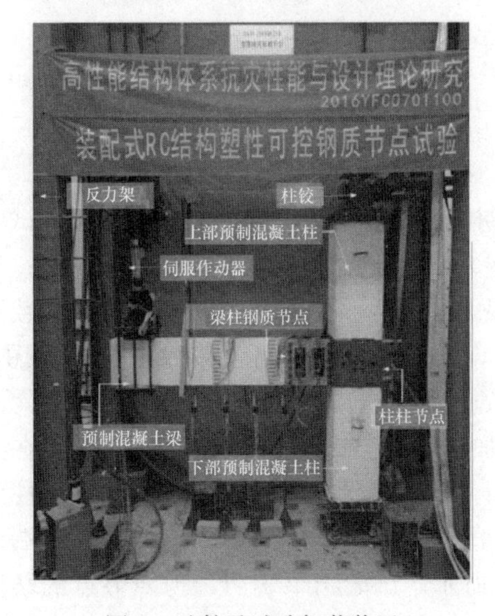

图 5　试件及试验加载装置

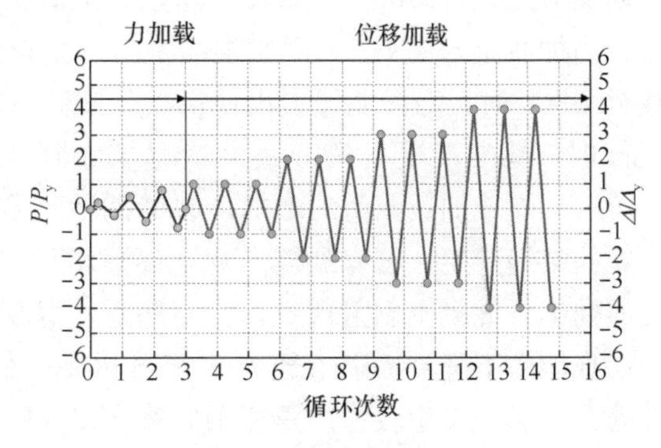

图 6　试验加载方式

为评估现浇节点和装配式节点的混凝土梁的状态，以及循环荷载下 MJ 试件和 PSJ 试件中钢筋的应变发展，分别对 MJ 试件和 PSJ 试件梁端布置一定数量的应变片，具体布置如图 7 所示。对于 PSJ 试件，在靠近钢板和距离钢板 330 mm 的预制梁端的顶部纵筋上布置应变片 A1～A4，在 A 系列应变片相对应的底部纵筋上布置应变片 B1～B4，在距钢板 100 mm 的箍筋上布置应变片 C1。对于 MJ 试件，在梁端、距梁端 350 mm 和 1000 mm 处的纵筋上布置应变片 D1～D6，在 D 系列应变片相对应的底部纵筋上布置应变片 E1～E6，在梁端和与柱相邻 100 mm 处的箍筋上布置应变片 F1。

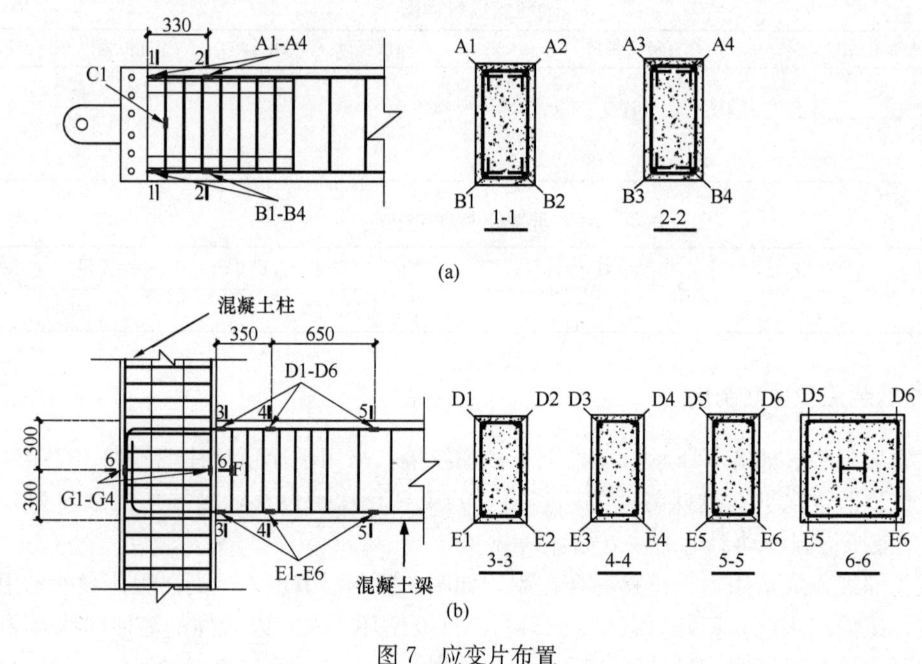

图 7　应变片布置

(a) 试件 PSJ；(b) 试件 MJ

4. 试验结果和分析

4.1　试验现象

现浇节点试件破坏形态如图 8 所示，加载初期，试件处于弹性状态，没有明显的试验现象。当加载至 60kN 时，梁端混凝土出现微裂缝，混凝土梁上下的微裂纹基本对称；当加载至 75kN 时，滞回曲线上出现拐点，试件屈服，试件屈服位移为 14 mm。位移加载到 28 mm 时，梁端混凝土裂缝最大缝宽扩大到 3mm，梁柱交界点出现较大通缝，如图 8 (a) 所示。随着加载幅值的增加，裂缝的数量和宽度逐渐增加，位移加载到 70 mm，不再有新的裂缝产生，已有裂缝的宽度逐渐增长。加载到 98 mm 时，梁端裂缝贯通，在梁柱交界处出现混凝土剥落现象，箍筋外露，混凝土被压溃。当加载到 110mm 时，梁柱节点的承载能力下降到极限承载力的 85％以下，停止加载，此时混凝土剥落严重，梁端纵筋外露，且纵筋发生较为严重变形，如图 8 (b) 所示。试验结果表明，现浇节点的塑性铰位置在梁端，混凝土梁的破坏集中于塑性铰位置。

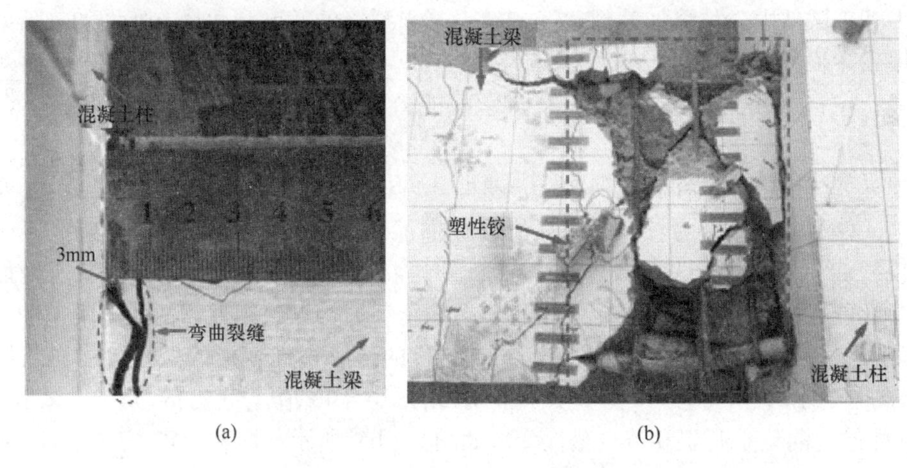

(a)　　　　(b)

图 8　现浇节点试件破坏形态

（a）弯曲裂缝；（b）塑性铰

装配式节点试件破坏形态如图 9 所示。从加载开始，混凝土梁下部开始出现微小裂缝，后期裂缝上下对称发展。阻尼器耗能段发生拉压变形，在加载过程中发挥耗能作用。当加载到 45kN 时，滞回曲线出现较为明显的拐点，试件屈服，试件屈服位移为 13mm。

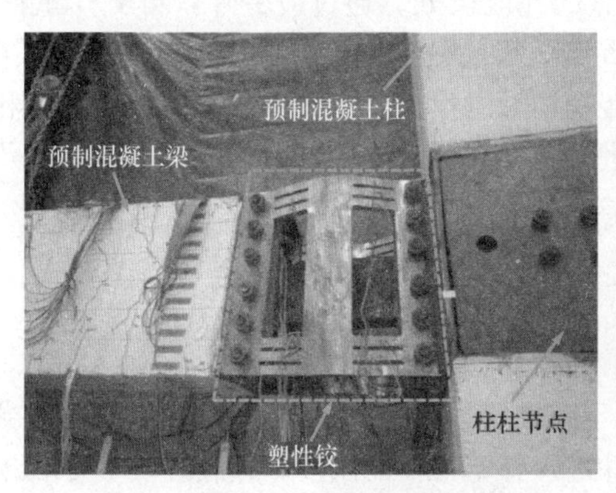

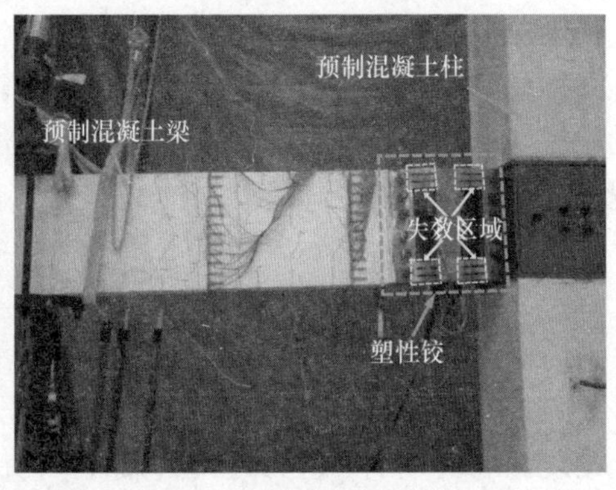

(a)　　　　(b)

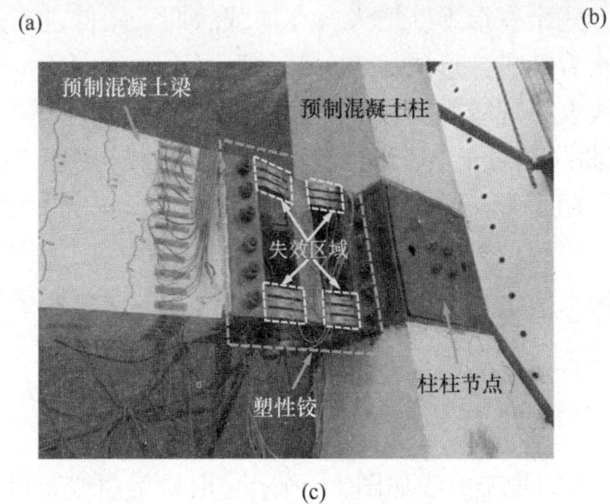

(c)

图 9　装配式节点试件破坏形式

（a）PSJ-1；（b）PSJ-2；（c）PSJ-3

在 PSJ-1 试件中，位移加载到 52 mm，混凝土裂缝的宽度和数量有所增加，阻尼器的非耗能段发生面外变形，最大面外位移为 32 mm；位移加载到 91 mm，阻尼器面外变形非常明显；位移加载到 104 mm 时，节点承载能力下降到极限承载力的 85%以下，停止加载。

在 PSJ-2 试件中，位移加载到 52 mm 时，节点最大平面外变形为 12 mm，面外变形较 PSJ-1 试件小，这是因为较厚的非耗能段减小了节点面外变形。位移加载到 65 mm 后不再出现裂缝。与 PSJ-1 试件相比，PSJ-2 试件的混凝土裂缝宽度和数量均有所增加，这是因为非耗能段厚度的增加提高了装配式节点的承载力，导致阻尼器的耗能段塑性变形较小，混凝土梁裂缝增多。PSJ-1 试件和 PSJ-2 试件的最终塑性变形均处于阻尼器的耗能段中。位移加载到 94 mm 时，PSJ-2 试件的承载力小于峰值承载力的 85%，试验结束。PSJ-2 试件最终混凝土开裂和节点塑性变形如图 9（b）所示。

在 PSJ-3 试件中，位移加载到 52 mm，混凝土裂缝开展的宽度和数量有所增加；PSJ-3 试件的裂缝扩展速率低于 PSJ-2 试件，这是由于耗能段的增加有效增大节点的塑性变形，从而减小混凝土梁产生的裂缝；PSJ-3 试件最终塑性变形处于阻尼器的耗能段中；位移加载到 94 mm 时，试验结束。PSJ-3 试件最终混凝土裂缝和节点塑性变形如图 9（c）所示。

在 PSJ 试件中，塑性变形集中于钢质节点的耗能段处。通过应变片获得 PSJ 试件混凝土梁的纵筋和箍筋应变，可知 PSJ 试件的混凝土梁均处于弹性范围内。试验结果表明，装配式梁柱塑钢质节点由于阻尼器模块和可转动销轴的设计，将梁端塑性铰限制在了阻尼器的位置，低屈服点阻尼器较现浇试件更早进入塑性并耗散能量，有效控制了混凝土梁的裂缝发展和损伤。

4.2 滞回曲线

MJ 试件和 PSJ 试件的滞回曲线见图 10，可以看出：

1）对于 MJ 试件，在加载初期，构件处于弹性阶段，滞回曲线呈线性发展。梁端混凝土出现微裂缝，钢筋应力达到屈服后，构件进入弹塑性耗能阶段。卸载时残余变形开始增大，随着荷载增加，残余变形不断增大。在加载后期，构件进入大变形阶段，滞回曲线开始出现弓形。混凝土裂缝贯穿形成通缝，部分混凝土压碎剥落，钢筋混凝土之间发生粘结滑移，导致滞回曲线上出现较为明显的捏缩现象。

2）PSJ 试件滞回曲线相对于 MJ 试件较为饱满。加载初期，阻尼器未出现塑性变形，试件处于弹性阶段。当阻尼器耗能段应力均达到屈服应力，开始进入塑性耗能。加载后期，由于梁柱节点的塑性铰位置限制在了阻尼器的位置，集中塑性耗能，减小混凝土梁发生破坏，滞回曲线表现出较饱满的梭形。可见，装配式节点相比于现浇节点具有较好的耗能能力。

4.3 骨架曲线

MJ 试件和 PSJ 试件的骨架曲线如图 11 所示。由骨架曲线获得的试验结果见表 5。其中，P_y 和 Δ_y 为试件屈服荷载和相应屈服位移，P_u 和 Δ_u 为试件峰值荷载和相应峰值位移，$P_{ult}=0.85P_u$ 和 Δ_{ult} 为试件极限荷载和相应极限位移。

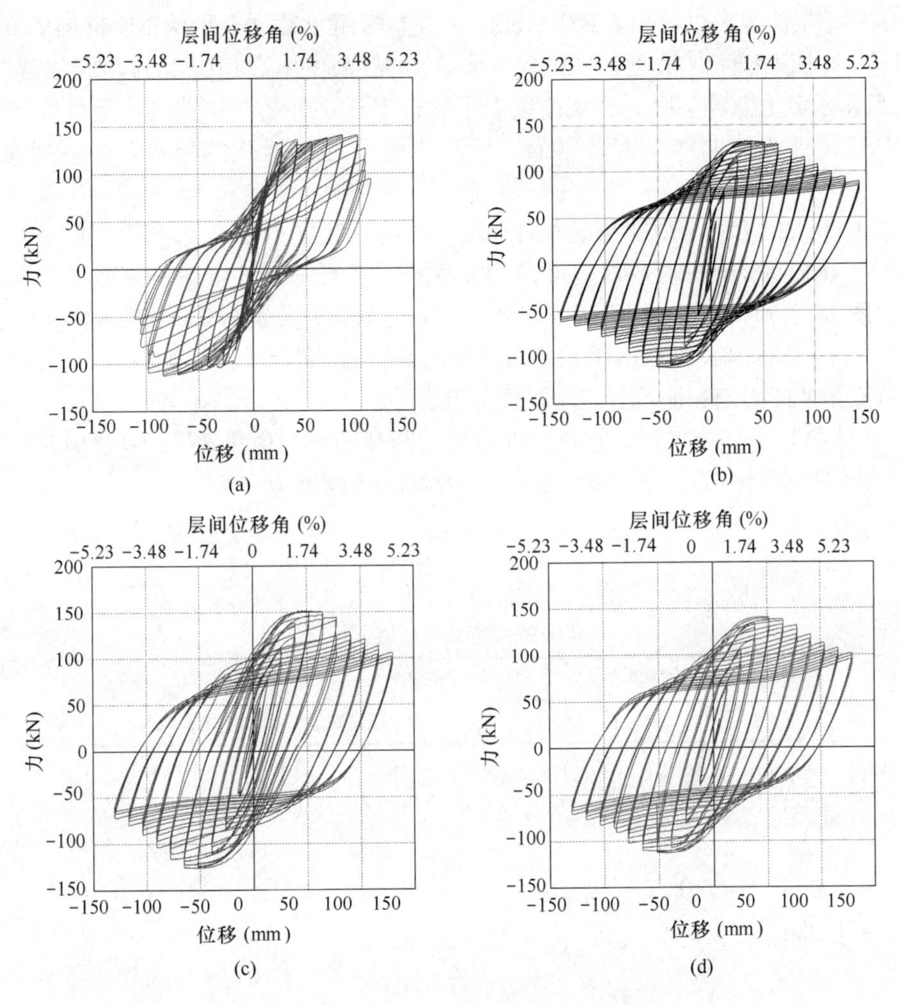

图 10　节点试件滞回曲线

(a) MJ；(b) PSJ-1；(c) PSJ-2；(d) PSJ-3

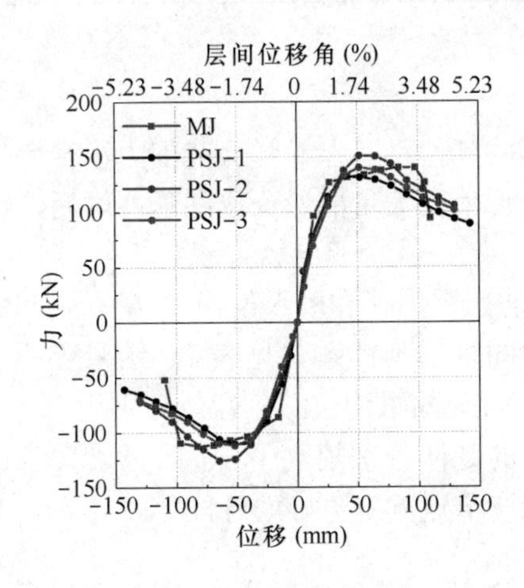

图 11　节点试件骨架曲线

由试件骨架曲线可以看出，PSJ-1 试件正负向峰值承载力分别为 MJ 试件的 94.22% 和 98.18%，峰值承载力相近；MJ 试件承载力在屈服后虽有所上升，但位移加载至 100mm 后，承载力急剧下降，骨架曲线向下穿过 PSJ 试件的骨架曲线，呈脆性破坏的趋势；而 PSJ 试件骨架曲线下降段平直，承载力没有加速退化的趋势，延性远高于 MJ 试件。

由图 10 可知，PSJ-2 试件在正方向和负方向的承载能力分别比 PSJ-1 试件高 14.2% 和 13.9%。由此可见，非耗能段厚度的增加减小了节点的面外变形，提高了节点的承载能力。由图 10 可知，PSJ-3 试件正方向和负方向的承载力分别为 140.25kN 和 112.05kN，PSJ-3 试件的正方向和负方向的承载能力分别比 PSJ-2 试件低 7.1%和 11.0%。由此可见，装配式节点承载能力随耗能段长度增加而降低。

MJ 试件和 PSJ 试件在正方向和负方向初始刚度的平均值见表 5。可以看出，PSJ 试件比 MJ 试件具有略高的初始刚度。节点初始刚度可以表示[28,29]：

$$K_0 = \frac{P_y}{\Delta_y} \tag{1}$$

式中 P_y 和 Δ_y 是由试验分析得到的骨架曲线给出的屈服载荷和屈服位移。

试件主要特征值　　表 5

试件类型	加载方向	屈服点		峰值点		破坏点		延性系数	初始刚度
		Δ_y	P_y (kN)	Δ_u	P_u (kN)	Δ_{ult} (mm)	P_{ult} (kN)		
MJ	正向	27.59	96.67	96.84	140.46	106.31	119.39	3.33	3.13
	负向	35.90	99.27	83.02	112.56	100.98	95.67		
PSJ-1	正向	22.10	107.70	39.04	132.22	95.76	112.39	3.79	4.19
	负向	24.67	86.41	50.91	110.51	80.19	93.94		
PSJ-2	正向	23.20	111.03	51.80	151.03	95.77	128.38	3.67	3.99
	负向	26.94	85.90	64.88	125.89	86.67	107.01		
PSJ-3	正向	21.80	94.10	51.80	140.25	95.76	119.22	3.69	3.84
	负向	28.31	95.13	52.15	112.05	84.60	95.24		

4.4 位移延性

节点在地震荷载作用下的延性是其在非弹性范围内承受大幅度循环变形的能力。节点在地震期间通过这些循环变形消耗大量能量。位移延性可用延性系数表示，可表示为[30]：

$$\mu = \Delta_{ult} / \Delta_y \tag{2}$$

式中，Δ_y 和 Δ_{ult} 分别为节点的屈服位移和极限位移。

试件在正方向和负方向的平均延性系数见表 5。结果表明，MJ 试件的平均延性系数为 3.33，PSJ-1、PSJ-2 和 PSJ-3 的平均位移延性值分别为 3.79、3.67 和 3.69。与 MJ 相比，PSJ-1、PSJ-2 和 PSJ-3 的延性分别提高了 13.81%、10.21%和 10.81%。由此可见，提出的装配式节点相比于现浇节点具有更好的延性。

4.5 承载力退化

装配式节点和现浇节点承载力退化是指节点承载力随着加载循环次数的递增而逐渐降

低，采用承载力退化系数 表征试件承载力退化特征，即[31]

$$\lambda_j = \frac{F_{3,j}}{F_{1,j}} \tag{3}$$

式中，$F_{3,j}$为第 j 级加载时第 3 次循环时峰值点荷载；$F_{1,j}$ 为第 j 级加载时第 1 次循环时峰值点荷载。

由图 12 可知，对于 MJ 试件，正向和负向加载过程中，MJ 试件一直存在承载力退化现象，MJ 试件承载能力退化系数在 0.71 到 1.00 之间，在层间位移角 0.5%到 2.9%之间，MJ 试件承载力退化较慢，但在层间位移角 3.65%以后时，MJ 试件强度退化系数突然快速下降，MJ 在正负方向的强度衰减分别为 19%和 29%，主要由于塑性铰区混凝土发生严重破坏导致节点承载力急剧下降。

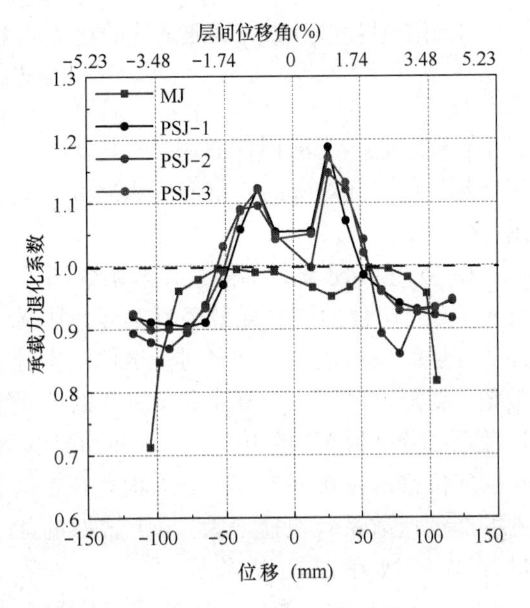

图 12　节点试件承载力退化

对于 PSJ 试件，在层间位移角 1.8%以后，承载力退化系数在 0.86～1.00 之间，即 PSJ 试件强度退化小于 14%。在层间位移角 3.48%以后，PSJ 试件承载力退化系数高于 MJ 试件。此外，由于 LYP100 低屈服点钢存在循环硬化现象，PSJ 试件在层间位移角为 0.5%和 1.8%之间存在承载力强化现象。

4.6　耗能能力

试件滞回曲线中每个循环的耗能面积 E 可以衡量试件在不同荷载下的耗能能量，如图 13 所示。节点能量耗散能力通常由累积滞回线面积评估[30]，如图 14 所示。PSJ 试件的

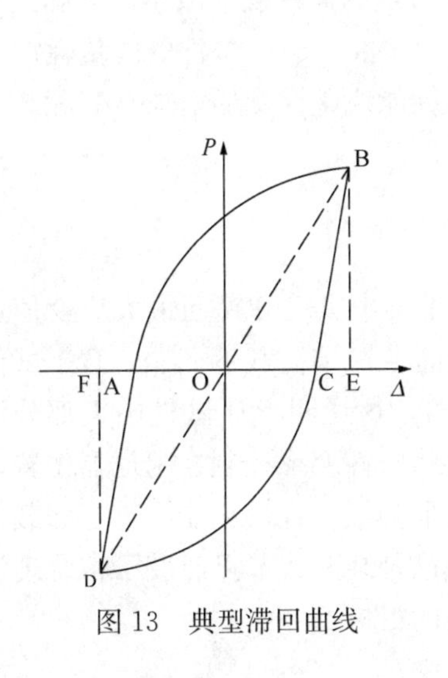

图 13　典型滞回曲线

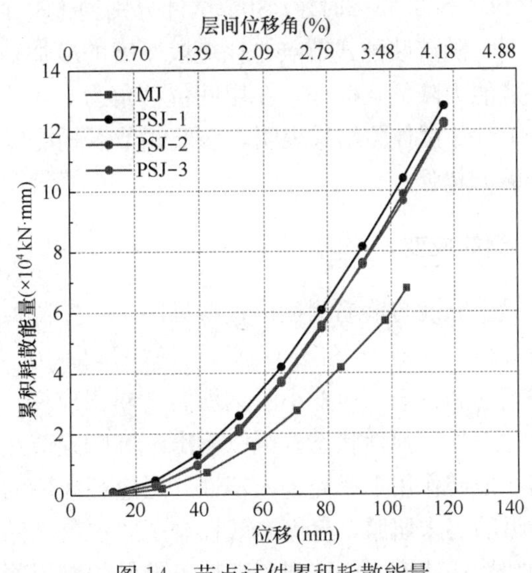

图 14　节点试件累积耗散能量

累积能量耗散高于 MJ 试件。当层间位移角为 3.33%时，PSJ-1、PSJ-2 和 PSJ-3 试件的累积能量耗散分别比 MJ 试件高 64%、55%和 52%。结果表明，装配式节点具有较好耗能能力。

采用能量耗散系数 η 表征试件的耗能能力，表达式为[31]：

$$\eta = \frac{S_{(ABC+CAD)}}{S_{(OBE+ODF)}} \tag{4}$$

式中：$S_{(ABC+CAD)}$ 为滞回环面积，$S_{(OBE+ODF)}$ 为滞回环峰值荷载对应的三角形面积之和，如图 12 所示。

由 MJ 和 PSJ 试件的能量耗散系数图 15 可见，PSJ 试件的能量耗散系数高于 MJ 试件。在层间位移角为 0.49%时，试件发生轻微塑性变形，PSJ-1、PSJ-2 和 PSJ-3 的能量耗散系数分别为 MJ 的 10.6、5.2 和 5.8 倍。在层间位移角为 0.49%～3.17%时，PSJ 试件的能量耗散系数快速增大。MJ 试件的能量耗散系数在层间位移角为 0.49% 和 1.95%之间时快速增大，而在层间位移角 1.95%后增加缓慢。当层间位移角为 3.33%时，PSJ-1、PSJ-2 和 PSJ-3 的能量耗散系数分别比 MJ 大 77%、62%和 60%。在层间位移角为 3.33%时，PSJ 试件的能量耗散系数大于 2，而 MJ 试件的能量耗散系数小于 1.5。结果表明，由于低屈服点钢的应用及阻尼器耗能段的设计，使得装配式节点具有良好的耗能能力。

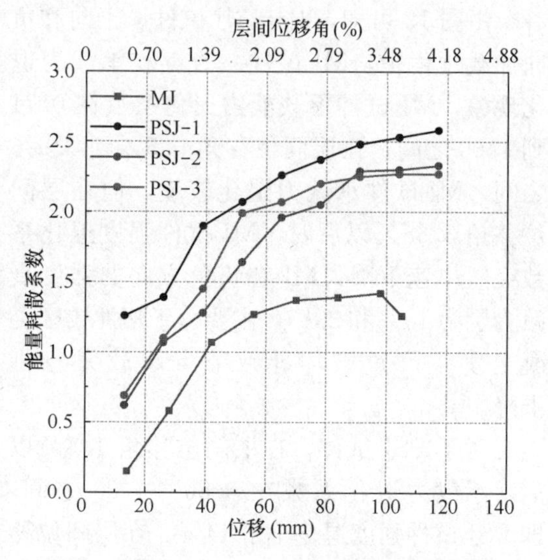

图 15　节点试件能量耗散系数

与 PSJ-1 试件相比，PSJ-2 试件的能量耗散系数要小很多。具体来说，在层间位移角为 0.49%和 3.33%时，PSJ-2 试件分别比 PSJ-1 试件的能量耗散系数小 50.8%和 8.2%。这表明，随着装配式节点非耗能段厚度的增加，装配式节点的面外变形减小，装配式节点的耗能能力降低。此外，在层间位移角为 0.49%～3.17%时，PSJ-3 试件的能量耗散系数高于 PSJ-2 试件。结果表明，装配式节点耗能段越长，耗能段塑性变形增加，装配式节点耗能能力越强。

4.7　钢筋应变

梁中纵筋与箍筋测点位置参见图 7，选取试件混凝土梁中纵筋和箍筋最大应变值进行比较分析，如图 16 和图 17 所示。MJ 试件中的纵向钢筋在初始荷载下屈服，PSJ 试件中纵向钢筋和箍筋的应变小于钢筋名义屈服应变（2000$\mu\varepsilon$）。这说明 PSJ 试件中大部分混凝土梁柱仍处于弹性状态。可以看出，装配式节点梁中能将塑性变形控制在阻尼器位置，阻尼器在地震作用下率先进入塑性并通过耗能段的拉压变形为结构提供耗能，直至加载结束梁中纵筋仍未屈服，从而有效控制了梁混凝土的损伤，该种装配式节点改变了传统现浇节点的塑性耗能模式，实现了塑性可控及控制混凝土损伤的目标。

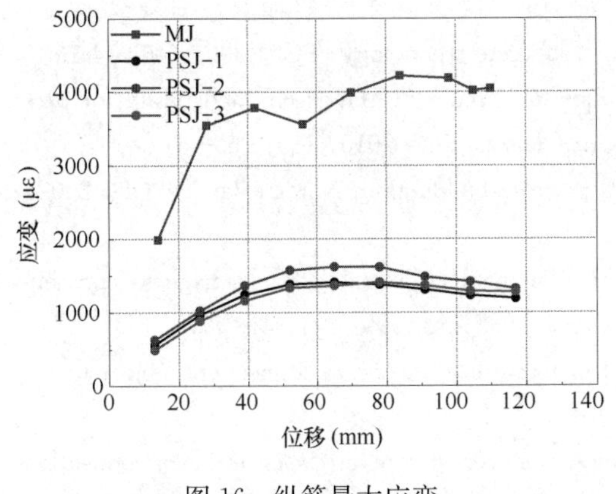

图 16　纵筋最大应变

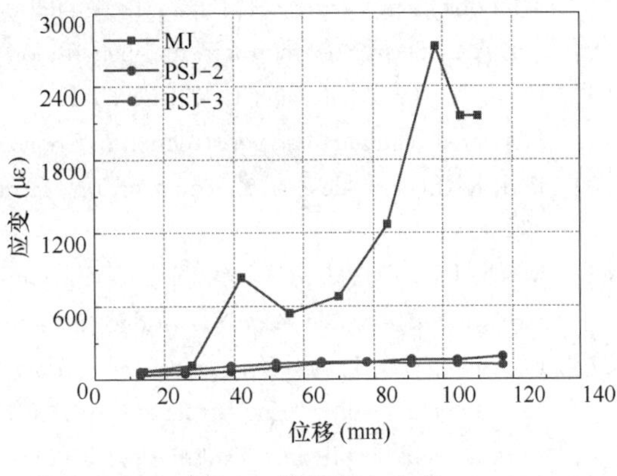

图 17　箍筋最大应变

5. 结论

本文提出了一种适用于预制混凝土结构的新型装配式梁柱钢节点。与现浇混凝土节点相比，该节点具有施工方便、梁内塑性铰位置可控等优点。为研究装配式节点抗震性能，对三个新型装配式节点和一个现浇混凝土节点进行了循环加载试验。同时，研究了装配式节点非耗能段厚度和耗能段长度对节点滞回性能的影响。从本研究的结果可以得出以下结论：

（1）现浇节点滞回曲线出现较为明显的捏缩现象，装配式梁柱钢质节点滞回曲线则表现出较饱满的梭形。

（2）装配式梁柱钢质节点与现浇节点初始刚度基本一致。随着荷载的增大，该节点较现浇节点更早进入屈服耗能阶段，低屈服点钢阻尼器塑性耗能，控制混凝土梁端不发生弯曲破坏。该节点能够满足梁端“塑性可控”原则，提高节点抗震性能，更好的保护混凝土梁避免发生破坏。

（3）装配式梁柱钢质节点具有较好的延性。与 MJ 试样相比，PSJ 试样的延性系数提高了 10%以上。

（4）装配式梁柱钢质节点承载能力退化速度均较现浇节点慢，退化系数保持在 0.86 以上。当层间位移角为 2.3%时，非耗能段厚度的增加和耗能段长度的减小导致钢节点强度退化更为严重。

（5）与现浇节点相比，装配式梁柱钢质节点具有更好的耗能能力。PSJ 试件的耗能系数大于 2，MJ 试件的耗能系数小于 1.5。装配式梁柱钢质节点中非耗能段厚度的减小和耗能段长度的增加能够提高节点耗能能力。

参考文献

［1］ Wang H，Marino，Marino EM，Pan P，Huang L，Xin N. Experimental study of a novel precast prestressed reinforced concrete beam-to-column joint. Eng Struct 2018；156：68-81.

［2］ Breccolotti M，Gentile S，Tommasini M，Materazzi AL，Bonfigli MF，Pasqualini B，Colone V，Gianesini M. Beam-column joints in continuous RC frames：Comparison between cast-in-situ and precast solutions. Eng Struct 2016; 127: 129-144.

[3] Yee AA. Social and environmental benefits of precast concrete technology. PCI J 2001; 46: 14-19.

[4] Chang Y, Li X, Masanet E, Zhang L, Huang Z, Ries R. Unlocking the green opportunity for prefabricated buildings and construction in China. Resour Conserv Recy 2018; 139: 259-261.

[5] Park R. Seismic design and construction of precast concrete buildings in New Zealand. PCI J 2002; 47: 60-75.

[6] Mitchell D, DeVall RH, Saatcioglu M, Simpson R, Tinawi R, Tremblay R. Damage to concrete structures due to the 1994 Northridge earthquake. Can J Civil Eng 1995; 22: 361-377.

[7] Korkmaz HH, Tankut T. Performance of a precast concrete beam-to-beam connection subject to reversed cyclic loading. Eng Struct 2005; 27: 1392-1407.

[8] Korkmaz Hasan Husnu, Tankut Tugrul. Performance of a precast concrete beam-to-beam connection subject to reversed cyclic loading [J]. Engineering Structures, 2005, 27(9): 1392-1407.

[9] Parastesh Hossein, Hajirasouliha Iman, Ramezani Reza. A new ductile moment-resisting connection for precast concrete frames in seismic regions: An experimental investigation [J]. Engineering Structures, 2014, 70(9): 144-157.

[10] 赵作周，韩文龙，钱稼茹，等．钢筋套筒挤压连接装配整体式梁柱中节点抗震性能试验研究 [J]. 建筑结构学报，2017，38(4)：45-53.

[11] Huang Y, Mazzarolo E, Briseghella B, Zordan T, Chen A. Experimental and numerical investigation of the cyclic behaviour of an innovative prefabricated beam-to-column joint. Eng Struct 2017; 150: 373-389.

[12] Yuksel E, Karadogan HF, Bal IE, Ilki A, Bal A, Inci P. Seismic behavior of two exterior beam-column connections made of normal-strength concrete developed for precast construction. Eng Struct 2015; 99: 157-172.

[13] Priestley M. J. Nigel. Overview of PRESSS research program [J]. Pci Journal, 1991, 36(4): 50-57.

[14] 李灿军，周臻，谢钦．摩擦耗能型 SMA 杆自复位梁柱节点滞回性能分析 [J]. 工程力学，2018，35(4)：115-123.

[15] Morgen BG, Kurama YC. Seismic design of friction-damped precast concrete frame structures. J Struct Eng 2007; 133: 1501-1511.

[16] Morgen B, Kurama Y. A friction damper for post-tensioned precast concrete beam-to-column joints. PCI J 2004; 49: 112-133.

[17] Wang C, Liu Y, Zheng X. Experimental investigation of a precast concrete connection with all-steel bamboo-shaped energy dissipaters. Eng Struct 2019; 178: 298-308.

[18] Ersoy U, Tankut T. Precast concrete members with welded plate connections under reversed cyclic loading. PCI J 1993; 38: 94-100.

[19] French CW, Amu O, Tarzikhan C. Connections between precast elements—failure outside connection region. J Struct Eng 1989; 115: 316-340.

[20] Magliulo G, Ercolino M, Cimmino M, Capozzi V, Manfredi G. FEM analysis of the strength of RC beam-to-column dowel connections under monotonic actions. Constr Build Mater 2014; 69: 271-284.

[21] Magliulo G, Cimmino M, Ercolino M, Manfredi G. Cyclic shear tests on RC precast beam-to-column connections retrofitted with a three-hinged steel device. Bull Earthquake Eng 2017; 1-21.

[22] Eom T, Park H, Hwang H, Kang S. Plastic hinge relocation methods for emulative PC beam-column connections. J Struct Eng 2015; 142.

[23] Yang K, Seo E, Hong S. Cyclic flexural tests of hybrid steel-precast concrete beams with simple connection elements. Eng Struct 2016; 118: 344-356.

[24] Choi HK, Choi YC, Choi CS. Development and testing of precast concrete beam-to-column connections. Eng Struct 2013; 56: 1820-1835.

[25] 中华人民共和国住房和城乡建设部，中华人民共和国国家质量监督检验检疫总局．建筑抗震设计规范 GB 50010—2010. 北京：中国建筑工业出版社；2010.

[26] 中华人民共和国国家质量监督检验检疫总局，中华人民共和国标准化管理局．金属材料　室温拉伸试验 GB/T 228— 2010. 北京：中国建筑工业出版社；2010.

[27] 中华人民共和国住房和城乡建设部，中华人民共和国国家质量监督检验检疫总局．混凝土物理力学性能试验方法标准 GB/T 50081—2002. 北京：中国建筑工业出版社；2003.

[28] Park R. Ductility evaluation from laboratory and analytical testing. In: Proc. of the 9th World Conference on Earthquake Engineering; 1988, pp. 605-616.

[29] 邓小芳，李治，翁运昊，等．预应力混凝土梁-柱子结构抗连续倒塌性能试验研究[J]. 建筑结构学报，2019，40(8)：71-78.

[30] Ghayeb HH, Razak HA, Sulong NHR. Development and testing of hybrid precast concrete beam-to-column connections under cyclic loading. Constr Build Mater 2017; 151: 258-278.

[31] 中华人民共和国住房和城乡建设部．建筑抗震试验规程 JGJ/T 101—2015. 北京：中国建筑工业出版社，2015.

14 组合隔震技术在高烈度区复杂高层结构的应用研究

吴小宾，彭志桢，秦　攀，曹　莉，韩克良，姜　雪
（中国建筑西南设计研究院有限公司，成都）

摘　要：结合某9度设防、近断层且体型复杂的高层医院建筑结构隔震设计实例，研究隔震技术对强地震输入、特别不规则和高预期抗震性能目标的复杂高层建筑结构的适应性及对应措施，提出发挥不同隔震器件或装置性能特点的组合隔震技术方法，结果表明组合隔震可减小隔震橡胶支座面压和水平变形，减少塔楼偏置引起的结构扭转效应、竖向高位收进引起的高振型效应，结构构件损伤程度和范围大幅降低，屈服耗能机制更为合理，有效提升了建筑结构的抗震能力；分析比较近场速度脉冲型地震波与考虑1.5近场系数的常规地震波作用下复杂高层隔震结构的地震响应差别，结果表明在速度脉冲型近场地震作用下的隔震层位移，层间位移角及层剪力均大于考虑1.5近场系数的常规地震波，即1.5增大系数不能包络近场地震影响，隔震设计时应慎重考虑。

关键词：9度设防；近断层；复杂高层建筑；组合隔震技术

Applied Research of Combined Isolation Technology in Complex High-rise Buliding Near Faults in 9 Degree Zone

Wu Xiaobin, Peng Zhizhen, Qin Pan, Cao Li, Han Keliang, Jiang Xue

Abstract: A hospital building with complex shape located in the 9-degree zone near the fault is presented; The adaptability and corresponding measures of isolation technology for complex high-rise building structures with strong earthquake input, especially irregular and high expected seismic performance targets are studied; The combined isolation technology method which can bring into play the performance characteristics of different isolation devices is put forward; The results show that it can reduce the surface pressure and horizontal deformation of rubber bearings, reduce the torsion effect caused by tower offset and high mode effect caused by vertical high-level retraction, and greatly reduce the damage degree and range of structural components, effectively improves the seismic capacity of the building structure; The difference of seismic response between near-field velocity pulse seismic wave and conventional seismic wave with 1.5 near-field coefficient is analyzed and compared; The results show that the displacement, inter story displacement angle and floor shear of isolation layer under the near-field seismic action of velocity pulse type are larger than that of conventional seismic wave with 1.5 near-field coefficient, the 1.5 near-field coefficient can not cover the near-field seismic effect, so the isolation design should be carefully considered.

Keywords: 9 degree seismic fortification, near fault, complex high rise, combined isolation technology

1. 引言

四川省属于地震多发区，而西昌市更属于9度设防且有活跃断裂带分布地区。由于城市发展，大量高层建筑以及复杂高层建筑出现，应对巨大的地震作用而保证较好的建筑使用功能成为建筑结构设计需要解决的问题。隔震技术在汶川、芦山等地震经受住了实际考验，得到广泛认可并获得政府的政策推广，住房和城乡建设部印发了《关于房屋建筑工程推广应用减隔震技术的若干意见（暂行）》[1]的通知，鼓励重点设防类、特殊设防类和位于8度以上（含8度）地震高烈度区的建筑采用减隔震技术。隔震结构通过降低上部结构输入地震作用方式，舒缓上部结构负担，能够较好解决抗震能力与建筑功能的矛盾问题。复杂高层建筑的隔震设计有其特殊性，包括结构体形不规则性对隔震效果的影响；强地震输入下超高、塔楼偏置导致的抗倾覆、抗扭转需求加大；上部结构薄弱部位较多，加强措施更多更严等。近几年，近断层地震动的运动特征及其对工程结构的影响不断受到关注。近断层地震动具有明显的长周期速度和位移脉冲[2]，对隔震建筑等长周期结构的抗震性能和设计带来不利影响。由于研究不足，我国抗震规范设计反应谱没有纳入近场条件，而通过近场放大系数考虑近场地震动影响。

本文结合某9度设防、近断层且体型复杂的高层医院建筑结构隔震设计实例，通过设置由橡胶隔震支座、滑板支座及黏滞阻尼器组合的隔震层的参数化计算分析，探讨组合隔震技术方法下复杂高层建筑结构设计思路及控制要点。并研究组合隔震技术在近场速度脉冲型地震波作用下复杂高层隔震结构地震响应特点，为今后类似建筑结构设计提供参考。

2. 工程概况

川投西昌综合医院，位于西昌市北城新区，抗震设防类别为9度，场地类别为II类，设计地震分组为第三组。由塔楼及裙房组成，其中塔楼地上共17层，建筑高度70.5m，Z字形平面，长度134.4m，宽度47.8m，单肢宽度23.9 m；裙房地上4层，建筑高度21.3m，为矩形平面，裙房长度193.2 m，宽度109.9m；塔楼与裙房不设置抗震缝，共用两层地下室。建筑效果见图1，塔楼结构布置平面见图2，裙房结构布置平面见图3，典型剖面见图4。

本项目具有以下特点：1）大型医疗建筑，属于重点设防类别；2）场地距离活跃断裂带小于2.5km，近断层的地震影响必须重视；3）建筑体型复杂，属高位收进的大底盘单塔结构，但塔楼偏置；塔楼平面不规则；4）建筑高度70.5m，大于9度区钢筋混凝土结构适用高度[3]。属于“超过高度的房屋，应进行专门研究和论证，采取有效的加强措施”范围[4]，为超限高层建筑工程。

图1　建筑效果图

Fig. 1　Architectural Renderings

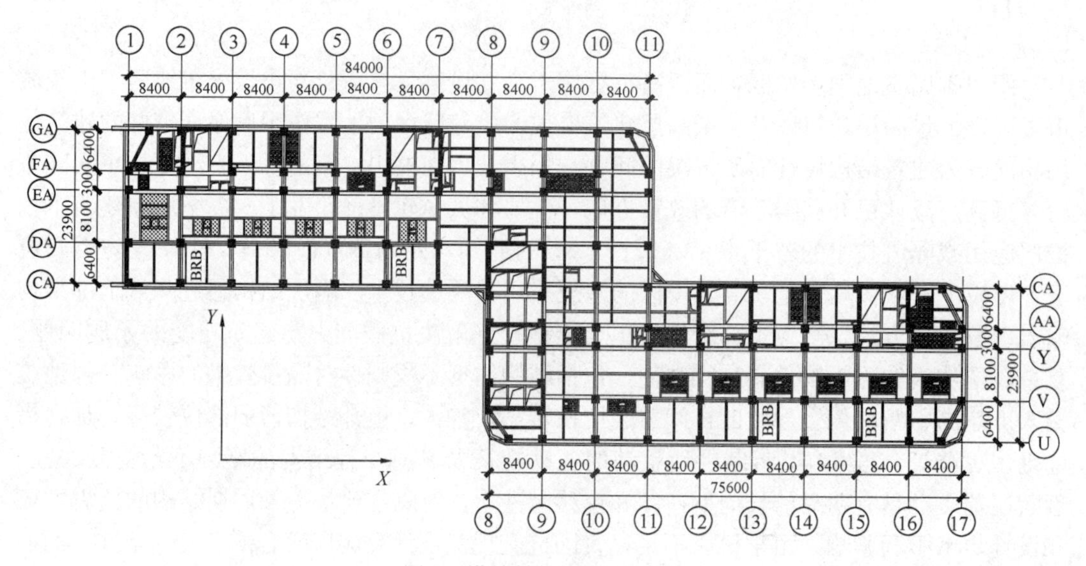

图 2　塔楼结构平面布置图

Fig. 2　Layout plan of tower structure

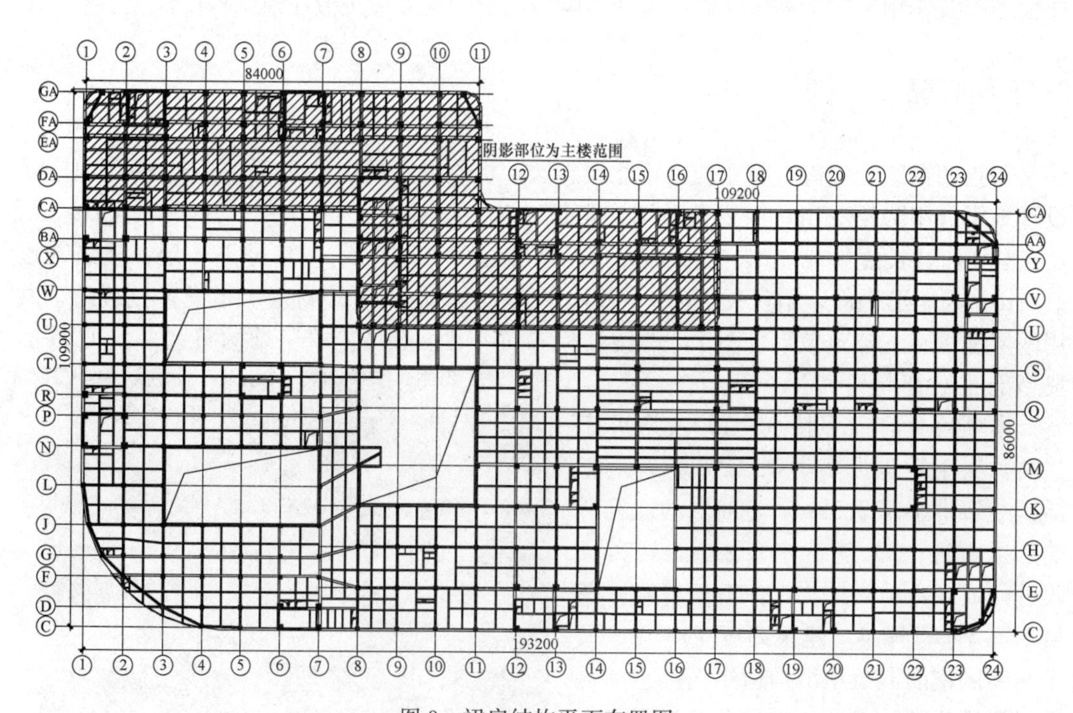

图 3　裙房结构平面布置图

Fig. 3　Layout plan of podium structure

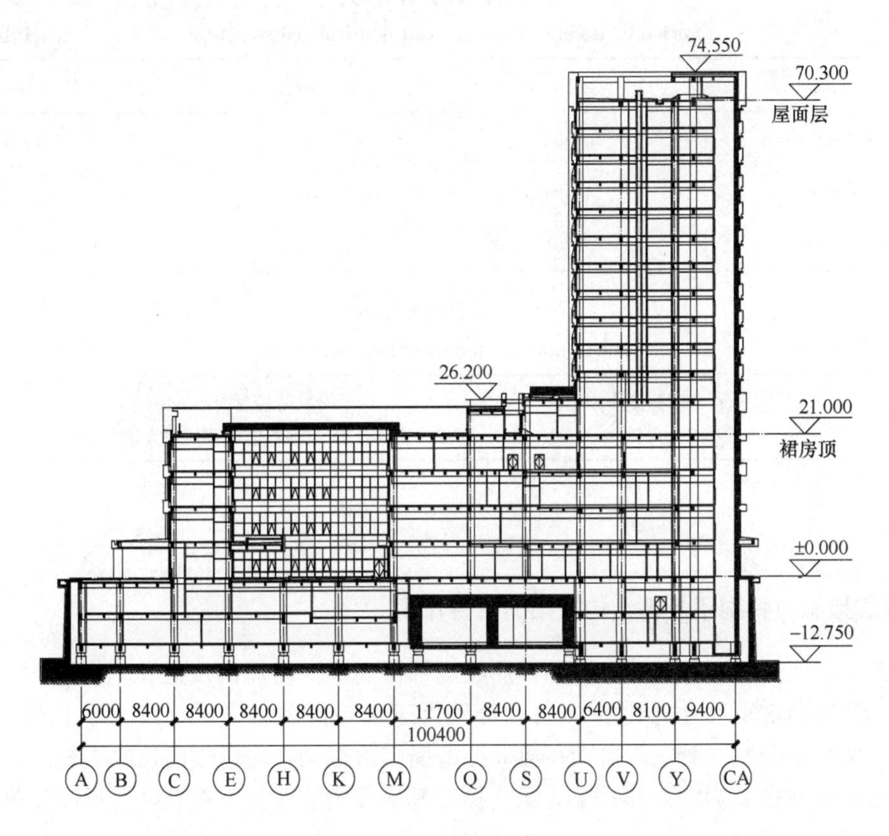

图4　典型剖面图

Fig. 4　Typical sectional drawing

3. 隔震技术的适应性研究

3.1　隔震效果受结构体型影响评价

一般来说，多层结构由于自振周期较短，隔震后自振周期大幅延长，动力响应减弱而产生较好的隔震效果。但高层结构的上述效应受到抑制，一是本身自振周期较长，隔震后延长比例较小且位于反应谱下降相对平缓段；二是由于橡胶支座面压控制原因，高层结构的支座较大进一步影响隔震效果，而高烈度的强地震输入进一步加大上述影响，导致隔震效果的降低。

本项目采用基础隔震，隔震层以上结构高度82m，按塔楼与裙房间设永久缝分离则塔楼高宽比3.5，而按塔楼与裙房一体则计裙房以上塔楼高宽比为2.1，两者的非隔震模型基本自振周期分别为1.323s、1.2481s；尽管属于高层结构，由于位于高烈度区，导致结构自身刚度较大，基本周期较小，采用隔震技术的效果较好，隔震前后周期比约为3，见表1；减震系数最大为0.369，小于0.38，见表2。隔震层大底盘有利于整体抗倾覆，防止隔震支座压屈或拉应力过大，因而隔震技术在本结构具有很好的适用性。

隔震前后结构周期 表 1

Period of isolated structure and non-isolated structure Table 1

周期比	隔震	非隔震	隔震/非隔震
T_1（Y 向平动）	3.6564	1.2483	2.93
T_2（X 向平动）	3.6553	1.1761	3.11
T_3（扭转）	3.0364	0.8922	3.40

水平地震减震系数 表 2

Horizontal seismic damping coefficient Table 2

地震作用	隔震结构与非隔震结构楼层剪力比值最大值	隔震结构与非隔震结构楼层倾覆弯矩比值最大值	减震系数
X 向	0.352	0.367	0.367
Y 向	0.348	0.353	0.353

3.2 隔震措施对特别不规则结构缺陷的改善作用

从图 4 剖面图可知，本结构上部楼层收进部位高度与房屋高度之比为 0.4，收进水平尺寸与下部楼层水平尺寸之比为 57%，分别大于规范限值 0.2 和 25%[5]。由图 5 可知，非隔震模型第 1 振型主属一阶振动；第 5 振型属多阶振动，与塔楼高位收进相关。图 6 表明，非隔震模型在 Y 向地震作用下，第 5 振型基底剪力大于第 1 振型，可见高位收进对非隔震结构的抗震不利影响较大，容易使结构出现薄弱部位。图 6 表明，通过合理设置隔震层，隔震模型 Y 向基底剪力主要由第 1 振型引起，其他振型对 Y 向基底剪力几乎无贡献。由图 7 不同振型的质量参与系数对比可知，隔震模型的振型质量参与系数集中在前三振型，而非隔震模型在 15～35 高阶振型均有质量参与。因此，隔震上部结构的振动形式由非隔震结构的摆动，变为主要变形集中于隔震层的整体移动后，有效减少了由高位收进引起的高阶振型效应的影响。

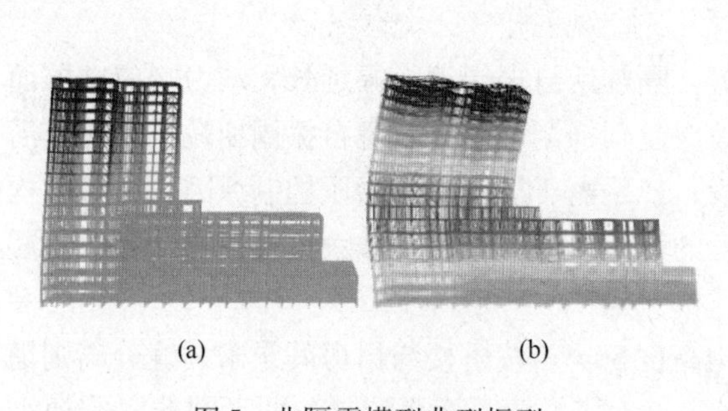

图 5 非隔震模型典型振型

Fig. 5 Typical modes of non-isolated model

(a) 第一阶（T_1=1.248s）；(b) 第五阶（T_5=0.395s）

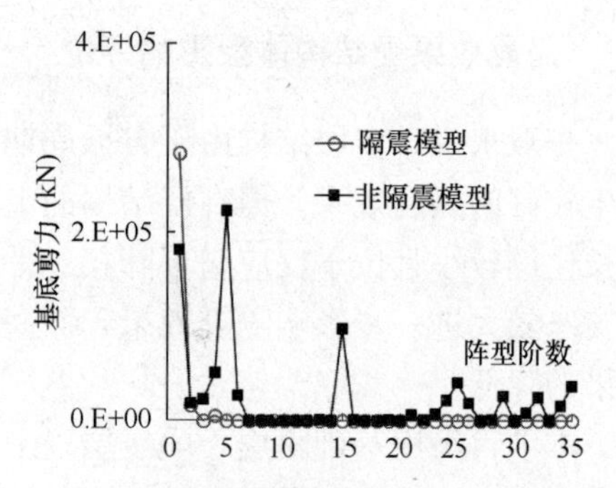

图 6 隔震与非隔震模型前 35 阶振型 Y 向的基底剪力

Fig. 6 Base shear in Y direction of the first 35 modes of isolated and non-isolated models

收进塔楼偏置于平面左上侧，其综合质心与裙房层质心偏离 25.8%，大于规范限值 20%[5]。图 8 表明，Y 向多遇地震下，降度后（按 8 度，并考虑近场增大系数）非隔震模型的层间位移角及扭转位移比接近或超过规范限值，特别是裙房扭转位移比达到 1.58，进一步调整难度大，代价高，且增加抗侧力构件影响建筑功能。而按 9 度且考虑 1.5 的近场系数后，在地震作用增大近一倍的情况下，隔震模型的最大扭转位移比仅为 1.38，小于规范限值 1.4[5]，且层间位移角沿高度分布较为均匀，远小于非隔震模型的层间移角，更小于规范限值［1/800］[5]。因此，通过选择适宜的隔震支座形式并合理布置，可以改善非隔震结构由于塔楼偏置不规则引起扭转不利情况，有利于上部结构抗震设计。

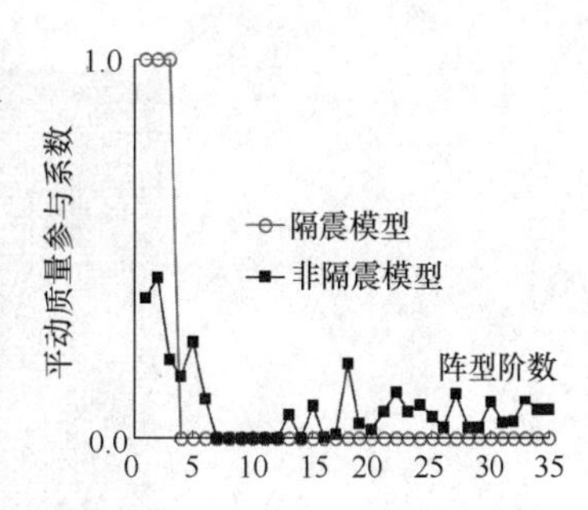

图 7　隔震与非隔震模型前 35 阶振型质量参与系数

Fig. 7　Participating Mass Ratios of the first 35 modes of isolated and non-isolated models

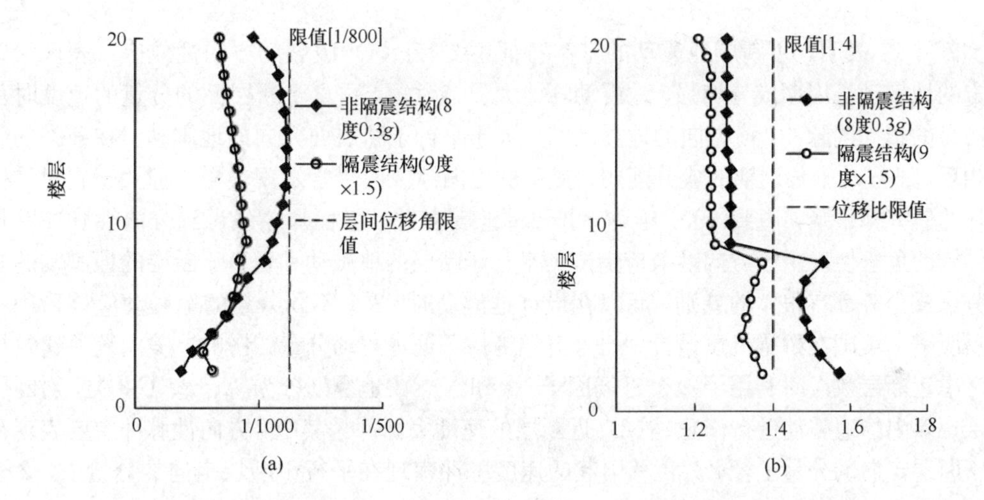

图 8　Y 向多遇地震下，隔震与非隔震模型的层间位移角及扭转位移比

Fig. 8　The Story drift angle and torsional displacement ratio of isolated and non isolated models under Y-direction frequent earthquakes

（a）层间位移角；（b）扭转位移比

3.3　隔震措施改善超长混凝土结构温度效应影响

本工程裙房长 193.2m、宽 109.9m，地下室长 200m、宽 124.3m，属于超长混凝土结构。为方便比较，按地上结构取 15℃的降温工况，比较地下室隔震上层的温度应力，未设置橡胶支座时拉应力普遍为 1.8～2.7MPa，最大拉应力 3.5MPa；设置橡胶支座隔震后，绝大部分拉应力为 1.14～1.78MPa，小于混凝土抗拉强度标准值 f_{tk}＝2.2MPa，仅少许楼板拉应力较大为 2.49MPa，见图 9。其余隔震层以上楼层受温度影响相对较小，情况也是类似。即设置橡胶支座隔震后，隔震上层楼板混凝土拉应力减少约 35%，温度效应显著降低。

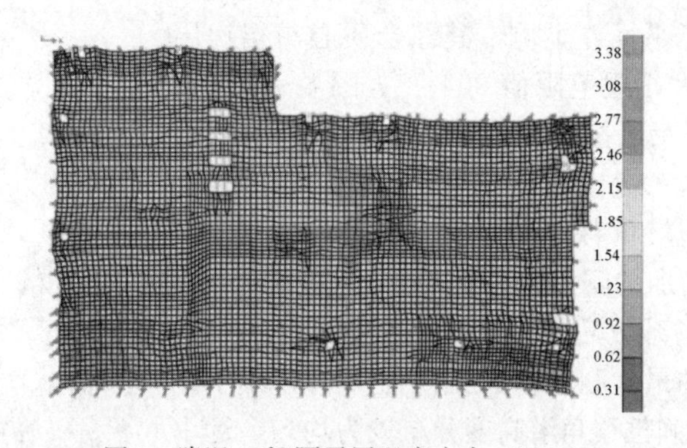

图 9　降温 15℃隔震层温度应力（MPa）

Fig. 9　Temperature principal stress of isolation layer at 15℃ cooling（MPa）

4. 近断层影响下的隔震设计研究

文献［6］表明，近断层地震动的基本特征主要有：（1）破裂的方向性效应：表现为破裂前方的地震动长周期成分加强，地震动峰值大、持时短，垂直于断层面的分量的速度时程中有明显的长周期脉冲，该方向的地震动明显大于平行于断层面方向的地震动。在破裂的后方则相反，地震动中长周期的成分削弱。破裂的方向性效应尽管发现很早，但关于它对结构的破坏的研究却很晚，直到 1994 年 Northridge 地震之后，工程师才认识到方向性脉冲对长周期结构的危害性。由于方向性效应造成的平行于断层的地震动和垂直于断层的地震动的幅值、频谱成分等都有很大的差别，所以在设计地震动时，2 个不同分量的地震动应该采用不同的反应谱。美国在桥梁抗震设计中已经开始考虑不同地震动分量的影响。美国统一建筑规范在考虑近断层效应时利用了 2 个近场因子 Na 和 Nv，将地震划分为 A、B、C 3 类，对断层距 15km 以内的地震动进行修正；（2）近断层的速度大脉冲：其中，方向性脉冲主要表现在垂直于断层走向的分量上，永久位移引起的速度脉冲表现在平行于断层走向的分量上，2 个速度脉冲是叠加在一起的。目前，关于速度脉冲的研究仅限于根据强震记录统计回归分析，而且大部分未区分方向性速度脉冲和永久位移引起的脉冲；（3）上盘效应：产生上盘效应主要是由于到断层在地表的迹线距离（断层距 ）相同的 2 点，上盘到发震断层面的距离（震源距）要小于下盘，而近断层区域地震动的几何衰减很快，导致上盘地震动要大于下盘地震动。另外，从断层面上辐射出去的地震波到达地表后会反射回断层面，再从断层面反射到地表，多次反射的地震波也放大了上盘的地震动。上盘效应提示我们在近断层抗震设计时，要对上下盘分别对待。文献［7］认为考虑近断层地震作用下的基础隔震结构设计时，PGV 与 PGD 应优于 PGA 成为地震动输入的控制参数。美国 UBC97 规范考虑了近断层地震效应，以 50 年超越概率 10％ 把全美国划分为 5 个地震区（对应的有效峰值加速度分别为 0.075g、0.15g、0.20g、0.30g 和 0.40g），其中需要考虑近场效应的 Zone 4 区（0.40g）主要分布在西海岸加州境内。该规范根据近断层的距离和断层类型，给出了近场因子 Na 和 Nv 的取值，通过地面反应系数 Ca 和 Cv 对设计谱进行调整，详图 10。我国抗震规范 GB 50011 的设计谱曲线中没有反应速度脉冲效应对结构破坏的影响，而提出统一的加速度幅值增大系数方法考虑近断层地震动对隔震结构影响，本节通过选取近断层脉冲波进行动力时程分析，讨论抗震规范方法的合理性。

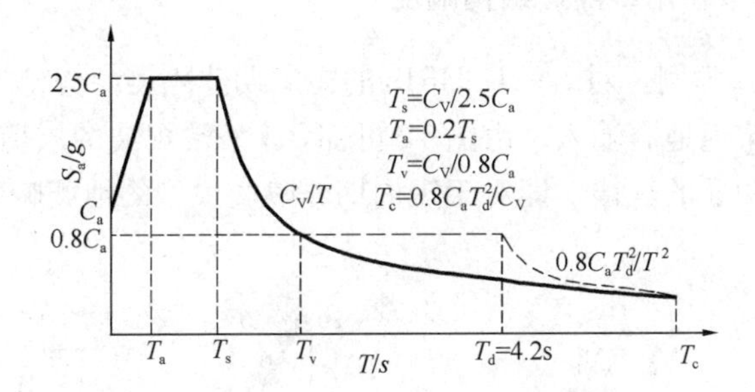

图 10 UBC97 标准谱（实线）与近场谱（虚线）对比（5%阻尼比）

Fig. 10 Comparison between standard spectrum (solid line) and near fault spectrum (dashed line) of UBC97 Code (5% damping ratio)

4.1 近断层脉冲波的特征及地震波选取

与远场地震相比，近场地震具有明显的速度脉冲特性，其显著特点是含有明显的速度脉冲波形、较长的脉冲周期和丰富的低频成分[8]。图 11 为典型近场脉冲型地震波 RSN1165 的加速度、速度时程曲线。本文选取 3 组常规波及 3 组近场脉冲波进行分析，且均满足抗震规范[3]选波要求。其中 3 条近场脉冲波，从 2007 年 BAKER[9]得到的 91 组速度脉冲时程中选取，其距离断层距离均在 10km 以内；而选用的常规地震波均为无速度脉冲效应的远场波。选用地震波信息如表 3 所示。以下针对隔震结构的分析，常规地震波均考虑 1.5 的近场影响系数，而脉冲波不考虑近场影响系数。

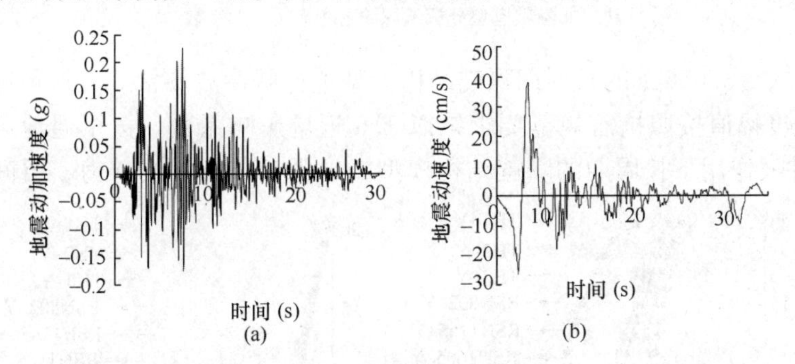

图 11 近场脉冲型地震波 RSN1165 加速度、速度时程曲线

Fig. 11 Acceleration and velocity time history curve of Rsn1165 with velocity pulse near fault

(a) 加速度时程；(b) 速度时程

选用地震波信息 **表 3**

Information of earthquake waves **Table 3**

地震波		地震事件	速度脉冲周期 s	PGV（cm/s）	震中距（km）
近场脉冲波	RSN802	Loma Prieta	4.57	45.93	7.58
	RSN1165	Kocaeli Turkey	5.37	38.26	7.21
	RSN1605	Duzce Turkey	4.93	71.05	6.58
常规波	TH1	Imperial Valley	—	26.3	22.03
	TH2	IWATE	—	5.28	66.6
	TH3	Northridge	—	9.8	41.1

4.2 近断层脉冲地震波作用下隔震结构响应

以 Y 方向地震输入为例，对 $t_1=1.2481s$ 的短周期非隔震结构模型进行脉冲波和常规波地震响应比较，按罕遇地震输入。由图 12 可知，3 条常规波的层剪力普遍比脉冲波大，而层间位移角则没有明显的规律，即对于短周期非隔震模型的地震响应，脉冲波没有呈现增强影响。

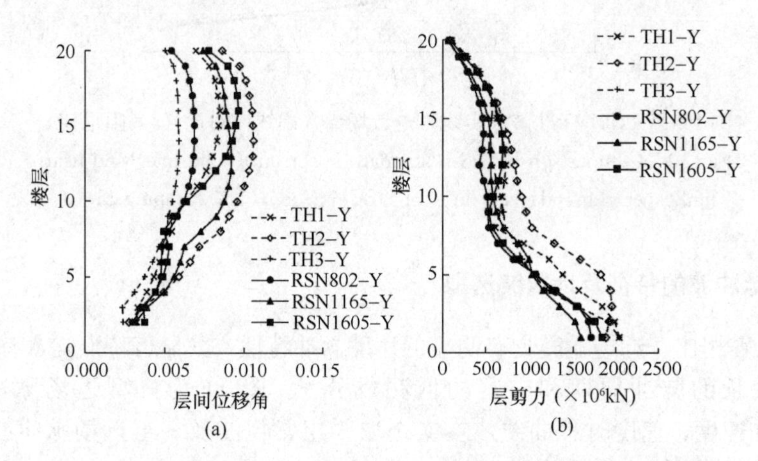

图 12　非隔震模型在常规波和脉冲波作用下的地震响应

Fig. 12　Seismic response of non isolated model under the action of conventional wave and pulse wave

(a) 层间位移角；(b) 层剪力

注：非隔震模型分析常规波未考虑 1.5 的系数

同样，对 $t_1=3.6564s$ 的长周期隔震结构模型进行脉冲波和常规波地震响应比较，其中常规波加速度幅值按照抗震规范要求乘以 1.5 近场影响系数，按罕遇地震输入。由图 13 可知，脉冲波作用下长周期的隔震结构模型的地震反应，如层剪力、层间位移角等都

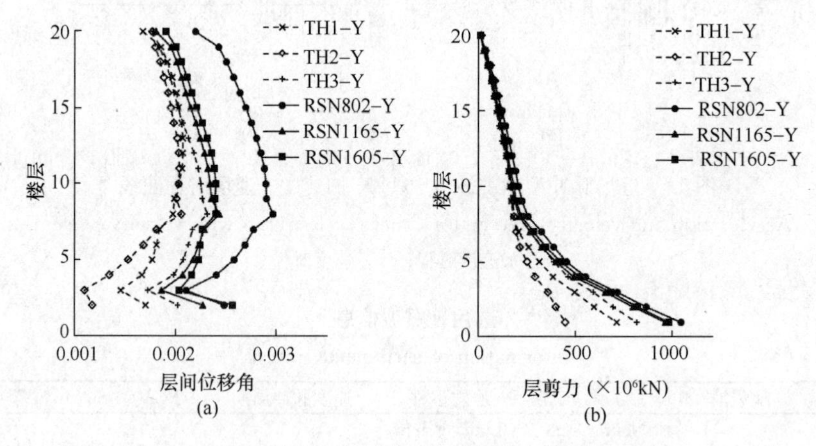

图 13　隔震模型在常规波和脉冲波作用下的地震响应

Fig. 13　Seismic response of isolated model under the action of conventional wave and pulse wave

(a) 层间位移角；(b) 层剪力

明显大于常规波的计算结果。

对比典型常规波 TH2 和脉冲波 RSN802 作用下隔震结构模型的瞬时输入能、隔震层位移时程曲线，按罕遇地震输入，见图 14。可知，尽管考虑 1.5 的近场影响系数[3]后常规波的瞬时地震输入能大于脉冲波，但由于速度脉冲效应的影响，脉冲波作用下的隔震层质心位移幅值明显大于常规波。

综上，由于抗震规范统一增大的近场影响系数仅考虑了地震波幅值的放大，而不能考虑近场脉冲地震的长周期频谱成分影响，不足以包络近断层脉冲波引起的地震响应，按此进行隔震结构设计不安全。

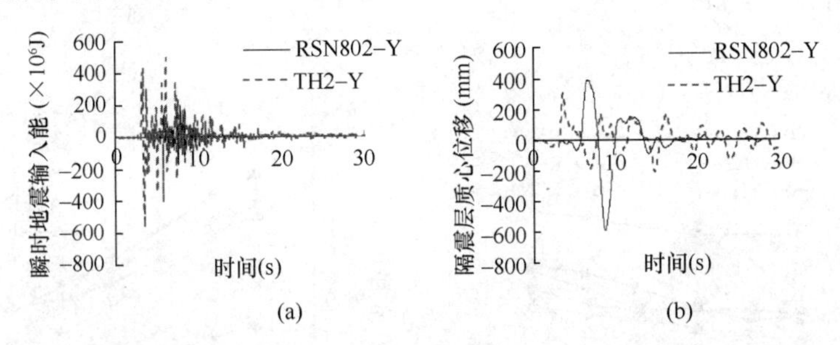

图 14　隔震模型在常规波与脉冲波作用下瞬时输入能及隔震层位移时程曲线

Fig. 14　Time history curve of Instantaneous input energy and displacement of isolation layer in isolated model under the action of conventional wave and pulse wave

(a) 地震瞬时输入能；(b) 隔震层位移

4.3　近断层脉冲波对隔震层设计影响

文献［10］表明，屈重比（隔震层支座总屈服力与上部结构重力比值）是决定隔震层力学性能、整体减震效果和结构抗震能力的重要参数。本节比较在设置不同屈重比时，脉冲波和常规波作用下隔震结构模型的计算减震系数和隔震层位移的大小，其中常规波均考虑 1.5 的近场影响系数。

从图 15 可知，不同屈重比下脉冲波的水平减震系数、基底剪力、最大层间位移角和隔震层质心位移均大于常规波，说明近断层脉冲波下隔震效果降低，考虑 1.5 的近场影响系数并不足够，上部结构设计需要进一步加强。此外，脉冲波和常规波作用下，随着屈重比的增加，隔震层质心位移明显减小，说明屈重比的增加导致隔震层刚度增加明显；但图 15（a）显示屈重比增加对水平减震系数影响相对较小。

5. 组合隔震技术的应用研究

5.1　设置黏滞阻尼器的组合隔震技术应用

近场脉冲波下最大的隔震层水平位移最大可达到 800mm，要满足罕遇地震时限值 Max（$0.55D=605$mm，$3r=696$mm）[3]＝696mm 的要求，屈重比需要达到 3.5%以上，意味着 90%以上的隔震支座需要采用铅芯支座，隔震层刚度相应增加，隔震效果下降。在隔震层 X，Y 向各设置 40 个黏滞阻尼器，形成橡胶隔震支座与黏滞阻尼器的组合隔震

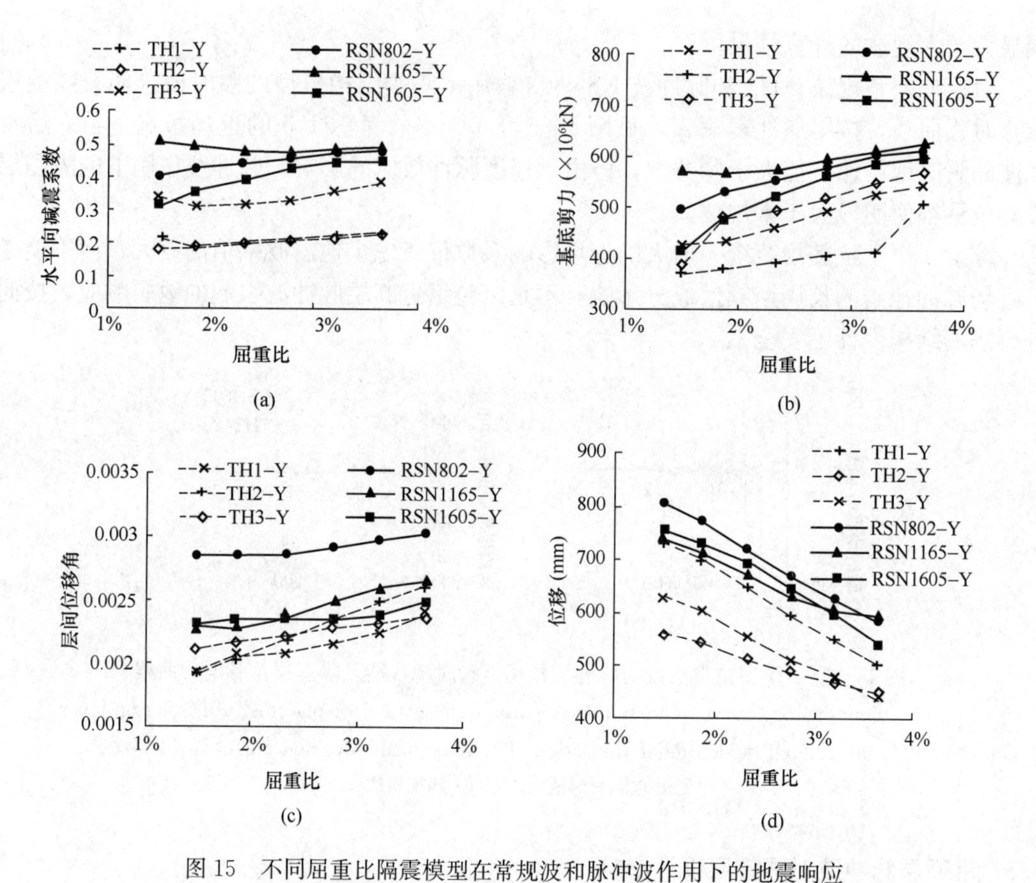

图 15　不同屈重比隔震模型在常规波和脉冲波作用下的地震响应

Fig. 15　Seismic response of isolated model with different yield to weight ratio under the action of conventional wave and pulse wave

(a) 减震系数（中震）；(b) 基底剪力（中震）；(c) 层间位移角（大震）；(d) 隔震层位移（大震）

方案，与单独橡胶隔震支座的隔震方案对比分析，比较考虑 1.5 近场系数的常规波以及近场脉冲波作用下隔震效果与限位情况。

以 RSN802 地震波为例，对比罕遇地震下两种隔震方案的典型能量耗散分布图 16 可知，两种方案总的输入能差别不大，且隔震层的消能器均耗散大部分能量。其中无黏滞阻尼的隔震方案，位移型的隔震支座耗能占比达到 51.3%；而组合隔震方案的隔震支座耗能占比约 34%，其速度型的黏滞阻尼器则消耗约 33%的能量，使得隔震支座位移减少，体现在隔震支座能量消耗有所减少。但组合隔震方案隔震层总的消耗能量比达到 67%、大于隔震方案的 51.3%，说明组合隔震不仅有效控制隔震支座位移，且能消耗更多的地震能量，使得结构构件塑性耗能占比不到 0.4%，屈服耗能机制更为合理，可显著提高结构抗震性能。

比较远场地震波 TH3 和近场脉冲地震波 RSN1605、罕遇地震作用下的上部结构楼层位移角和楼层地震剪力见图 17。从图可知，在 RSN1605 脉冲波与考虑 1.5 倍近场系数 TH3 常规波作用下，组合隔震方案的上部结构楼层位移角、楼层剪力小于隔震方案；且前述近场脉冲波作用下楼层位移角、楼层剪力大于远场波的情况发生改变，部分楼层近场脉冲波的地震响应小于远场波。

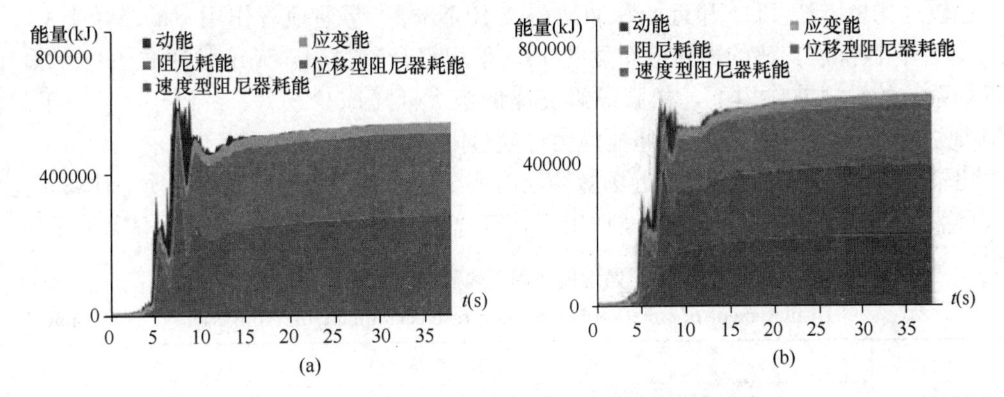

图 16　罕遇作用下，两种隔震方案的能量耗散分布

Fig. 16　Energy dissipation distribution of two isolation schemes under rare action

（a）隔震方案；（b）组合隔震方案

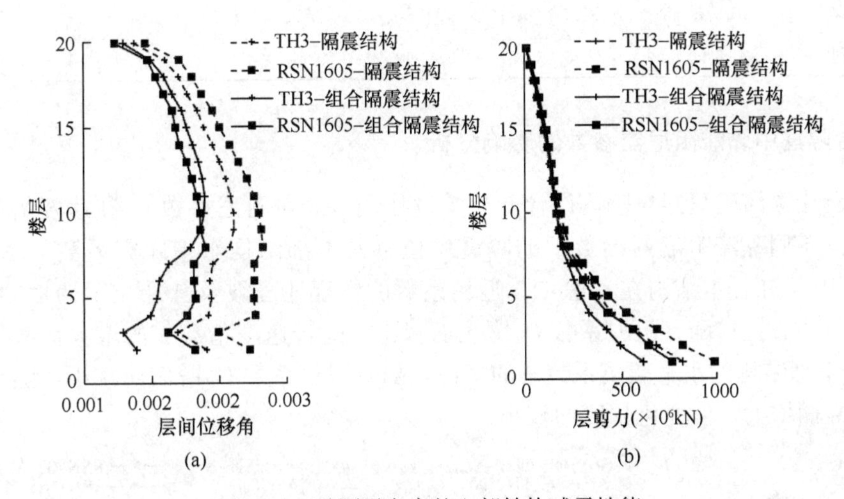

图 17　两种隔震方案的上部结构减震性能

Fig. 17　Seismic performance of superstructure with two isolation schemes

（a）层间位移角；（b）楼层剪力

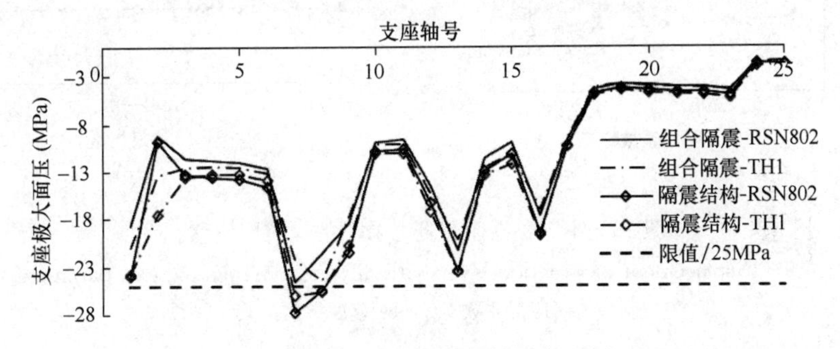

图 18　罕遇地震下，隔震支座的极大面压（MPa）

Fig. 18　Maximum pressure of seismic isolation bearings under rare action（MPa）

比较远场地震波 TH1 和近场脉冲地震波 RSN802、罕遇地震作用下的塔楼上 CA 轴线上 25 个橡胶隔震支座的面压分布见图 18。采用组合隔震后，无论脉冲波还是考虑 1.5 倍近场系数的常规波作用下，橡胶隔震支座的极大面压减少约 8%～15%，且小于限值 25MPa[4]；而隔震方案的个别支座压应力已超过限值 25MPa。

比较 3 条远场地震波和 3 条近场脉冲地震波罕遇地震作用下的隔震层质心位移，见表 4。不论远场常规波还是近场脉冲波作用下，组合隔震方案的隔震层位移均显著减小。

Y 向罕遇地震作用下隔震层质心位移 **表 4**

Displacement of isolation layer under rare earthquake in Y direction **Table 4**

隔震方案	隔震层位移（mm）					
	常规波（考虑 1.5 近场系数）			脉冲地震波		
	TH1	TH2	TH3	RSN802	RSN1165	RSN1605
组合隔震	483.5	411.3	569.6	587.1	596.8	531.7
隔震结构	702.5	545.8	604.8	778.1	715.2	735.8
降低比例	31.2%	24.6%	5.8%	24.5%	16.6%	27.7%

5.2 组合隔震中黏滞阻尼器参数的影响分析

本工程中黏滞阻尼器对隔震设计起到关键作用，故对阻尼系数 C_d 和速度指数 α 进行优化分析。不同黏滞阻尼器参数下的隔震层位移及上部结构水平减震系数见图 19 及图 20；由图 19 可知，本结构在考虑 1.5 近场系数的常规地震波及脉冲波罕遇地震作用下，结构隔震层质心位移随着阻尼系数 C_d 增大而减小，随着速度指数 α 的增大而减小；由图 20 可知，上部结构的水平减震系数总的来说，随着阻尼系数 C_d 增大而减小，随着速度指数 α 的增大而增大，但变化幅度均较小。

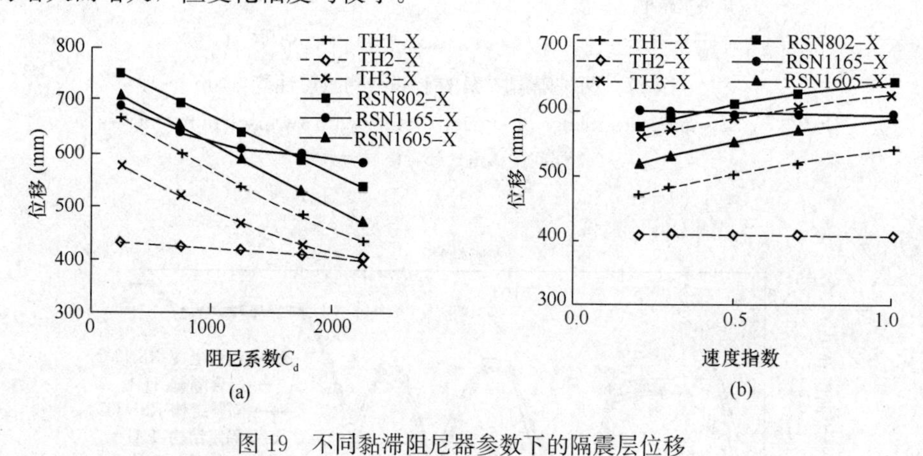

图 19 不同黏滞阻尼器参数下的隔震层位移

Fig. 19 Displacement of isolation layer with different viscous damper parameters

(a) 阻尼系数 C_d；(b) 速度指数 α

综上阻尼系数 C_d 与速度指数 α 对隔震层变形与上部结构反应的影响分析且考虑厂家的实际生产情况，黏滞阻尼器的阻尼系数取为 $1700\text{kN/(m/s)}^{\alpha}$，速度指数取为 0.3。

最终，组合基础隔震系统由铅芯橡胶支座（LRB）、无铅芯橡胶支座（LNR）、黏滞

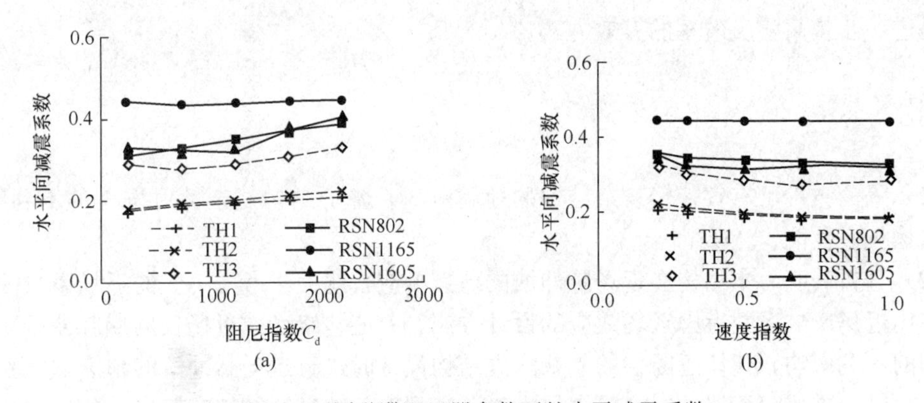

图 20 不同黏滞阻尼器参数下的水平减震系数

Fig. 20 Horizontal damping coefficient with different viscous damper parameters

(a) 阻尼系数 C_d；(b) 速度指数 α

阻尼器（VFD）和弹性滑板支座（ESB）组成，隔震层布置如图 21 所示。其中铅锌橡胶支座：LRB1400 支座 77 个，LRB1300 支座 41 个，LRB1100 支座 123 个；橡胶支座：LNR1400 支座 29 个，LNR1300 支座 27 个，LNR1100 支座 205 个；弹性滑板支座 ESB600 15 个，用于调整隔震层刚度中心与上部结构质心的偏心率。黏滞阻尼器 X 向和 Y 向各 40 个，最大输出力 2000kN，设计允许位移 720mm。整个隔震层屈重比约为 2%。

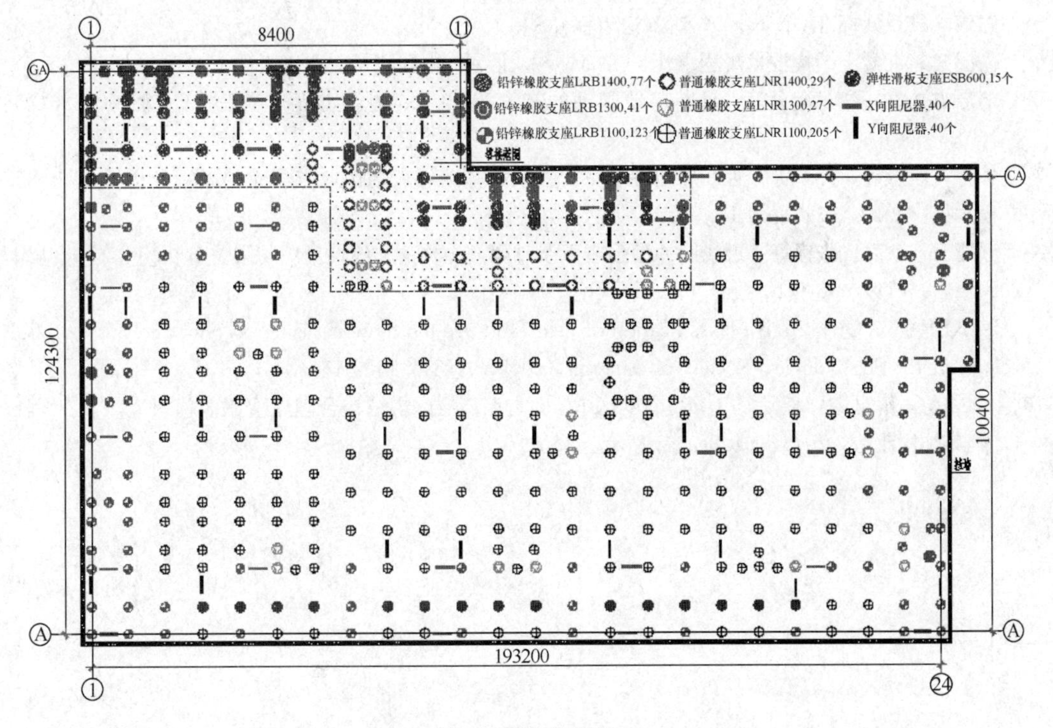

图 21 隔震层布置

Fig. 21 Layout of isolation layer

综上所述，隔震层设置速度型黏滞阻尼器不增加隔震层的静刚度、不影响上部结构隔震效果同时，耗散大量地震能量，有效降低了地震作用下的隔震支座受力及隔震层的位移，是进一步提高隔震效率的有效方式。

6. 结论

（1）隔震技术可有效减轻复杂结构的地震响应，降低结构复杂性对抗震的不利影响，提高结构抗震性能。

（2）分析表明，本结构在近场脉冲波的地震响应普遍大于考虑 1.5 的近场系数的常规波。采用近场影响系数考虑近场效应的设计方法，不能包络考虑近场长周期脉冲运动对隔震结构的不利影响，建议近断层的复杂隔震结构设计时进行近场脉冲波的地震响应验算。

（3）采用黏滞阻尼器的组合隔震技术对于高烈度区的复杂结构有着良好的适用性，是提高隔震效率、增加复杂结构抗震性能的有效方式。

参考文献

[1] 中华人民共和国住房和城乡建设部．关于房屋建筑工程推广应用减隔震技术的若干意见(暂行)[EB/OL](2014－0221). http：//www. mohurd. gov. cn/wjfb/201402/t20140226 _ 217207. html.

[2] 火明譞，赵亚敏，陆 鸣．近断层地震作用隔震结构研究现状综述[J]. 世界地震工程，2012，28(3)：161-170.

[3] 建筑抗震设计规范 GB 50011—2010(2016 年版)[S]. 北京：中国建筑工业出版社，2016.

[4] 建筑隔震设计标准 GB ×××××(送审稿)[S].

[5] 高层建筑混凝土结构技术规程 JGJ 3—2010[S]. 北京：中国建筑工业出版社，2010.

[6] 刘启方，袁一凡，金星等．近断层地震动的基本特征[J]. 地震工程与工程振动，2006，26(1)：1-10.

[7] 韩森，段燕玲，孙欢等．近断层地震动特征参数对基础隔震结构地震响应的影响分析[J]. 土木工程学报，2013，46(6)：8-13.

[8] 颜桂云，项洪，张铮等．近场脉冲型强震下基础隔震结构的非线性反应与限位分析[J]. 建筑结构，2016，46(11)：96-101.

[9] BAKER JW. Quantitative classification of near-fault ground motions using wavelet analysis[J]. Bulletin of the Seismological Society of America，2007，97(5)：1486-1501.

[10] 刘立德，解琳琳，李爱群等．高烈度区 RC 剪力墙高层隔震结构合理屈重比取值研究[J]. 工程抗震与加固改造，2018，40(5)：82-87.

Industry-Academia Forum on Advances in Structural Engineering（2020）

第九届结构工程新进展论坛简介

本届论坛主题：韧性结构与结构减隔震技术

会议时间：2020 年 12 月 3 日～4 日

会议地点：中国　广州

主办单位

中国建筑工业出版社

同济大学《建筑钢结构进展》编辑部

香港理工大学《结构工程进展》编委会

承办单位

广州大学

广东省工程勘察设计行业协会

协办单位

中国地震学会工程隔震与减震控制专业委员会

中国建筑学会抗震防灾分会结构减震控制委员会

中国灾害防御协会城乡韧性与防灾减灾专业委员会

☞ 关于论坛

“结构工程新进展论坛”自2006年首次举办以来，十余年来已打造成为行业内颇有影响的交流平台。论坛旨在促进我国结构工程界对学术成果总结及交流，汇集国内外结构工程各方面的最新科研信息，提高专业学术水平，推动我国建筑行业科技发展；同时，论坛面向工程领域，面向结构工程师，针对工程技术人员当前最关心、最有争议的技术难点，积极探索，相互切磋，寻求更好的解决方案和措施，以服务于更广泛的工程领域和更多的工程技术人员。

论坛原则上以两年一个主题的形式轮流呈现，历届论坛主题分别为：

- **新型结构材料与体系**（第一届，2006，北京）
- **结构防灾、监测与控制**（第二届，2008，大连）
- **钢结构研究和应用的新进展**（第三届，2009，上海）
- **混凝土结构与材料新进展**（第四届，2010，南京）
- **钢结构研究和应用的新进展**（第五届，2012，深圳）
- **结构抗震、减震新技术与设计方法**（第六届，2014，合肥）
- **工业建筑与特种结构新进展**（第七届，2016，西安）
- **可持续结构与材料**（第八届，2018，上海）

☞ 本届论坛特邀报告人

周福霖　院士（广州大学）

欧进萍　院士（哈尔滨工业大学）

陈政清　院士（湖南大学）

徐　建　院士（中国机械工业集团有限公司）

薛松涛　日本工程院院士（日本东北大学）

葛汉彬　日本工程院院士（日本名城大学）

李国强　教授（同济大学）

徐幼麟　教授（香港理工大学）

郁银泉　教高，勘察设计大师（中国建筑标准设计研究院）

丁洁民　教授，勘察设计大师（同济大学）

杜修力　教授（北京工业大学）

孙柏涛　教授（中国地震局工程力学研究所）

李宏男　教授（沈阳建筑大学）

李忠献　教授（天津建筑大学）

李爱群　教授（北京建筑大学）

滕　军　教授（哈尔滨工业大学）

陈　星　教高（广东省勘察设计协会）

朱忠义　教高（北京市建筑设计研究院）

吴小宾　教高（西南建筑设计院）

周　云　教授（广州大学）

☞ **本届论坛组织机构**

指导委员会：

顾　问：周福霖

主　任：咸大庆　滕锦光　李国强

委　员：韩林海　李宏男　吴智深　徐正安　任伟新　苏三庆　史庆轩

肖建庄　周　云　赵梦梅　刘婷婷

组织委员会：

主　任：谭　平　陈　星

委　员：黄襄云　马玉宏　刘彦辉　罗振城　文小菲　龙耀球　林菲菲

徐　丽　王玉梅　陈洋洋　郝霖霏　张　颖　罗俊杰　和雪峰

第九届论坛特邀报告论文作者简介

（按本书论文顺序）

周福霖 中国工程院院士，教授，现任广州大学工程抗震研究中心主任兼总工程师，曾任中国工程院土木、水利与建筑工程学部主任，广州市科协主席。是世界著名的工程结构与工程抗震、隔震与减震控制领域的专家，对我国工程结构隔震减震控制技术体系的建立、应用与发展做出了奠基性、开拓性的贡献。1963年湖南大学土木工程学院工民建本科毕业，1983年加拿大不列颠哥伦比亚大学研究生毕业得到硕士学位。曾任机械工业部第四设计研究院副总工程师、联合国工发组织（UNIDO）隔震技术顾问和国际减震学会（ASSISI）主席。目前为中国建筑学会抗震防灾研究会结构减震与控制专业委员会主任委员、中国土木学会防震减灾技术推广委员会主任委员、中国工程标准化协会工程振动专业委员会副主任委员、中国地震学会工程隔震与减震控制专委会名誉主任。曾主持过中国国家科学基金研究项目、美国国家科学基金项目、中美科技合作项目、联合国工发组织（UNIDO）科学技术研究开发项目等10余项。主编我国第一本《叠层橡胶支座隔震技术规程》、国家标准《建筑隔震设计标准》和《橡胶支座》，参编《建筑抗震设计规范》。出版学术著作4部，发表论文近200篇，研究成果获国家、省、部科技进步奖8项，其中国家科技进步二等奖2项。曾获得全国“五一劳动奖章”、“广州市十大优秀留学回国人员”、广东省“首届南粤创新奖”等个人奖励获得者。广东省2015年“科技创新突出贡献奖”和全国“第七届优秀科技工作者”等奖励等10多个光荣称号。

欧进萍 哈尔滨工业大学（深圳）土木与环境工程学院教授，结构监测、控制与防灾减灾专业领域专家。曾兼任中国振动工程学会理事长、中国建筑学会副理事长、国际结构控制与监测学会（IASCM）执行理事、国际智能基础设施结构健康监测学会（ISHMII）副理事长。2003年当选中国工程院院士。在结构抗震抗风及其动力可靠性、结构振动控制、结构健康监测、海洋平台结构及其安全保障技术、纤维复合（FRP）材料制品与结构等方面取得系统的研究与应用成果。发表学术

论文300余篇、出版著作4部；获得国家科技进步二等奖4项(3项排名第一、1项排名第三)、省部级科技进步一等奖5项、国家发明专利20余项；主持编制国家相关规范标准2部；培养毕业博士60余人、毕业硕士100余人、博士后出站30余人。

Youlin Xu Professor Youlin Xu is Dean of the Faculty of Construction and Environment, Chair Professor of Structural Engineering, and Yim, Mak, Kwok & Chung Professor in Smart Structures at The Hong Kong Polytechnic University. Professor Xu has conducted researches and served as a consultant in structural engineering for over three decades, with particular interests in wind effects on long-span bridges and tall buildings, structural health monitoring of mega infrastructure, structural vibration control and smart structures. He has published over 290 SCI journal papers, delivered over 100 keynote/invited lectures at conferences or workshops. He is on the Civil Engineering list of the Most Cited Researchers developed for Shanghai Ranking's Global Ranking of Academic Subjects 2016 by Elsevier. In recognition of his outstanding research achievements, he received several prestigious awards, including the Guanghua Engineering Science and Technology Award in 2018, the IAWE Davenport Medal in 2018, the ASCE Robert H. Scanlan Medal in 2012, the Qian Ling Xi Computational Mechanics Award in 2010, and Croucher Award in 2006. Professor Xu has published three books: Structural health monitoring of long-span suspension bridges, Wind effects on cable-supported bridges, and Smart civil structures. He has been engaged in many high-impact knowledge-transfer projects, including the health monitoring projects on the Tsing Ma Bridge and the Stonecutters Bridge in Hong Kong, the CCTV Tower in Beijing and the Shanghai Tower in Shanghai. He is a Fellow of The Hong Kong Institution of Engineers, the American Society of Civil Engineers, the Engineering Mechanics Institute of the U.S.A, the Institution of Structural Engineers of the U.K., and Hong Kong Academy of Engineering Science.

Hanbin Ge Dr. Hanbin Ge is a full Professor of Structural Engineering at Meijo University, Japan since 2008, Member of Engineering Academy of Japan since 2018, Fellow of Japan Society of Civil Engineers (JSCE) since 2013, and Changjiang Scholar (Chair Professor), MOE, China. He graduated from Huazhong University of Science and Technology, China in 1986, and then studied and worked at Nagoya University, Japan for 20 years. Professor Ge is co-author of 18 books including seismic design codes and more than 380 journal papers. His research interests include behavior of structural members and systems, with particular emphasis on developing high-performance seismic dampers, seismic and damage control design, seismic performance evaluation, fractural behavior and retrofit of steel and steel-concrete composite structures. Professor Ge received The JSCE (The Japan Society of Civil Engineers) Thesis Award in 1995, The JSCE Structural Engineering Thesis Award in 1999, The JSCE Bridge Engineering Thesis Award (Tanaka Award) in 2016 and 2020.

李国强 同济大学教授，国家土建结构预制装配化工程技术研究中心主任，建筑钢结构教育部工程研究中心主任，中国钢结构协会副会长，中国工程建设标准化协会副理事长。长期从事多高层建筑钢结构、钢结构抗震与抗火的研究工作。出版中英文著作 17 部；发表学术期刊论文 700 多篇，其中 SCI 收录英文论文 233 篇，EI 收录中文论文 221 篇；应邀作国际学术会议大会主题报告或邀请报告 60 多次；主持编写国家标准《建筑钢结构防火技术规范》等 12 部国家、行业或地方工程建设标准；研究成果取得美国专利 2 项，日本专利 1 项，中国发明专利 45 项，并在国家会展中心、国家会议中心、上海虹桥交通枢纽、广州新电视塔、天津 117 大厦、中国大飞机总装厂等一大批国家重大工程中应用，获国家技术发明二等奖 1 项（排第 1）、国家科技进步二等奖 2 项（排第 1、第 9），上海市、教育部等省部科学技术奖一等奖 9 项（5 项排第 1、2 项排第 2、2 项排第 6），二等奖 6 项（均排名第 1）。

杜修力　1962年生，教授级高级工程师，博士生导师，北京工业大学副校长，土木工程学科带头人，国家自然科学基金创新研究群体带头人。主要从事工程结构抗震研究，主持完成两项国家级重大项目，发表SCI、EI收录论文400余篇，获国家授权发明专利30余项，出版著作6部，获得国家科技进步二等奖5项、省部级科技进步奖10余项。目前兼任国际生命线与基础设施地震工程学会理事长、中国地震学会基础设施抗震减灾专业委员会主任、国际防护工程学会理事、国务院学位委员会学科评议组成员、国家自然科学基金委员会工程与材料学部咨询专家和学科评议组成等。结合交通、能源等基础设施工程抗震问题开展研究，在基础理论与方法、关键技术和规范编制、工程应用等方面取得了突出成就。

薛松涛　1963年12月15日生于上海，1985年毕业于同济大学固体力学系，之后赴日本东北大学工学部建筑学科留学，主攻结构抗震。1989年获东北大学硕士学位，1991年获博士学位，同年留任于东北大学助教并于1995年升任副教授。1996年回母校同济大学担任教授，至今。2010年应日本东北工业大学邀请，兼任教授并且任制振工学研究所长。主要研究方向是建筑抗震，制振以及健康监测，至今发表了150篇以上论文。主持了多项纵向科研项目，包括日本文部省科研项目4项，已完成的国际合作项目2项，主持完成国家自然科学基金项目3项，上海浦东国际机场科研项目1项和国家教委留学回国人员重点资助项目。主持国家杰出青年科学基金1项，主持科技部重点项目1项。2018年被评为日本工程院外籍院士。

李爱群　国家杰出青年基金获得者，国家级教学名师，北京学者，东南大学特聘教授，现任北京建筑大学副校长。中国建筑学会抗震防灾分会常务理事，中国振动工程学会结构控制与健康控制分会常务理事，中国勘察设计协会结构隔减震专家委员会副主任委员，国家住建部抗震设防专业委员会副主任委员。长期从事工程抗震减振隔震和健康监测研究工作。主持获国家科技进步二等奖3项、省部级科技进步一等奖6项。主编出版著作教材9部，发表SCI、EI收录论文180余篇。获国家发明专利50余项。研发了结构减振隔震和健康监测系列技术，研究成果成功应用于北京国家会议中心（国内第一个大跨楼盖减振工程）、四川成都绿地高层减震楼群（国内最大高层减震住宅群）、北京奥运转播塔、北京奥林匹克塔和南京博物院老大殿整体顶升工程（顶升点数、顶升高度、顶升面积为国际之

最，国家住建部科技示范工程）。

郁银泉 1962年生，教授级高级工程师，国家勘察设计大师，中国建筑科技集团副总工程师，中国建筑标准设计研究院副院长、总工程师。兼任住房和城乡建设部科学技术委员会超限高层建筑工程技术专业委员会副主任委员；中国建筑学会会员；中国钢结构协会常务理事；《钢结构》及《建筑结构》编委会委员等职位。研究方向主要为建筑钢结构、建筑抗震与减隔震技术。主编了《高层民用建筑钢结构技术规程》《建筑隔震设计标准》等多部国家及行业标准，主持了多项国家、省部级课题项目。

丁洁民 同济大学教授、博士生导师，全国工程勘察设计大师，国家一级注册结构工程师，英国皇家资深注册结构工程师；现任同济大学建筑设计研究院（集团）有限公司总工程师。1990年毕业于同济大学结构工程系，获工学博士学位。长期从事复杂结构的研究与设计咨询工作，在超高层结构、复杂结构和大跨度钢结构等方面取得许多研究成果，荣获建设部科技进步一等奖，国家科技进步二等奖，教育部科技进步一等奖，中国建筑学会科技进步奖特等奖、一等奖等，并参与编制了《建筑抗震设计规范》GB 50011—2010、《空间结构设计规程》DG/TJ 08—52—2004等国家和上海市的有关设计规程。已完成高层、超高层、大跨度体育场馆、会展中心、大型影剧院、高铁交通枢纽等百余项工程项目，荣获全国优秀工程勘察设计行业建筑工程一等奖、二等奖，设计银奖，全国优秀建筑结构设计一等奖、二等奖，英国结构工程师协会全球奖，英国结构工程师协会金质奖章，IABSE杰出结构奖，CTBUH世界最佳高层建筑奖等。出版书籍5本，包括《减隔震建筑结构设计指南与工程应用》《黏滞阻尼技术工程设计与应用》等。

周云 广州大学土木工程学院教授，博士生导师，广州大学副校长。先后主持完成国家重点研发计划课题、国家自然科学基金项目和羊城学者首席科学家项目等三十余项；发表学术论文300余篇；获国家发明、实用新型专利80余项；主编国家行业标准《建筑消能减震技术规程》，标准化协会标准《屈曲约束支撑应用技术规程》，参编国家、标准化协会等6个规范或规程；出版著作8本，教材8本；获省（部）、市科技进

步奖 6 项。兼任中国土木工程学会防震减灾工程技术推广委员会副主任（兼秘书长）及多个协会或专业委员会委员、《土木工程学报》、《建筑结构学报》等学术刊物编委。被评为南粤优秀教师、省“千百十工程”省级培养对象先进个人、广州市劳动模范、市优秀专家、市“121 人才梯队工程”后备人才、高层次人才。

朱忠义　教授级高工，工学博士，现任北京市建筑设计研究院有限公司总工程师、住建部全国超限工程审查委员会委员，入选北京学者、国家百千万人才工程、享受国务院政府特殊津贴、首届中国钢结构协会“钢结构大师”，荣获 2017 年国家“有突出贡献中青年专家”、2018 年全国“杰出工程师奖”等称号。长期从事大跨度空间结构、建筑减隔震的设计和研究。主持了中国“天眼”FAST 主动反射面主体支承结构、北京大兴国际机场航站楼、北京首都国际机场 T3 航站楼、2022 年冬奥会国家速滑馆、2022 年卡塔尔世界杯主体育场、中国科学院江门中微子探测器、国家体育馆、昆明新机场等多项国家重大工程的设计与研究，曾获国家科学技术进步奖二等奖 1 项、北京市科学技术奖一等奖等省部级奖 10 项，国际桥梁与结构工程协会（IABSE）杰出结构大奖和英国结构工程师学会（ISE）杰出结构奖等国际奖两项。

滕军　哈尔滨工业大学（深圳）土木与环境工程学院教授，国务院特殊津贴专家，主持国家科技支撑计划重点课题 1 项、国家研发计划重点专项课题 1 项，主持国家自然科学基金重大项目 1 项、重点项目 2 项、面上项目 7 项，参与国家自然科学基金重大国际（中美）合作研究项目 1 项。主要社会兼职包括中国振动工程学会结构健康监测和抗震控制专业委员会常务理事、副主任委员；国家自然科学基金会评议专家组成员等。滕军提出基于最优失效模式的抗灾结构体系及设计理论，搭建了成套的结构抗震分析和评价平台，建立了系统的结构灾害监测和控制的理论、技术和方法，并进行了重大工程的实践，其中结构健康监测成套技术应用于国家游泳中心“水立方”、深圳市民中心、万科中心、深圳湾体育中心等；结构振动控制成套技术应用于广州塔、深圳京基 100、深圳梧桐山电视塔等，结构抗震理论和方法应用于数十项超限建筑的抗震性能化设计中。获得国家科技进步二等奖 2 项，省部级一等奖 5 项。出版专著 2 部，参编国家标准 3 本，发明专利 10 余项，发表论文 300 余篇，SCI、EI 收录 270 篇。

吴小宾 教授级高级工程师，国家一级注册结构工程师。中国建筑西南设计研究院有限公司总工程师。担任中国建筑学会建筑结构分会第七届理事会副主任委员、中国勘察设计协会结构设计分会第一届理事会常务理事；中国勘察设计协会抗震防灾分会全国隔震减震专家工作部委员等。从事建筑结构设计及建筑抗震、减隔震技术研究和应用。在大型既有建筑改造加固、复杂结构和超高层结构、大型索膜结构、高烈度地区隔震和消能减震结构等领域业绩斐然，主持数十项大型工程，如：四川省科技馆（四川展览馆改造）、贵州铜仁奥体中心、乐山奥体中心、临沂奥体中心、领地环球金融中心、成都天府新区中海 489m 超高层项目、川投西昌综合医院、西昌人民医院等。多次获得四川省工程勘察设计“四优”一等奖、中国勘察设计协会全国优秀工程勘察设计行业奖优秀建筑工程设计一等奖和优秀建筑结构专业一等奖、优秀抗震防灾一等奖。